Religious Tourism

This thought-provoking book embarks into the captivating world of religious tourism ar
cate relationship between travel and spirituality, shedding light on whether tourists are
meaning or simply pleasure-seekers.

In first section of the book, the chapters delve into the complex dynamics of pilgrim
significance of sacred heritage sites, and the evolving role of tourism in religious experi
volume digs deeper into the spiritual dimensions of religious tourism, examining the res
spiritual retreats, the anthropology of spirituality and tourism, and the motivations of mo

Written in an accessible style, this volume invites readers from all backgrounds to exp
impact of religious tourism on individuals and societies alike. Engaging and enlightenin
must-read for anyone intrigued by the intersection of travel and spirituality.

The chapters in this book were originally published in *Tourism Recreation Research.*

Tzung-Cheng (TC) Huan is Professor at National Chiayi University, Taiwan. He is also Edito
ism Recreation Research and a member of the UNWTO (United Nations World Tourism Orga
Tourism Experts. He has been honoured by the 2011–2012 Outstanding Reviewer Award fr
tality Quarterly; the 2013, 2017, and 2018 Emerald Literati Awards; and the 2018 Publons Pee

Anestis K. Fotiadis is Full Professor at Zayed University, College of Business, Department
Abu Dhabi, UAE. His main area of research is associated with tourism management, even
hospitality management, and sustainable development. He has published more than 40 res
international academic journals such as *Annals of Tourism Research*, *Tourism Management*, *Jou*
Research, and others.

Nikolaos Stylos is Associate Professor in Marketing and Digital Innovation at the University
ness School, UK, and Honorary Professor of Hospitality Management at the Tainan University
Taiwan. He is an internationally recognised and awarded scholar specialising in marketing i
digital transformation in the service industries.

Religious Tourism

An Essential Reader

Edited by
Tzung-Cheng (TC) Huan, Anestis K. Fotiadis and Nikolaos Stylos

Routledge
Taylor & Francis Group
LONDON AND NEW YORK

First published 2025
by Routledge
4 Park Square, Milton Park, Abingdon, Oxon, OX14 4RN

and by Routledge
605 Third Avenue, New York, NY 10158

Routledge is an imprint of the Taylor & Francis Group, an informa business

British Library Cataloguing-in-Publication Data
A catalogue record for this book is available from the British Library

ISBN13: 978-1-032-87161-5 (hbk)
ISBN13: 978-1-032-87162-2 (pbk)
ISBN13: 978-1-003-53120-3 (ebk)

DOI: 10.4324/9781003531203

Typeset in Myriad Pro
by codeMantra

Publisher's Note
The material in this volume has been reproduced in facsimile. This means we can retain the original pagination to facilitate easy and correct citation of the original essays. It also explains the variety of typefaces, page layouts and numbering.
The publisher accepts responsibility for any inconsistencies that may have arisen during the conversion of this book from journal articles to book chapters, namely the inclusion of journal terminology.

Contents

Citation Information

The following chapters were originally published in various issues and volumes of *Tourism Recreation Research*. When citing this material, please use the original page numbering for each article, as follows:

Chapter 1
Is Tourist a Secular Pilgrim or a Hedonist in Search of Pleasure?
Dan Knox, Kevin Hannam, Peter Jan Margry, Daniel H. Olsen and Noel B. Salazar
Tourism Recreation Research, volume 39, issue 2 (2014), pp. 235–267

Chapter 2
Self-expression and play: can religious tourism be hedonistic?
Sean Lochrie, Ian W. F. Baxter, Elaine Collinson, Ross Curran, Martin Joseph, Gannon, Babak Taheri, Jamie Thompson and Ozge Yalinay
Tourism Recreation Research, volume 44, issue 1 (2019), pp. 2–16

Chapter 3
Pilgrimage: Journeying Beyond Self
Chris Devereux and Elizabeth Carnegie
Tourism Recreation Research, volume 31, issue 1 (2006), pp. 47–56

Chapter 4
Writing on the Graceland Wall: On the Importance of Authorship in Pilgrimage Landscapes
Derek H. Alderman
Tourism Recreation Research, volume 27, issue 2 (2002), pp. 27–33

Chapter 5
Sacred World Heritage Sites: Balancing Meaning With Management
Myra Shackley
Tourism Recreation Research, volume 26, issue 1 (2001), pp. 5–10

Chapter 6
Complementarity: bridging the tourism academic/religion divide
Stephen Schweinsberg
Tourism Recreation Research, volume 48, issue 4 (2023), pp. 627–629

Chapter 7
Sacred Journeys: Religious Heritage and Tourism
Dallen J. Timothy
Tourism Recreation Research, volume 27, issue 2 (2002), pp. 3–6

Chapter 8
Exploring the restorative benefits of spiritual retreats: the case of clergy retreats in Australia
Chelsea Gill, Jan Packer and Roy Ballantyne
Tourism Recreation Research, volume 43, issue 2 (2018), pp. 235–249

Chapter 9
Religion, spirituality, and the formation of tourism knowledge
Stephen Schweinsberg
Tourism Recreation Research, volume 48, issue 4 (2023), pp. 593–604

Chapter 10
Spiritual Hosting and Tourism: The Host-Guest Dimension
Yoel Mansfeld and Alison J. Mcintosh
Tourism Recreation Research, volume 34, issue 2 (2009), pp. 157–168

Chapter 11
Spirituality and Tourism: An Anthropologist's View
Sagar Singh
Tourism Recreation Research, volume 34, issue 2 (2009), pp. 143-155

Chapter 12
Pilgrimage and Tourism Anthropological Perspectives
Sagar Singh and J. Anthony Paredes
Tourism Recreation Research, volume 17, issue 2 (1992), pp. 84-88

Chapter 13
Eudaimonic benefits of tourism: the pilgrimage experience
Marc van Iwaarden and Jeroen Nawijn
Tourism Recreation Research, volume 49, issue 1 (2024), pp. 37–47

Chapter 14
Motivations of urban millennials for spiritual travel in India
Snigdha Kainthola, Nimit Chowdhary, Rahul Pratap Singh Kaurav and Pinaz Tiwari
Tourism Recreation Research, volume 49, issue 2 (2024), pp. 410–425

Notes on Contributors

Derek H. Alderman is Professor and Former Dept. Head of Geography at the University of Tennessee, USA, and Past President of American Association of Geographers (2017–2018).

Roy Ballantyne is Emeritus Professor at the University of Queensland Business School, Australia. His research focuses on environmental education, environmental interpretation, freechoice environmental learning, wildlife tourism, and ecotourism research. He is the joint author of *Contemporary Issues in Heritage and Environmental Interpretation* and *Designing Interpretive Signs: Principles in Practice* and joint editor of the *International Handbook on Ecotourism*.

Ian W. F. Baxter is Professor of Historic Environment Management at Heriot-Watt University, Edinburgh, UK. He originally trained as an archaeologist and has spent 25 years undertaking applied research and consultancy, working with national heritage and tourism organisations on historic site visitor experience management and organisational knowledge development.

Elizabeth Carnegie is Associate Professor and Section Group Head of Tourism and Events at Northumbria University, Newcastle upon Tyne, UK.

Nimit Chowdhary is Professor and Head of the Department of Tourism and Hospitality Management at Jamia Millia Islamia, New Delhi, India. With more than 25 years of experience in academics, he is a recipient of several awards, authored eight books, edited two books, and contributed more than 116 papers.

Elaine Collinson is Associate Professor in the School of Social Sciences – Director of Postgraduate Studies & Director of Corporate Executive Development at Heriot-Watt University, Edinburgh, UK. Over 25 years' experience in the Higher Education sector, with publications in the areas of marketing, entrepreneurship, education, and industry collaboration.

Ross Curran is Assistant Professor in Management at Heriot-Watt University, Dubai, UAE. His primary research interests focus on nonprofit marketing and volunteer management practices, as well as areas of authenticity and heritage in tourist consumption.

Chris Devereux was formerly Professor at Institute of Education at the University of London, UK.

Anestis K. Fotiadis is Full Professor at Zayed University, College of Business, Department of Management, Abu Dhabi, UAE. His main area of research is associated with tourism management, event management, hospitality management, and sustainable development. He has published more than 40 research papers in international academic journals such as *Annals of Tourism Research, Tourism Management, Journal of Business Research*, and others.

Martin Joseph Gannon is Lecturer in Entrepreneurship at the University of Edinburgh Business School, UK. He holds a variety of research interests including entrepreneurial philanthropy, family business governance and marketing, heritage marketing management, tourism management, and consumer behaviour. To this end, his research is available in the *Journal of Consumer Culture*, the *International Journal of Contemporary Hospitality Management, Current Issues in Tourism*, the *Journal of Travel and Tourism Marketing*, and the *Service Industries Journal*.

Chelsea Gill is Assistant Professor of Management at Bond University, Australia. Chelsea has a background in event and hospitality management and has previously completed a Bachelor of Business Management and Master of Arts (Leadership). Her research interests include religious tourism, mental restoration, clergy well-being, and visitor experience.

Kevin Hannam is Professor of Tourism at the City University of Macau, China.

Tzung-Cheng (TC) Huan is Professor at National Chiayi University, Taiwan. He is also Editor-in-Chief of *Tourism Recreation Research* and a member of the UNWTO (United Nations World Tourism Organization) Panel of Tourism Experts. He has been honoured by the 2011–2012 Outstanding Reviewer Award from *Cornell Hospitality Quarterly*; the 2013, 2017, and 2018 Emerald Literati Awards; and the 2018 Publons Peer Review Award.

Marc van Iwaarden is a graduate of the BSc Tourism from Wageningen University and Breda University of Applied Sciences, Netherlands. His interests lie in the field of positive psychology, spirituality, religion, and sustainable tourism. It is his passion to understand how people experience increased well-being in a lasting and sustainable way.

Snigdha Kainthola is a doctoral researcher in the Department of Tourism and Hospitality Management at Jamia Millia Islamia, New Delhi, India. Her core areas for research are spiritual tourism, over tourism, spirituality, and tourism.

Rahul Pratap Singh Kaurav is Associate Professor at the Fortune Institute of International Business (FIIB), New Delhi, India. With more than 12 years of academic experience, he is a prolific writer, has edited seven books, and contributed more than 40 papers. His areas of interest are tourism marketing, internal marketing, destination performance, and qualitative analysis.

Dan Knox is Associate Professor and Director of the Department of International Tourism Management at Surrey International Institute at the Dongbei University of Finance & Economics, Dalian, China.

Sean Lochrie is Assistant Professor of Marketing at Heriot-Watt University, Dubai, UAE. His primary research interest focuses on the creation of custodianship behaviours within World Heritage Site management. Recent publications have explored stewardship and local community engagement in World Heritage Site management.

Yoel Mansfeld is Associate Professor and former Head of the Centre for Tourism, Pilgrimage and Recreation Research at the Department of Geography and Environmental Studies at the University of Haifa, Israel.

Peter Jan Margry is emeritus professor of European ethnology at the University of Amsterdam. He is an ethnologist and historian, and his current focus is on contemporary religious cultures, cultural memory practices, and cultural heritage in the Netherlands and Europe. He also served as the director (executive vice-president) of the International Society for Ethnology and Folklore (SIEF).

Alison J. McIntosh is Professor at the Auckland University of Technology, School of Hospitality & Tourism, New Zealand. A central theme of her research is that experiential, qualitative and social justice analyses reveal subjective, emotional, spiritual, and neglected aspects of tourism experiences.

Jeroen Nawijn is Senior Lecturer at the Breda University of Applied Sciences, Netherlands. His work addresses issues of well-being, sustainable tourism, and cultural heritage. His main interest is to study areas or phenomena in which there is friction. This friction could relate to practical issues such as differing stakeholder objectives, tensions between visitors and residents, or conceptual and methodological differences in academic interpretations and applications of theories within the field of tourism.

Daniel H. Olsen is Professor in the Department of Geography at Brigham Young University in Provo, Utah, USA. His research interests revolve around religious and spiritual tourism, heritage tourism, and the management of sacred sites, with secondary research interests in tourism in peripheral areas and tourism and disabilities. He is co-editor of several books including Religion, Tourism, and Spiritual Journeys (Routledge, 2006), Encyclopedia of Religious and Spiritual Tourism (Routledge, 2021) and has published over 50 journal articles and book chapters.

Jan Packer is Associate Professor and Principal Research Fellow at the University of Queensland Business School, Australia. Jan's research focuses on applying principles from educational, environmental, and positive psychology to understand and improve visitor experiences at natural and cultural tourism attractions such as museums, zoos and aquariums, botanic gardens, national parks, ecotourism, and wildlife tourism attractions. She has a special interest in the quality-of-life outcomes of museum visits, and tourism and leisure more broadly.

J. Anthony Paredes was formerly Professor at the Department of Anthropology at Florida State University, Tallahassee, USA.

Noel B. Salazar is Professor in Social and Cultural Anthropology and Founder of the Cultural Mobilities Research (CuMoRe) cluster at KU Leuven. His research interests include anthropologies of (im)mobility and travel, heritage and tourism, discourses and imaginaries of Otherness, world anthropologies, and endurance locomotion. He has won numerous grants for his research projects, including from the National Science Foundation (NSF), the EU, the Wenner-Gren Foundation, the Research Foundation Flanders (FWO) and KU Leuven.

Stephen Schweinsberg is Senior Lecturer in Sustainable Management at the UTS Business School at the University of Technology Sydney, Australia. Stephen has an active research interest in academic knowledge formation and in the subjective and heterogeneous nature of an organisation's place-based setting. Recent research has been published in *Annals of Tourism Research*, *Tourism Management* and *Journal of Sustainable Tourism*. Stephen is on the Editorial Review Board of *Annals of Tourism Research and Tourism Review.*

Myra Shackley was formerly Professor of Culture Resource Management and Head of the Centre for Tourism and Visitor Management at Nottingham Trent University Business School, UK.

Sagar Singh is Researcher and Editor with the Centre for Tourism Research and Development, Lucknow, India.

Nikolaos Stylos is Associate Professor in Marketing and Digital Innovation at the University of Bristol Business School, UK, and Honorary Professor of Hospitality Management at the Tainan University of Technology, Taiwan. He is an internationally recognised and awarded scholar specialising in marketing innovation and digital transformation in the service industries.

Babak Taheri is a Full Professor at Texas A&M University in the Department of Hospitality, Hotel Management & Tourism. His research is characterised by a dual focus on 'consumer behaviour' and 'marketing management', with particular interest in the area of tourism, hospitality, and heritage. Babak has over 90 academic publications and currently serves as Associate Editor for the *Service Industries Journal* and *Tourism Management Perspectives*.

Jamie Thompson is Lecturer in Marketing at Edinburgh Napier University, UK. His primary research interests are in tourism management, particularly looking at international volunteer tourism from a host community perspective and the management of volunteers within host communities.

Dallen J. Timothy is Professor in the School of Community Resources and Development at Arizona State University, Tempe, USA. He holds a B.Sc. Degree in Geography from Brigham Young University, Provo, USA; an M.A. in Geography from the University of Western Ontario, Canada; and a Ph.D. in Geography from the University of Waterloo, Canada. His research interests in tourism include heritage, community-based planning, political boundaries and supranationalism, shopping and consumption, peripheral and rural regions, less developed countries, and religiously motivated travel.

Pinaz Tiwari is a research scholar in the Department of Tourism and Hospitality at Jamia Millia Islamia, New Delhi, India. Her interest areas are in tourism marketing, destination management, tourism education, and stakeholders capacity development.

Ozge Yalinay is Assistant Professor in Marketing at Heriot-Watt University, Edinburgh, UK. Her current research interests comprise contemporary cultural studies in a marketplace, with a particular focus on the Istanbul Bazaar.

Introduction

Tzung-Cheng (TC) Huan, Anestis K. Fotiadis and Nikolaos Stylos

In an era where the lines between the sacred and the secular often blur, religious tourism presents a fascinating lens through which we can examine the interplay between travel and spirituality. *Religious Tourism: An Essential Reader* ventures into this captivating domain, addressing the profound question of whether tourists embarking on spiritual journeys are genuine seekers of meaning or merely pursuing pleasure. Through a series of thought-provoking chapters, this book explores the intricate relationship between faith and travel, inviting readers to delve into the evolving landscape of religious tourism.

The concept of religious tourism is not new; for centuries, people have traveled to sacred sites and participated in pilgrimages as acts of devotion. However, in today's globalized world, this ancient practice is subject to new interpretations and transformations. The chapters in this volume, originally published in *Tourism Recreation Research*, provide a comprehensive exploration of how tourism intersects with religious and spiritual practices, offering insights into whether modern travelers are driven by profound spiritual quests or by more hedonistic desires.

Part I of the book, *Exploring the Intersection of Tourism and Religion*, sets the stage by examining the complex dynamics of pilgrimage landscapes and the role of tourism in religious experiences. The chapter by Dan Knox, Kevin Hannam, Peter Jan Margry, Daniel H. Olsen, and Noel B. Salazar provocatively questions whether tourists are secular pilgrims or hedonists in search of pleasure, setting the tone for a deeper inquiry into the motivations behind religious travel. Following this, Sean Lochrie and colleagues discuss the potential for religious tourism to embody elements of hedonism, challenging traditional notions of sacred travel.

Chris Devereux and Elizabeth Carnegie's chapter, *Pilgrimage: Journeying Beyond Self*, explores the transformative potential of pilgrimage, suggesting that such journeys often transcend personal gratification to achieve deeper spiritual fulfillment. Derek H. Alderman's analysis of authorship in pilgrimage landscapes, *Writing on the Graceland Wall*, underscores the importance of personal narratives and cultural expressions in shaping religious tourism experiences. Myra Shackley's discussion on managing Sacred World Heritage Sites highlights the delicate balance between preserving spiritual significance and addressing tourism demands.

Stephen Schweinsberg's contribution, *Complementarity: Bridging the Tourism Academic/Religion Divide*, offers a critical perspective on the academic discourse surrounding religious tourism, advocating for a more integrated approach. Dallen J. Timothy's exploration of religious heritage and tourism further contextualizes the broader implications of sacred travel, revealing how tourism can both enhance and complicate the preservation of religious sites.

Part II, *Spiritual Dimensions of Religious Tourism*, delves deeper into the restorative and transformative aspects of spiritual travel. Chelsea Gill, Jan Packer, and Roy Ballantyne examine the restorative benefits of spiritual retreats, using clergy retreats in Australia as a case study. Stephen Schweinsberg's discussion on the formation of tourism knowledge through religion and spirituality provides valuable insights into how spiritual experiences contribute to our understanding of tourism.

Yoel Mansfeld and Alison J. McIntosh's exploration of spiritual hosting and the host-guest dimension highlights the relational aspects of spiritual tourism, while Sagar Singh's anthropological perspectives offer a nuanced view of spirituality and tourism. In their joint chapter, Singh and J. Anthony Paredes explore the anthropological dimensions of pilgrimage and tourism, revealing the complex interplay between cultural practices and spiritual journeys.

Marc van Iwaarden and Jeroen Nawijn's study on the eudaimonic benefits of pilgrimage emphasizes the deep psychological and emotional rewards of spiritual travel, while Snigdha Kainthola and colleagues' research into the motivations of urban millennials for spiritual travel in India provides contemporary insights into how modern travelers engage with spirituality.

Religious Tourism: An Essential Reader is more than just an academic inquiry; it is an invitation to reflect on the profound ways in which travel intersects with spiritual quests. Written in an accessible style, this book is designed to engage a broad audience, from scholars and practitioners to enthusiasts of spiritual travel. By examining both historical and contemporary dimensions of religious tourism, this volume offers a rich tapestry of perspectives that illuminate the transformative power of sacred journeys.

We hope that this book will inspire readers to consider the deeper implications of their own travel experiences and foster a greater appreciation for the diverse motivations and meanings that shape religious tourism. As we navigate the complexities of spirituality and travel in the 21st century, *Religious Tourism: An Essential Reader* serves as a valuable resource for understanding the intricate and evolving relationship between faith and exploration.

Part I

Exploring the Intersection of Tourism and Religion

This exclusive department is aeated to include findings of special significance and to identify areas of subtle research nuances through mutual debates, discourse and discussions. Elenctic method is used wherein knowledge progresses through articulation, aoss-examination and rejection of spwious hypotheses. Thus, probe aims at encowaging scholars to think against the grain by unmasking the stereotype and dogmatic that has taken the mould of research conservatism. Contact the Editor-in-chief for *more detafls.*

Is Tourist a Secular Pilgrim or a Hedonist in Search of Pleasure?

Dan Knox, Kevin Hannam, Peter Jan Margry, Daniel H. Olsen and Noel B. Salazar

Context

The metaphor of the tourist as pilgrim and tourism as a pilgrimage has been an important idea in tourism studies, reproduced in both academic and popular accounts with varying degrees of criticality. This research probe considers a number of different ways of thinking through the degree to which tourists could be said to be either secular pilgrims or hedonists in search of pleasure. As such it considers the meanings, uses and potentialextensions of metaphors of pilgrimage and how these relate to religion, to tourism and to hedonism, as well as how all of these categories interconnect. There is no unity of approach to this question among the authors here and this on the whole makes for a lively and stimulating debate. Knox and Hannam extend the metaphor of the pilgrim into the realm of hedonistic tourism through an account of popular and mass tourist practice which considers the role of religion and spirituality as objects of tourist practice. Margry makes the case that secular pilgrim is an oxymoron and that more scholarly effort ought to be expended on identifying the limited but significant commonalities between tourism and pilgrimage. Olsen situates the discussion in relation to secularization and challenges Knox and Hannam's playful extension and multiplication of metaphors. Salazar undertakes an analysis of the emergence and development of metaphors in tourism studies to demonstrate their continued utility but also the ways in which they shape representations and understandings. The range of opinions here represents a sustained reconsideration of established terminologies.

The Secular Pilgrim: Are We Flogging a Dead Metaphor?

Introduction

It is possible to argue that the very act of engaging in the phenomenon of tourism is in itself an act of pilgrimage in which the 'faith' or 'doctrine' being reinforced is the notion of the holiday or trip as an escape, as a punctuation mark in a biography or as a rite of passage. People may disagree about which specific kinds of tourism or travel they wish to engage in, but they generally will not disagree about the more general need to engage in some form of going away whether it is motivated by a desire to engage with local cultures, climb mountains, learn to ski or lounge on a beach with a cocktail. In this probe, we seek to explore the distinction betweennotionsofpilgrimageandhedonism as well as the intersections between these definitions. In doing so, we seek to playfully extend definitions of both as well as to encourage further debate. Contemporary tourism is now so diverse that the original terms of the opposition set up between secular pilgrims and hedonists arguably no longer apply. There is hardly any difference between pilgrims and tourists (Santos 2002) and many pilgrimages are hedonistic (Norman 2011). It is no longer simply the case, if ever it were, that the tourist can lazily be assigned to either the category of seeking distracting pseudo-events or instead the intrepid traveller seeking adventure. This of course is not necessarily remarkable in that any time we set out to explore a dualism we go through a predictable process of arriving at a conclusion whereby we might reject the notion of the categories as mutually exclusive or discover instead an additional set of categories. Such is the stuff that tourist typologies are made of. What we want to do here is to suggest that it is not so much that the categories are no longer of relevance but that they have in some cases grown towards and overlap each other so that we can identify elements of the pilgrimage in hedonistic forms of tourism and hedonistic behaviour among supposed pilgrims.

Hedonistic Tourism

It is not necessary to look too hard in order to find manifestations of hedonistic behaviour in tourism. Traditionally, we would look to mass tourism in order to find examples of heavy drinking, casual sex, drug use and wild behaviour. Many authors have already done this and unsurprisingly found that British youths travelling to Mediterranean mass tourism destinations for two weeks to partake of all of these activities and more (Knox 2009; Andrews 2011). Equally, we might look to slightly older representatives of the same kind of tourist pursuit in exploring mass tourism more generally, and witness the apparent debauchery and lack of authentic cultural interest of the hordes that disport themselves on the beaches of Spain, Turkey, Greece, Bulgaria etc. each summer. Again, these groups are seen to be simply pursuing basic pleasures such as high levels of consumption of food and drink, relaxing by the pool, visiting the beach and other mild diversions. We can look elsewhere, too, and find the behaviour of extreme sports or winter sports enthusiasts, of backpackers, or of independent travellers and recognize a certain degree of risk-taking hedonism in their pursuits (Thorpe 2012). Holly Thorpe (2012: 325) for instance discusses transnational snowboarding cultures and notes that:

> *committed snowboarders often migrate to mountain towns where they find accommodation and employment, and spend several months to many years practicing, playing and perftrming in the various physicnl (for example, mountains) and social (such as, bars, cafes, shops) spaces and places.*

Note, however, that these kinds of accounts tend to lack the degree of disparagement that is reserved for the package holiday customers in the all inclusive resorts. *Apres-ski* may be as drunken, unregulated and wild as any other tourist night out, but it has never been viewed with the same disdain as what happens *apres* la plage. So far, then, we have established that some tourists very clearly fall into the camp of being hedonists in search of pleasure.

Tourists as Secular Pilgrims

That hedonists are in pursuit of pleasure suggests that the lot of the pilgrim, whether secular or otherwise, is somewhat more ascetic as they self-flagellate themselves from museum to gallery to heritage attraction. Contained within the original distinction between pilgrims and hedonists is a moral judgement about the relative worth of such choices which is made not from the position of religious authority but from both the privileged position of the academic and that of the well-travelled and self-styled enlightened commentator. Surely, there is significant pleasure to be derived from tourism as pilgrimage? Studies of tourist motivation attest to the pleasure that attaches to visiting heritage sites, sporting venues, wilderness landscapes, 'exotic' cultures or urban centres. It is simply that these are different kinds of pleasures, that they are imagined to be more refined or to represent a 'better' class of tourist, and that they may form part of a programme of education or self-improvement as in many forms of volunteer tourism (see Conran 2011). They are also, by and large, the pleasures that appeal to the academic or popular commentator on the tourism industry. The idea of the secular pilgrim is probably more reflective of the views of tourism academics rather than of tourists themselves. The same moral distinction noted above applies here and can be characterized as a sense of self satisfaction.

There is a certain irony that churches and other religious sites are such a focal point of cultural and heritage tourism practice in the Christian world. Cathedrals, basilicas, abbeys, minsters, priories and so on are not entirely stripped of their original purpose or meaning but rather re-oriented towards a different kind of pilgrimage of a secular nature. The act of pilgrimage is no longer always a component part of a Christian lifestyle but rather a pilgrimage to remember Christians and Christianity of the past (see, for example, Coleman 2007). This is not to belittle the faith or pilgrimages of practising Christians visiting such sites, but simply to point out that for the majority the practice of visiting such sites is as much about architectural wonder, historical interest or curiosity about the behaviour of adherents. The *Rough Guide to Vietnam* (2013) includes in an entry relating to Notre Dame Cathedral in Ho Chi Minh City the following account:

> *The attractive redbrick bulk of the late nineteenth-century Notre Dame Cathedral straddles the northern reach of Dong Khoi. Aside from the few stained-glass windows above and behind its altar, and its marble relief Stations of the Cross, the interior boasts only scant decoration. There's plenty of scope for people-watching, however, as a steady trickle of Catholics pass through in their best silk tunics and black pants, fingering rosary beads, their whispered prayers merging with the insistent murmur of the traffic outside.*

The value of the church as a tourist attraction is based upon not only physical appearance and decoration but the apparently mystical and spectacular behaviour of the faithful. The implicit peculiarity suggested in this travel guide in relation to the behaviour of other visitors to the site is indicative of a relatively dismissive attitude towards believers in contemporary society. Or, rather, it is indicative of an attitude towards particular kinds of 'believers' in particular contemporary societies. It is not that the readers of the *Rough Guide to Vietnam* do not have orienting beliefs, but that they believe in something else. That something else, or an element of that something else, might be characterized as a belief in the power and value of travelling, of undertaking secular pilgrimages, visiting exotic places and 'people-watching'.

Extending the pilgrimage metaphor further, there are undoubtedly sets of central beliefs that act as social rules, norms and conventions within specific tourist sectors.

These beliefs differ according to the nature of the tourist encounters, motivations and practices of ecotourism, mass tourism or adventure tourism. There is a sense in which such faith is more about the act of travelling or touring than about the objects, cultures, places or peoples toured or visited. Thus, for example, we would be mistaken to overstate the devotion of the average pilgrim to sites associated with Dali, Picasso, Shakespeare, Kafka, Joyce or any other representative of esteemed cultural forms. These high cultural forms potentially encourage us to make the mistake of elevating such practices above those where the nature of the engagement is with more popular or populist forms of culture such as the world of entertainment. The beliefs are that visiting sites associated with such culturalfigures is in some way self improving, ought to be a part of an itinerary (the guide book says so) and indicates a degree of cultural sophistication. This, in short, is Bourdieu's (1984) cultural capital thesis.

Now, if such acts are secular pilgrimages, to what extent are they characterized by a sense of devotion to the artist, political figure or historical person represented? Visiting a Picasso museum in Malaga or in Paris does not require a lifelong devotion to the oeuvre of Picasso or a deep understanding of his biography, the differing natures of the periods of his outputs or the underlying philosophies and contexts of the Cubist movement. It simply tells us that the individual is a member of a social, cultural, national or international group within which Picasso is considered a worthwhile artist. Equally, few visiting Prague's Franz Kafka Museum will have dedicated much of their lives to reading his novels or short stories and will instead be working from an understanding that Kafka is a major literary figure of international importance. Many of those visiting, indeed, probably do not want to visit at all but are simply being taken along by somebody else who does want to visit. Neither is this to say that there are not Kafka or Picasso devotees among the cultural tourism hordes: nor is it to denigrate the desire to learn as a part of the cultural experience of tourism. We make these observations simply in order to avoid the trap of elevating certain cultural forms over others and to demonstrate that cultural tourism is not necessarily any more thoughtful, reflexive or meaningful than beach tourism, sun seeking or cruising.

So, cultural tourists are not always particularly devoted to an artist, an historical period, an architectural movement or a physical place. A greater degree of devotion and dedication can perhaps be found in the realm of practice that has been termed creative tourism (Richards 2011; Hannam and Knox 2010). At the intersection of serious leisure, tourist practice and lifestyle, participation in acts of creativity during a vacation can be seen as the culmination of a broader set of practices lived out at home. In this context, we see something of the temporal aspect of pilgrimage in the irregular departure from home to participate in acts of great significance, to partake of ritualistic behaviour and to reaffirm a connection. Creative tourists or serious leisure participants engage in pilgrimage-like behaviour in relation to a very wide-ranging set of interests - golf, pottery, landscape photography, cookery, language learning etc. The breadth of these definitions and the relationship they establish between everyday and tourism means that we can even incorporate not only secular pilgrimage within them but also traditional faith-based pilgrimage. At the opposite end of this spectrum, it would also be possible to imagine pilgrimages based upon hedonistic behaviour and the opportunity to behave in a similarly hedonistic style as at home but for a longer period of time, with a different group of co participants or simply in a place with a hedonistic reputation. Indeed, Jansson (2002) has argued that through mediatization, a shift has come about from realistic hedonism to imaginative hedonism, where the former relies on bodily pleasures and the latter on emotional and spiritual pleasures. But, again, such a distinction can easily be overturned when we examine hedonism in pilgrimage more closely.

Hedonistic Pilgrimages

The distinction between the sacred and the profane becomes inherently less important as society becomes ever more secular in Europe, North America and elsewhere. We have argued elsewhere that cultural capital and the idea of serious leisure can apply in the setting of any kind of tourism including mass tourism, youth tourism or hedonistic tourism and we want to build on this understanding here to investigate the ways in which the polar opposites of hedonism and pilgrimage have grown together as the tourist market has diversified and become ever-deeper embedded in contemporary societies (Hannam and Knox 2010). In certain settings, the mass tourist is undertaking a pilgrimage of a sort to exactly the same extent as the cultural or any other kind of tourist might be (see Obrador et al. 2009). For example, the mediatization of tourist landscapes and particular resorts creates motivations, aspirations and desire to visit those resorts and live the lifestyle represented in popular television programmes, in magazines and newspapers, or on websites. The cultural image of somewhere like Benidorm, for example, is refracted through word of mouth reports from associates, television drama, holiday brochures, online reviews and photographs etc. In addition, we need to note that in particular social groups the desire to visit such a resort and to partake of mainstream tourist behaviours constitutes a secular pilgrimage.

Furthermore, it has been suggested elsewhere that this kind of relationship applies to a greater extent in the context of youth tourism, particularly where it is motivated by a special-interest in musical forms or clubbing culture (Knox 2009). Thus, visiting Ibiza to experience the annual clubbing season is a pilgrimage for the individual who usually visits similar sorts of clubs and consumes the same sorts of music at home each weekend. It is a temporal punctuation mark within the general flow of serious leisure. We can also ascribe similar sorts of motivations and connections for extreme sports enthusiasts who learn to snowboard at home and visit ski resorts each year on vacation (Thorpe 2012). Who are we to say that visiting a Manumission party or snowboarding a particular run is not as significant as visiting the birthplace of Mozart, the grave of Karl Marx or wandering around the Kremlin? The quest in this kind of pilgrimage might not be for anything recognizable as sacred within conventional definitions but we can certainly point to these activities being rites that carry significance within distinct, bounded cultural communities. There is certainly a search for authenticity within these activities in the same way that there is in the activities of the culture or creative tourist.

Cultural and Creative Tourism as Orgies of Consumption

Tourist typologies would tend to represent cultural tourists as being motivated by authenticity, by cultural engagement and the opportunity to learn something new as a part of their tourist encounters (Cohen 1979). The verbal dexterity demonstrated in applying, stretching and murdering half dead metaphors has not yet come to a conclusion and we now move to explore the similarities between cultural and mass tourism in terms of motivation, behaviour and pilgrimage. The above section has shown that we can see correspondences between the behaviour of hedonistic tourists and our understandings of pilgrimage, and we now move to look for manifestations of hedonism in other forms of tourism. We should perhaps take a moment here to reflect on the fact that cultural tourism is no longer clearly separate from mass forms of tourism in that it does not represent a very significant market that continues to grow (Obrador et al. 2009). It is probably more likely that a tourist would find himself in a crush of people when visiting the Louvre, the Vatican or other cultural attractions as in visiting Mediterranean sun destinations.

Manifestations of hedonism in the contexts of cultural and creative tourism are clearly visible even if we usually decide not to talk about them. They may not quite be 'the elephant in the room', but they certainly constitute a trunk sticking through the door. Firstly, the holiday as a period of increased activity and signification points to a rapid uptick in consumption over a relatively short period of time. While this may not be a wild or unrestrained process, there is a sense of hedonism about the content of packed itineraries and tours that provide more information, education and potential self improvement than most individuals would reasonably be able to absorb in that time period. Secondly, there is an increase in alcohol and food consumption as a part of holidays of all kinds and the sorts of situational dis-inhibition that this might lead to, whether or not the drunkenness that results is of the respectable middle-class kind. Even something as apparently benign as the tendency of cultural tourists visiting Amsterdam to engage in the smoking of cannabis, or visitors to Thailand to consume live sex or striptease acts point to a hedonistic pursuit of pleasure at the core of apparently pilgrim-like vacations. Thirdly, and possible most crucially, as flagged earlier, is the sense that the cultural, creative or any other kind of tourist is to a large extent motivated by a degree of expected pleasure regardless of which activities provide them with that pleasure. All are travelling abroad to partake of pleasures they probably would not participate in at home to the same extent - whether these pleasures are based around booze, beaches or Bauhaus architecture.

Widening the Debate

We do not pretend to have answered the question originally posed at the opening of this piece, but instead seek to participate in a broader debate about the utility and fit of the terms secular pilgrim and hedonist in 21•t century tourist studies. Undoubtedly, many will disagree with our particular take on this debate and we welcome that as much, if not more, than simple agreement. Indeed, we look forward to reading the responses to the issues we have addressed, and to the challenging of our conceptions.

We ought to recognize within this discussion, however, that this is a highly situated, contingent and Eurocentric debate that fails to take account of tourist practices outside of the Western World or to consider societies within which shared faith in the form of organized religion remains important or even shows signs of growth. Does this debate have the same resonance in Islamic cultures, or in China, India, Sub-Saharan Africa or indeed anywhere outside the historical heartlands of tourist studies? As tourist studies internationalizes, we would expect to see accounts of different practices and from different perspectives, or may be not. Does this debate make sense, for instance, among the masses of people on organized tours to view folk villages and ethnic cultures in China? Undoubtedly, participation in tourism in Chinese society relates to the accumulation of cultural capital but is this accrued on the same basis as it is in European societies? Do the same outlets for hedonistic tourism exist within other tourism marketplaces? Club 18–30-style booze and sex holidays simply are not an option for some consumers, although gambling, drinking and prostitution exist in all tourism markets and are consumed by some participants in those markets.

There remains much more to be unpacked in terms of the religious imagery of the metaphor of the pilgrim and questions as to the extent to which we believe that all acts of historical pilgrimage are motivated by a deep sense of devotion rather than a degree of self-interest, self-preservation or self improvement. Pilgrimage as an earlier form of mobility is very clearly an ancestral behaviour to the diverse tourism industry of the early 21st century but this does not mean that we need to remain fixated by the explanatory power of the concept itself if we should find that has become redundant. If society changes, behaviour changes and lifestyles change, we should accept that the ways we categorize and think about cultural phenomena should also change in response. So, we have argued that while the twin concepts of pilgrim and hedonist continue to have some utility they are not sufficiently nuanced or all-encompassing to describe, let alone explain, the totality of contemporary tourist practice.

References

ANDREWS H. J. (2011). Porkin Pig Goes to Magaluf. *Jownal* of *Material Culture* 16: 151–170.

BOURDIEU, P. (1984). *Distinction: A Social Critique* of *the Judgement* of *Taste.* London. Routledge.

COHEN, E. (1979). Rethinking the Sociology of Tourism. *Annals* of *Tourism Research* 6: 18–35.

COLEMAN, S. (2007). A Tale of Two Centres? Representing Palestine to the British in the Nineteenth Century. *Mobilities* 2(3): 331–345.

CONRAN, M. (2011). They Really Love Me! Intimacy in Volunteer Tourism. *Annals* of *Tourism Research* 38(4): 1454–1473.

HANNAM, K. and KNOX, D. (2010). *Understanding Tourism: A Critical Introduction.* London. Sage.

JANSSON, A. (2002). Spatial Phantasmagoria: The Mediatization of Tourism Experience. *European Journal* of *Communication* 17(4): 429–443.

KNOX, D. (2009). Mobile Practice and Youth Tourism. In Obrador, P., Crang, M. and Travlou, P. (Eds) *Cultures of Mass Tourism: Doing the Mediterranean in the Age of Banal Mobilities.* Farnham, Surrey, UK. Ashgate: 143–156.

NORMAN, A. (2011). *Spiritual Tourism: Travel and Religious Practice in Weslmt* Society. London. Continuum.

OBRADOR, P., CRANG, M. and TRAVLOU, P. (Eds) (2009). *Cultures* of *Mass Tourism.* London. Ashgate.

RICHARDS, G. (2011). Creativity and Tourism: The State of the Art. *Annals* of *Tourism Research* 38: 1225–1238.

ROUGH GUIDES (2012). *Rough Guide* to *Vietnam.* London. Rough Guides.

SANTOS, X. (2002). Pilgrimage and Tourism at Santiago de Compostela *Tourism Recreation Research* 27(2): 41–50.

THORPE, H. (2012). Transnational Mobilties in Snowboarding Culture: Travel, Tourism and Lifestyle Migration. *Mobilities* 7(2): 317–345.

Submitted: April 3, 2013
Accepted: May 1, 2014

Continued ....

Whiskey and Pilgrimage: Clearing Up Commonalities

Peter Jan Margry

I have previously argued that the notion of 'secular pilgrimage' is an oxymoron (Margry 2008: 1–2). The two elements which comprise this concept are contradictory: pilgrimage is inherently a religious/ spiritual activity. Postulating a secular pilgrimage (or pilgrim) therefore introduces a vague, unscholarly notion. By analogy, the word pilgrimage has become popular as a term for located practices in which people go somewhere without a religious or spiritual motivation. The notion is then usually applied in a metaphorical way. To clarify this confusing usage, we might also apply a profane metaphor: if we define whiskey as an alcoholic beverage, then secular pilgrimages can be described as alcohol-free whiskey. But can we then still speak of whiskey? Thus, for example, it is misleading to state as Knox and Hannam do in various phraseologies, that tourists who visit religious buildings and sites perform a pilgrimage of a secular nature while gazing at the architecture or peering at the devotional practices inside. Why call it pilgrimage if such behaviour cannot be accounted for as pilgrimage? Is it simply the lure of an intriguing word replete with expressive connotations, or is it an exercise in presenting superficial analogies, without demonstrating basic commonalities?

From a scholarly perspective both possibilities would imply an unsatisfactory process of analysis. This is also the case for the assignment for the research probe of this issue of Tourism Recreation Research: 'Is tourist a secular pilgrim or a hedonist in search of pleasure?' This double question, including the oxymoron, is even partly redundant on its face, as modem tourism is often equated with hedonistic endeavour for pleasure and relaxation. It remains unclear why the issue of hedonism is included in this probe, unless it conveys the moral stance of the authors which surfaces in the text.

To my mind the central issue actually at stake is the question of whether modem tourists are engaged in a religiously inspired journey. In other words: is tourism a sort of religion and does pilgrimage also express forms oftourism? Itis necessary then to define pilgrimage: a journey out of religious or spiritual inspiration, consciously undertaken by individuals or groups, to a place or along a track that is regarded as more sacred or salutary than the environment of everyday life, to seek a transcendental encounter with the sacred for the purpose of acquiring spiritual, emotional or physical healing or benefit.

With the rise of modern mobility and travel the notion of tourism has been connected to pilgrimage research for some decades now. It was thought that tourism, with its seemingly comparable characteristics, could generate new insights into the operation of pilgrimage, and vice versa. The French anthropologist Alphonse Dupront was the first scholar to put this theme on the research agenda. He sought to investigate how human collective psychology relates to both social mass phenomena and their capability to 'liberate' people and widen their horizons (Dupront 1969: 119–121). His comparison was not convincing, as he neither evaluated the presence of the 'alcohol' nor realized that in se pilgrimage is basically an individual practice (Margry 2012: 282).

It was the American anthropologists Victor Turner and Edith Turner who in their influential book Image and Pilgrimage remarked that 'a tourist is half a pilgrim, if a pilgrim is half a tourist' (1978: 20). This frequently quoted assumption has become proverbial and hence usually cited as the 'proof for the similar genes of tourism and pilgrimage. Along with a striking rise in mass tourism and renewed interest in the phenomenon of pilgrimage, this interesting issue was picked up by researchers all over the world. Cohen, for example, argued that secularization robbed pilgrimage from its symbolic and mystical power and therefore sees tourism as a modem metamorphosis of pilgrimage and travel (Cohen 1992: 58–59). The present day world shows that nothing is less true. Comparative research is however important and relevant, as long as the basic intrinsic differences - constitutive for both phenomena - are not overlooked or ignored, which otherwise, as a consequence, would subvert research results.

It would be short-sighted not to acknowledge that certain human behaviours can indeed commonly be identified in both tourism and pilgrimage. But it would also be inexpedient to allow derivative aspects of the phenomena to prevail over basic distinctions, as it usually happens. Throughout the centuries the repertoire of pilgrims' collateral behaviours has been comprehensive, and has included aspects of 'tourism' as well, notwithstanding that pilgrims always had one basic, existential, religious motif: arriving at the sacred (shrine). This quest did not exclude that underway they performed practices related to 'tourism', whether they were going to Jerusalem, Santiago or Lourdes, as modem research confirms.

In the past decades various pilgrimages - for example to Amsterdam, Lourdes, Santiago and Wittem (NL) - have been the subject of sociological research based on multiple-choice questionnaires aimed at collecting data on motivational repertoires (Post et al. 1998: 19–48, 173–203). One reason for choosing such a non-qualitative method was that most pilgrims find it quite difficult to formulate their motives. This difficulty characterizes not only traditional Catholic pilgrims to Lourdes or Fatima, but is equally applicable to more 'elitist' present-day journeys to Santiago de Compostela and other spiritual walking routes (Margry 2014). Nevertheless, in the research performed the presence of a wide variety of 'motives' mentioned by the pilgrims is striking. The number of motives rises as high as 20 when multiple-choice questions were used by social scientists. Applying this method proved to have strong steering effects and leads to facile checking in the line up of possible motives. All sorts of secondary touristic, cultural or social aspects of a pilgrimage were included by the researchers, and were subsequently dutifully ticked by the respondents. This fed the researcher's notion that pilgrims are actually more like tourists, as they simply expressed that they indeed liked the beauty of the scenery, bought commercially produced souvenirs, gazed at cultural heritage and admitted that they could also enjoy the sociability of a collective journey (cf. Post et al. 1998: 157–242). It is obvious that all sorts of secondary motives do come in the picture when they are placed next to the main objective, and thus obscure the central religious or spiritual incentive forgoing on a pilgrimage. Using the method of open questions, by which pilgrims themselves have to reflect and formulate their motivation, produces answers of a more unalloyed and reflective nature, focusing on the pilgrim's main objective (Margry 2008, 2014). These answers reflect the principal motives and objectives people have whenthey do a pilgrimage.

Elaborating on Turner and Turner (1978), the anthropologists Badone and Roseman tried to reconcile the conceptual dichotomy between religious travel and secular journeying or tourism. They rightly observe, 'Rigid dichotomies between pilgrimage and tourism, or pilgrims and tourists, no longer seem tenable in the shifting world of postmodem travel' (Badone and Roseman 2004: 2). This is part of what Badone and Roseman call 'intersecting journeys'. This concept does not imply that tourism and pilgrimage are interchangeable. For Badone and Roseman intersections only come to the fore when tourists allow themselves to be carried away - intentionally or unintentionally- by sacred, transformati.ve experiences of the pilgrimage or at the shrine, and actually become pilgrims.

The presentation of Badone and Roseman's altered perspective on pilgrimage was stimulated by the fact that the new phenomenon of inspirational travelling on old pilgrimage roads had come into being. Since somewhere in the 1960s the traditional pilgrimage model has been re-enacted in a seemingly recreational manner, its re invention of the Camino to Santiago created a new format, a geme of which it is both the precursor and example par excellence. This development created a new, non institutionalized manner of pilgrimage, to one degree or another separate from formal religion. The Camino is now partly a Catholic pilgrimage and partly a 'new-agey' or spirituality pilgrimage. The revival and popularity of walking the Camino to Santiago has resulted in the creation of hundreds of 'historical' and religious/spiritual roads and tracks all over the world. Did this change then result in a 'touristic pilgrimage' or, if we recall my earlier metaphorical comparison, in 'whiskey light'? What is actually the case?

Apart from the walking and sport enthusiasts, the fans of tranquility and culture and (art)history lovers, Santiago was discovered and reinvented by spiritual seekers. For many walkers the journey along the Camino has become an individual rite of passage, or 'a pilgrimage to one's self'. Trekking along the European Camino network can thus be described as the practice, inspired by history and tradition, of creating a religiously open ritual format for oneself for reflecting on one's life and possibly, depending on a variety of spiritual values, for experiencing transcendence in a certain manner. For such individuals, it functions as a pilgrimage. In this way, walking the pilgrim trails has become an inward-orientated activity for questioning oneself or giving meaning to one's life in relation to the world. These tracks created a possibility to move, to walk, but not so much for arriving at a destination where one would find a shrine. Modem, 'unchurched' times have created an alternative for those who do not want to participate in pilgrimages of institutionalized religions. For those who operate within an eclectic religious paradigm, an open manner of overcoming personal problems is to do a pilgrimage on such spiritual tracks.

It is not only the Camino network that is populated by groups which exhibit different practices and discourses next to one another. Also in other seemingly 'secular' or mixed places one should always take into account that there are multiple motif structures present. The graves of Elvis Presley and Jim Morrison, for example, are both tourist attractions where mass tourism is manifestedly present. The visits to Graceland and Pere Lachaise are however much more diversified than it might appear at first sight. Next to the millions of fans and tourists there exist specific groups of visitors for whom a religious (pilgrimage) factor is in fact the leading motive, employing narratives which are very distinct from those of the fans and tourists. This then makes it clear that ethnography is necessary to clarify whether people who visit specific locations have a religious motive or 'just' follow the secular or tourist circuits, and are thus not involved with the 'essence' of pilgrimage. Often things are not what they seem. For example, despite the fact that the Morrison fan-tribe which was camped out around his grave consumed many bottles of highly spiritous 86 proof Jack Daniels and were depicted as pilgrims, they proved in fact to be very distinct from a group of individuals who had a religious or spiritual motive, who invoked Morrison in a transcendental way and actually constituted a pilgrimage (Margry 2008). Graceland or Pere Lachaise are thus tourist destinations, but also represent-if only for a specific group - a shrine which is not secular at all, and to which they thus make a pilgrimage.

And so let us look again at the stance of Knox and Hannam. I think it is problematic to start a scholarly debate on concepts by 'playfully extend[ing] definitions', as they write. In this way, they suggestively launch into the metaphorical mode by equaling the escapism of holiday with 'faith' (their quotation marks!), thus ignoring the aspect of religion: 'there is little difference between pilgrims and tourists', they bluntly state. Just as easily one could argue that there is no greater difference than between a pilgrim and tourist, maybe apart from the fact that both categories have to come a certain distance to arrive at their destination. Even travelling is a relative concept in this context, as although among the public pilgrimage is best known for its renowned long-distance sanctuaries, quantitatively speaking most pilgrimages are at short or middle-long distance from the pilgrims'point of origin. Tourism too is varied, local, international and intermediate. In both cases, the exotic and far away often dominate the spotlight of the international media. Thus, when dealing with behaviour, functions and meanings the differences between the two categories become obvious. The basic mental predispositions are quite opposite. The pilgrim goes to deepening his or her faith or intensify his or her relation to god/ the sacred, and/ or (s)he is engaged in an existential endeavour related to physical or mental problems or support In that regard-contrary to what both authors suggest - I would stress the importance to keep up categories as much as is useful and possible, certainly related to the domain of religion.

But, of course, it is also not that easy. A major difficulty is that hardly any expression of human activity is unambiguous. As having a job in daily life also means drinking coffee, having lunch, buying objects, experiencing the urban or rural environment, socializing, gossiping, private emailing and checking web news, similarly pilgrims experience all kinds of secular encounters under way or after arrival. They indeed can enjoy the natural and cultural beauties they see while under way, or engage in recreational and socializing activities, aspects that explain why so many boxes on secular motives were checked on the multiple-choice forms. And if they do that, what can then be distilled from that knowledge? Various scholars call them both pilgrim and tourist. However, to keep to clear definitions it is preferable to describe them as pilgrims with additional interests. They do not have a touristic motive, but while on pilgrimage they participate in collateral experiences of a social, cultural, recreational or touristic nature.

Allowing various categories to overlap makes analysis problematic in general. Without hardly any definition of terms, in their probe Knox and Hannam produce a dysfunctional bricolage, and not only on pilgrimage. How, for example, is hedonism to be understood: in a philosophical way or as a practical container concept for people who like to drink and have sex during holidays? The latter could become quite a category if these are the two criteria. The two week holidays of British or Dutch youth to mass tourist destinations are neither form of hedonism; they merely engage in a (yearly) holiday within the reach of their budget and social peers, in which all the experiences of a vacation must take place. The moral tone in Knox and Hannam's text is another questionable way of addressing the topics. How can one make a neutral analysis by using descriptive cliches and derogatory terminology like 'debaucheries' or 'lack of authentic cultural interest of the hordes'? Or did I miss the irony?

In conclusion, I must reiterate that without defining and delineating, research cannot be done in a proper way. Therefore Knox and Hannam's stance is a mistaken one in which basically differing categories are mixed up, resulting in an obfuscating bricolage instead of providing better insight, which should be the major endeavour of scholarly research.

References

BADONE, E. and ROSEMAN, S. R. (Eds) (2004). *Intersecting Journeys.The Anthropology of Pilgrimage and Tourism*. Urbana University of Dlinois Press.

COHEN, E. (1992). Pilgrimage and Tourism: Convergence and Divergence. In Morinis, A (Ed) *Sacred Journeys: The Anthropology of Pilgrimage*. Westport. Greenwood: 47–61.

DUPRONT, A (1967). Tourisme et Pèlerinage: Refléxion de Psychologie Collective. *Communications* 10: 97–120. MARGRY, P. J. (2008). *Shrines and Pi1grimage in the Modern World: New Itineraries into the Sacred*. Amsterdam. Amsterdam University Press.

MARGRY, P. J. (2012). European Religious Fragmentation and the Rise of Civil Religion. In Kockel, U., Frykman, J. and Craith, M. (Eds) *A Companion to the Anthropology of Europe*. Malden/Oxford. Blackwell Wiley: 275–294.

MARGRY, P. J. (2014). To Be or Not to Be, ... a Pilgrim? Spiritual Pluralism on the Finisterre Trail. In Groschwitz, H and ScMne, A (Eds) *Religiosität und Spiritualität. Fragen, Kompetenzen, Ergebnisse*. Miinster. Waxmann.

POST, P., PIEPER, J. and UDEN, M. V. (1998). *The Modern Pilgrim: Multidisciplinary Explorations of Christian Pflgrimage*. Leuven. Peeters.

TURNER, V. W. and TURNER, E. (1978). *Image and Pflgrimage in Christian Culture*. Anthropological Perspectives. New York/Oxford: Columbia UP/Blackwell.

Submitted: June 11, 2013
Accepted: May 5, 2014

Continued....

Metaphors, Typologies, Secularization, and Pilgrim as Hedonist: A Response

Daniel H. Olsen

Introduction

In recent years academics have been drawn to the multiple crossroads between tourism and pilgrimage (read: religion), which has led to a growing body of research regarding this tourism niche area from a variety of disciplinary lenses. To this rapidly growing literature base we can add the provocative research probe by Knox and Hannam which is in essence a re-visitation of discussions related to the similarities and differences between pilgrims/tourists and pilgrimage/ tourism through attempting to "playfully" broaden the definition of pilgrimage and pilgrims into the realms of both hedonism and secular travel.

While the authors touch upon a number of points in their research probe that warrant comment, I wish to focus on four critical issues that arise from this paper. I first discuss the way in which religious metaphors are used in this paper, and then address the authors' discussion of typologies in the context pilgrim/ tourist and pilgrimage/ tourism dichotomies. I then critique the implications of the implicit use of the secularization theory in this research probe, and finally address the contention that, in my words, 'the hedonist is half a pilgrim, if a pilgrim is half a hedonist', before concluding.

Metaphors

Scholars have long used metaphors, analogies and similes in tourism studies as a 248 part of the 'language of tourism' (Dann 2002, 2005). For example, terms such as 'tourism as play', 'tourism as sacred journey', and the 'tourist as child' are used to highlight what tourism 'is like'and how tourists and tourism can act as metaphors of the social world (Cohen 1985; Graburn 1989; Dann 2002, 2005; Sharpley and Sundaram 2005). As a part of this 'language of tourism', some scholars have borrowed religious terminology to metaphorically compare tourism to a 'sacred quest', sacred journey' or a religious pilgrimage; where tourists, like religious pilgrims, travel in search of quasi-religious authenticity, truth, and self-actualization through participation in the 'ritual' of sightseeing and in the process of sacralizing tourist attractions and places (MacCannell 1976; Grabum 1989). This view of tourism and pilgrimage-what Cohen (1992) terms the convergence perspective-has come about in part because of the contention that as travel solely for religious reasons (i.e., medieval pilgrimage) began to give way to more multi-functional and multi motivational trips (i.e., modem tourism) after World War II there began to be both a structural and a theoretical 'blurring of the lines' between pilgrimage and tourism (Kaelber 2002, 2006). The convergence of these two 'travel styles' (Adler 1989) has encouraged scholars to metaphorically transpose the image of the religious pilgrim onto the modem tourist and, subsequently, the act of religious pilgrimage onto the act of travel.

Throughout their research probe Knox and Hannam attempt to utilize a number of additional religious terms-such as 'quest', 'ritualistic behaviour', 'faith', 'belief, 'rite of passage', and 'doctrine'-as metaphors in an attempt to convince readers of the close relationship between hedonistic tourism and pilgrimage. For example, they state that '...the very act of engaging in the phenomenon of tourism is in itself an act of pilgrimage in which the 'faith' or 'doctrine' being reinforced is the notion of the holiday or trip as an escape...a rite of passage' (p. 236). Another example is found on page 237, where the authors state: 'That hedonists are in pursuit of pleasure suggests that the lot of the pilgrim, whether secular or otherwise, is somewhat more ascetic as they self-flagellate themselves from museum to gallery to heritage attraction'.

Unfortunately, the authors seem to feel that the 'playful' use of religious terminology throughout their research probe will lend credence to their arguments regarding the close ties between pilgrimage and hedonistic forms of tourism. This, however, is not the case. For example, based on these two quotations one could rightfully ask: How is the notion of the holiday a part of someone's 'faith' or 'doctrine'? What 'faith' or 'doctrine' is being taught, who teaches this 'doctrine', and how is this 'doctrine' taught? Whose 'faith' is being reinforced? What is it exactly people who travel should have 'faith' in? In what ways is the holiday considered a 'rite of passage'? What is the 'rite' that is being performed? What are the rewards for completing this 'rite'? How do pilgrims, 'secular or otherwise', 'self-flagellate themselves', and do they really do this as they travel to museums, galleries, and heritage attractions?

The problem with the uncritical use of religious terminology as metaphors to describe tourism-related activities and experiences stems from the fact that 'religion has its own specialized language' to help believers 'talk...about the world and about life as [it is] really experienced, beyond the prejudices of the modern, secular life' (Heintzman 2011: 5–6). As such, one must consider the difficulties that can arise in merging the 'language of tourism' with the 'language of religion' and how terms translate from one 'language' to another. For example, can all religious terminology be equally utilized as metaphors in describing the relationships between tourism and pilgrimage? Will readers automatically see the metaphorical connection between pilgrimage and tourism when using such terminology? Will the terminology used be synonymous in the minds of readers?

To me, if religious terms are going to be used to justify a particular blurring of tourism and pilgrimage, some sort of explanation and/ or definition must accompany these terms if they are to be appropriately used as metaphors. For example, is 'faith', as understood in a religious context, really the same as the holiday as an act of 'faith' or as a part of someone's 'faith'? Even if the metaphor is extended, as some scholars have done, to include'tourism as religion' (Selwyn 1994; Burns 1999; see Sharpley 2009) or even 'religion as tourism' (Grimshaw 2001), this does not mean that all religious terms are inherently appropriate to use or are transferable when comparing tourism andpilgrimage travel.

Typologies

In their research probe Knox and Hannam critique the way in which typologies differentiate the pilgrim from the hedonist.They argue that the traditional differentiation between pilgrims and hedonists stems in part from tourism scholars who, from their privileged position within academia, have (purposefully?) succeeded in labelling certain types of tourism behaviours and activities as more 'refined' than others, and that certain classes and types of tourists are 'better' than others. As such Knox and Hannam argue that'original distinctions between pilgrims and hedonists [tourists?]' was based on the'moral judgment[s] about the relative worth of such choices' by academics and 'well travelled and self-styled enlightened commentator[s]' (p. 237).

Knox and Hannam suggest that this moral interference by academics regarding the'original distinctions between pilgrims and hedonists' is outdated in the modem world where dualisms in general do not apply. They argue this is the case because '...any time we set out to explore a dualism we go through a predictable process of arriving at a conclusion whereby we might reject the notion of the categories as mutually exclusive or discover instead an additional set of categories' (p. 236). Typologies, they argue, tend to 'grow...towards and overlap each other' (p. 236), and as such it should be no surprise that 'we can identify elements of the pilgrimage in hedonistic forms of tourism and hedonistic behaviour among supposed pilgrims' (p. 236).

Knox and Hamman are correct in their view that when it comes to pilgrim-tourist comparisons that 'the tourist [has been] lazily...assigned' to either one or the other category (p. 236). As I have more extensively outlined elsewhere (Olsen 2010), past studies comparing pilgrims and tourists have typically been done through the lens of Max Weber's 'ideal types', contrasting the 'ideal pilgrim' (the medieval aesthetic Christian pilgrim) with the 'ideal tourist' (described by Cohen's (1979) as a diversionary tourist or a 'superficial nitwit'). In comparing the 'ideal pilgrim' with the 'ideal tourist' scholars have ignored the fact that 'individuals rarely fit into tidy categories such as "pilgrim" and "tourist"' (Olsen 2010: 849). As Maoz and Bekerman (2010) note from their study of Israeli backpackers in India, there are many aspects of travel that cannot be classified as either 'pure' travel or as typical pilgrimage or religious tourism. Therefore they question the analytical value of typologies as fixed categories. Indeed, any postmodemist or post-structuralist scholar would argue that typologies can no longer be viewed from their modernist perspective as bounded and fixed categories; rather, the boundaries between types should be seen as malleable and contingent on changing cultural and personal circumstances (Tilley 2006). As such these scholars would have difficulty accepting anachronistic and abstract universal travel typologies that place types of travel and travellers into static 'everything/nothing' categories (see Adler 1989; Bauman 1996; Uriely 2005, 2009).

Knox and Hannam, however, take their criticism of typologies a step further and suggest that the terms 'pilgrimage' and 'pilgrim' may not be suitable metaphors for describing tourists and tourism in the modem era (see p. 241), and that 'while [the] twin concepts of pilgrim and hedonist continue to have some utility they are not sufficiently nuanced or all-encompassing to describe, let alone explain, the totality of contemporary tourist practice' (p. 241). As such, Knox and Hannam argue that the terms 'pilgrimage' and 'pilgrim' as metaphors may be seemingly'redundant'.

Their concern that 'pilgrimage' is an outdated metaphor for modem tourism is not theirs alone, however. Bauman (1996) for example, has argued that the pilgrim is more appropriately a metaphor for modernity rather than for post-modernity, in part because in the context of identity building the pilgrim represents identity-maintenance rather than identity-building. He argues that the pilgrim has already elected his/her point of arrival, their life-course is set, and present matters are sacrificed for future promises; whereas postmodem consumers refuse to have their life-course set and mortgage their present for future rewards. Therefore, Bauman argues that the term 'pilgrim' is no longer a metaphor for identity-building in modern life. Instead, he suggests four different metaphors that suggest lifestyles that the majority of people today practice: the stroller, the vagabond, the tourist and the player. As such, one might wonder which of Bauman's metaphors Knox and Hannam would consider being the most appropriate for the hedonist tourist, however they are defined. In addition, Maoz and Bekerman (2010: 424) have also suggested that 'pilgrimage' and 'religious tourism' are 'no longer suitable for examining the touristic experience, which we believe has now become complex, full of contradictions and in other words-postmodem'.

One needs to keep in mind, however, that objections over the use of the terms 'pilgrimage' and 'pilgrim' as metaphors for tourism and tourists are intertwined with the more general tensions between modernist and postmodernist theories of tourism and market segmentation (i.e., differentiation vs. de-differentiation of tourist experiences). For example, tourist typologies can be an invaluable marketing tool for tourism destinations, organizations, and operators due to the recognition that tourists are heterogeneous in nature and therefore tourists are segmented in terms of motivations, activities, travel experiences, socio-demographic characteristics, lifestyles and values. In fact, Knox and Hannam themselves acknowledge that the categorizing of tourists is still an important part of tourist studies! However, the use of typologies or categories to differentiate different types of tourists are seen, as viewed from a postmodern perspective, to have limited predictive ability (Uriely 2009; Moscardo et al. 2012) and can only identify regularities between homogeneous groups of tourists and describe 'different tourist practice rather than types of people' (Edensor 2001: 59–60).

As such, even though Knox and Hannam imply that the postmodern approach to typologies is the better approach, it would be interesting to have them articulate why they think modernist views of typologies still seem to be a mainstay in current tourism studies, and, in this context, whether or not the modernist use the 'pilgrimage' and 'pilgrim' metaphors might actually have some sort of utility, particularly when considering the 'post-tourist' who is able to create and participate in a wide range of touristic experiences when they travel.

Secularization

As noted above, Knox and Hannam suggest that pilgrimage as a metaphor is 'redundant'. They express the view that 'Pilgrimage as an earlier form of mobility is very clearly an ancestral behaviour to the diverse tourism industry of the early 21st century but this does not mean that we need to remain fixated by the explanatory power of the concept itself if we should find that has become redundant' (p. 241). The reason for this potential redundancy, they argue, is because 'If society changes, behaviour changes and lifestyles change, we should accept that the ways we categorize and think about cultural phenomena should also change in response' (p. 241). The question, however, is that what is causing society, behaviours, and lifestyles to change that warrants pilgrimage as a metaphor to become 'redundant'? Their answer to this question would be because 'The distinction between the sacred and the profane becomes inherently less important as society becomes ever more secular in Europe, North America and elsewhere' (p. 239). Their argument implicitly seems to be that as Europe and North America become more secular these societies become less religious, and therefore pilgrimage, as a dying form of travel, will eventually no longer have the metaphorical 'explanatory power' to describe what tourism'is like'.

This line of reasoning is recognizable to those who are familiar with the secularization thesis or theory, which has been discussed at length by scholars in the fields of history, sociology, and religious studies. While classical secularization theory can be traced back to the nineteenth-century, it was during the 1960s that scholars began to note the waning public influence of religion. This lead some academics to argue that as societies continued to industrialize, modernize, and become more rational, complex, and individualistic (read: less religious), that religion was destined to (indeed, must!) decline and eventually disappear on both a personal and a communal level (Luckmann 1963; Berger 1967; Gorski 2003). As Wallace (1966: 265) confidently declared, 'The evolutionary future of religion is extinction'!

However, a number of scholars have pointed out the limitations of this secularization theory. For example, opponents of the secularization theory argue that secularization and modernization have been components of every society and culture historically, and that this trend in modem society is nothing new. As Stark and Bainbridge (1985: 1–2) point out, rather than secularization and modernization leading to the end of religion, societies see 'merely a shift in fortunes among religions as faiths that have become too worldly are supplanted by more vigorous and less worldly religions.' As well, there seems to be a geography to secularization (Wilford2010), in that'religion has declined in some sectors and levels of society, but not all' (Demerath 1995: 110). As Stark and Bainbridge (1985:1–2) note, 'While secularization progresses in some parts of a society, a countervailing intensification of religion goes on in other parts'. Indeed, not only have both the decline and resilience of religion been illustrated in a European context (e.g., Green 2011; Kaufmann et al. 2012), but around the world there has been an intensification of religion and religiosity with the rise of Islamic fundamentalism, global Pentecostalism, and the growing strength of the American religious right.

While these and other criticisms, as well as 'religion's stubborn refusal to disappear' (Chaves 1994: 749), have led some scholars to either revise the secularization theory to make it less linear (Martin 2005, 2011; see Pickel 2011) or to suggest a middle ground argument where societies can be more secular without becoming less religious (Gorski 2000), Knox and Hannam seem to be using the secularization thesis in an uncritical manner in sweepingly suggesting that the 'distinction between the sacred and the profane' is less important in secularizing societies such as Europe and North America, which, by extension, implies a lessening of the influence of religion in those societies. Indeed, they seem implicitly follow early proponents of the secularization theory who viewed this theory more as a 'doctrine'; a 'taken-for-granted ideology' which would come to fruition even though there has been a lack of empirical evidence suggesting the demise of religion as well as the lessening importance of the sacred/ profane divide (Hadden 1987: 588, in Swatos Jr. and Christiano 1999; see also Stark 1999).

As such, I would like to know from Knox and Hanamm who exactly has suggested that the sacred/ profane distinction is becoming less important in Europe and North America? In what ways and to whom has this distinction become less important? It is important to note that any arguments related to the secularizing tendencies of any society needs to be contextualized in terms of cultural change, political surroundings, and identity building processes (Pickel2011), and as such the generalized use of the secularization theory, considering the criticisms noted above, is an inadequate basis upon which to claim that 'pilgrimage' is a poor and possibly 'redundant' metaphor.

Pilgrim as Hedonist

I think that the arguments Knox and Hannam make regarding their questioning of typologies and who gets to place moral judgment on tourist motivations and activities are, based on my own understanding of their lead probe, made mainly to bolster their argument that 'hedonism tourism' (pp. 239, 241) is not tourism just related to fu1filling sensual pleasures and self-centered and self indulgent lifestyles; participation in hedonistic activities can also be seen as a pursuit after both 'emotional and spiritual pleasures' (p. 239). As such, 'hedonism tourism', like cultural tourism or religious tourism, is also about a 'search for authenticity', at least 'within [the] distinct, bounded cultural communities' of hedonistic tourists (p. 240). However, considering the conventional understandings of the hedonist tourist being 'motivated by a desire for sensual pleasure' (Swarbrooke and Homer 1999: 36) within an 'enclosed, exceptional, duty-free zone' (Diken and Laustsen 2004:102) where they are 'licensed to participate in all kinds of otherwise forbidden pleasures' (Caruana and Crane 2011: 1504), the only way in which Knox and Hannam can make this argument persuasively is to make 'hedonism tourism' co-equal with -or at least akin to- more respected and 'refined' market segments like cultural tourism and religious tourism.

In arguing for the blurring of the lines between 'serious pilgrims and frivolous tourists' (Pfaffenberger 1983), Knox and Hannam attempt to shift the position of pilgrimage and pilgrims on the pilgrim-tourist continuum (Smith 1992) so that 'the original terms of the opposition set up between secular pilgrims and hedonists arguably no longer apply' (page 1). For example, Knox and Hannam suggest that pilgrimage is 'clearly an ancestral behaviour' which 'is no longer always a component part of a Christian lifestyle' (p. 237). They argue that the majority of people who travel to religious sites are not pilgrims in a true devotional sense; rather, the majority who visit these sites are more interested in the architectural or historical aspects of the sites as well as 'to remember Christians and Christianity of the past' (p. 237). In addition, Knox and Hannam put forward the view that 'there is little difference between pilgrims and tourists', and that 'many pilgrimages are hedonistic' in nature (p. 236). They question 'the extent to which...all acts of historical pilgrimage are motivated by a deep sense of devotion rather than a degree of self-interest, self-preservation or self-improvement' (p. 240). As such, 'we can identify elements of the pilgrimage in hedonistic forms of tourism and hedonistic behaviour among supposed pilgrims' (p. 236).

While one might agree that pilgrimage in the Christian world has waned over time in terms the importance of pilgrimage for social cohesiveness, Knox and Hannam seem to forget that Christian pilgrimage is alive and well While overall statistics are difficult to come by with regards to the religious tourism market, the total market for religious travel in the United States alone is estimated to be approximately 16.6 million people (Saltzman 2010), and North American religious tourists supposedly account for $10 billion of an $18 billion worldwide faith tourism industry Gordan 2007). While granted these numbers do not suggest that everyone travelling to a religious sites is a 'pilgrim' in a truer sense of the word, these numbers also include more 'pilgrim-like' travel including religious conferences, attendance at World Youth Day (in which over 70,000 Catholic youth participate) (Rymarz 2007; Norman and Johnson 2011), and Vatican events which attracted over 2.3 million people in 2012 (Catholic News Service 2013).

Also, while Knox and Hannam do not discuss, give evidence of, or identify hedonistic behaviour among 'supposed pilgrims' (whoever 'supposed pilgrims' are), their view mirrors Turner and Turner's (1978: 20) classic phrase 'a tourist is half a pilgrim, if a pilgrim is half a tourist', with 'hedonist' being substituted for 'tourist'. This argument that both historical and modem pilgrimage contains elements of hedonism, however, is not a new idea. Anyone who has read Chaucer's classic work Canterbury Tales (Furnivall 1967) would note that many medieval Christian pilgrims were more akin to modem tourists in their behaviours rather than true pilgrims (Olsen 2008). In fact, logically, would not suggest that pilgrimage contains no hedonistic elements run counter to Knox and Hannam's concerns regarding the pilgrim-tourist dualism as an ahistorical and abstract attempt to universalize travel typologies (Adler 1989)? Where the pilgrim being discussed is, again, the medieval aesthetic Christian pilgrim motivated by devotional and initiatory rituals?

While Knox and Hannam's argument for 'hedonistic behaviour among supposed pilgrims' is correct, their view that there are 'elements of the pilgrimage in hedonistic forms of tourism' is a bit more problematic. Along with their view that hedonism is about a 'search for authenticity' and involves 'emotional and spiritual pleasures', Knox and Hannam suggest that 'Surely there is significant pleasure to be derived from [hedonism] as pilgrimage' (p. 236). In stating this they implicitly seem to equate hedonistic 'pleasure' with religious experiences. Such a suggestion would, of course, offend the religious sensibilities of many, as this would make pilgrimage and pilgrim activity equal with more hedonistic forms of tourism activity such as wine tourism or sex tourism (Ostrowski 2000). In fact, it is for this very reason that religious commentators have had difficulty using the pilgrim as a metaphor for the secular traveller (Vukonic 2002), as well as with the use of terms such as *pilgrimage tourism, spiritual tourism, religious tourism,* and *tourism pilgrimage* to describe modem pilgrimage travel.

As well, one could question whether the 'authenticity' and the 'emotional and spiritual pleasures'and maybe experience are akin to the same as those sought for and experienced by pilgrims or even cultural tourists. As such, while scholars have noted the structural similarities between pilgrimage and tourism (e.g., MacCannell1976; Graburn 1989; Gupta 1999; Fleischer 2000), when it comes to motivation it is difficult to argue that pilgrims and hedonists share much in common, as they are vastly different in terms of their goals, motivations, and mindsets while traveling.

Conclusion

Knox and Hannam cover additional topics and issues in their paper, but I have limited myself to four key matters related to religious metaphors, typologies, the secularization thesis, and the blurring of the pilgrim/hedonist dichotomy. While my comments up to this point have been critical, I wish to note that Knox and Hannam do make a couple of valuable contributions to some of the current research debates related to tourism and pilgrimage.

For example, the authors, echoing Towner (1995) and Olsen (2010), are correct in reminding readers that 'We ought to recognize within this discussion, however, that this is a highly situated, contingent and Eurocentric debate that fails to take account of tourist practices outside of the Western World or to consider societies within which shared faith in the form of organized religion remains important or even shows signs of growth. Does this debate have the same resonance in Islamic cultures, or in China, India, Sub-Saharan Africa or indeed anywhere outside the historical heartlands of tourist studies?' (p. 241). I wonder if the themes in this research probe, as discussed below, would even be considered by scholars in a Middle Eastern or an African context. In fact, instead of asking a North American (like myself) or European scholar to respond to the probe by Knox and Hannam, it may have been much more interesting to have a scholar from one of the world regions listed above to comment on their arguments in order to engage in more inter-cultural dialogue and understanding regarding typologies, metaphors, and secularization.

In addition, I appreciated the concern raised by Knox and Hannam, as noted above, regarding who defines certain tourist motivations and performances as morally superior or inferior. As the privileged domain of academics and 'well-travelled and self styled enlightened commentator[s]', how do these definitions relate to how certain tourism market segments are represented, and how these representations affect those who are a part of those market segments? How is it that we can avoid 'the mistake of elevating [certain tourist] practices above those where the nature of the engagement is with more popular or populist forms of culture such as the world of entertainment' (p. 238) rather than with the world of high culture, if it is indeed a mistake to do so? I expect that Knox and Hannam would agree with other scholars (e.g., Aziz 1987, Collins Kreiner and Kilot 2000, Wickens 2002, and Moscardo et al. 2012) who have all argued that instead of imposing pre-selected unitary types on tourists, scholars should instead seek to examine the self-perceptions and experiences of the tourists who are being typologized in order to create more grounded typologies and increase their predictive abilities.

However, for all their 'playful extension[s]' of the definitions of pilgrimage and hedonism, Knox and Hannam never did define the terms they suggest have been conflated together in the modem world. For example, how exactly would Knox and Hannam define 'pilgrimage' and 'pilgrim', considering they are being compared to hedonism and the hedonist tourist? What is 'hedonism tourism', and what are the characteristics of the 'hedonistic tourist'? These are important distinctions to make, considering that Knox and Hannam in essence are comparing pilgrimage as a 'travel style' with hedonism which is more of a 'lifestyle'. As well, what is meant by 'vacation-as-doctrine'? What is meant by 'authenticity'? What is 'secular pilgrimage' as opposed to 'religious pilgrimage'?

After reading this paper, I note that there is an increasingly evident need for definitional consistency when it comes to the way scholars blur pilgrimage and tourism, particularly when it comes to the translation and interpretation of words from the 'language of religion' to the 'language of tourism'. While Knox and Hannam may be correct in stating that 'The verbal dexterity demonstrated in applying, stretching and murdering half-dead metaphors has not yet come to a conclusion...' (p. 240), without definitional consistency regarding metaphors us to descnbe the blurring of tourists/ pilgrims and tourism/ pilgrimage there may never be a conclusion. Thus, to paraphrase Voltaire, 'Define your terms, then we can talk'.

References

ADLER, J. (1989). Travel as Performed Art. *American Journal of Sociology* 94: 1366–1391.

AZIZ, B. N. (1987). Personal Dimensions of the Sacred Journey: What Pilgrims Say. *Religious Studies* 23: 247–261.

BAUMAN, Z. (1996). From Pilgrim to Tourist- Ora Short History ofldentity. In Hall, S. and Gay, P. du (Eds) *Questions of Cultural Identity*. London. Sage: 18–36.

BERGER, P. L. (1967). *The Sacred Canopy: Elements of a Sociological Theory of Religion*. New York. Anchor Books.

BURNS, P. (1999). *An Introduction to Tourism and Anthropology*. New York. Routledge.

CARUANA, R. and CRANE, A. (2011). Getting Away From It All: Exploring Freedom in Tourism. *Annals of Tourism Research* 38(4): 1495–1515.

CATHOLIC NEWSSERVICE (2013). VaticanSays 2.3 Million People Joined Pope for Vatican Events in 2012. Availabale at http://www.catholicnews.com/datafstories/cns/1300050.htm- Accessed on 15 May 2013.

CHAVES, M. (1994). Secularization as Declining Religious Authority. *Social Forces* 72(3): 749–744.

COHEN, E. (1979). Rethinking the Sociology of Tourism. *Annals of Tourism Research* 6(1): 18–35.

COHEN, E. (1985). Tourism as Play. *Religion* 15(3): 291–304.

COHEN, E. (1992). Pilgrimage and Tourism: Convergence and Divergence. In Morinis, A. (Ed.) *Sacred Journeys: The Anthropology of Pr1grimage*. Westport, CT. Greenwood Press: 18–35.

COLLINS-KREINER, N. and KUOT, N. (2000). Pilgrimage Tourism in the Holy Land: The Behavioural Characteristics of Christian Pilgrims. *Geojournal* 50: 55–67.

DANN, G. M.S. (2002). The Tourist as a Metaphor of the Social World. In Dann, G. M.S. (Ed) *The Tourist as a Metaphor of the Social World*. Wallingford. CABI: 1–17.

DANN, G. M. S. (2005). The Theoretical State of the Art in the Sociology and Anthropology of Tourism. *Tourism Analysis* 10: 3–15.

DEMERATH III, NJ. (1995). Rational Paradigms, A-Rational Religion, and the Debate over Secularization *Journal for the Scientific Study of Religion* 34(1): 105–112.

DIKEN, B. and LAUSTEN, C. B. (2004). Sea, Sun, Sex and the Discontents of Pleasure. *Tourist Studies* 4(2): 99–114.

EDENSOR, T. (2001). Performing Tourism, Staging Tourism: (Re) producing Tourist Space and Practice. *Tourist Studies* 1(1): 59–81.

FLEISCHER, A. (2000). The Tourist behind the Pilgrim in the Holy Land. *International Journal of Hospitality Management* 19: 311–326.

FURNNALL, F. J. (1967). The *Cambridge Ms. of Chaucer's Canterbury Tales*. New York. Johnson Reprint Corporation.

GORSKI, P. S. (2000). Historicizing the Secularization Debate: Church, State, and Society in Late Medieval and Early Modem Europe, ca.1300 to 1700. *American Sociological Review* 65(1): 138–167.

GORSKI, P. S. (2003). Historicizing the Secularization Debate: An Agenda for Research. In Dillon, M. (Ed) *Handbook of the Sociology of Religion*. Cambridge, UK Cambridge University Press.

GRABURN, N. H. H. (1989). Tourism: The Sacred Journey. In Smith, V. L. (Ed) *Hosts and Guests:* The *Anthropology of Tourism*. Philadelphia. University of Pennsylvania Press: 21–36.

GREEN, T. (2011). Secularization of European Society. The *Encyclopedia of European Society*. Oxford, UK Blackwell: 1–3.

GRIMSHAW, M. (2001). Tourist, Traveler, or Exile: Redefining the Theological Endeavour. The *Journal of Religion* 81(2): 249–270.

GUPTA, V. (1999).Sustainable Tourism: Learning from Indian ReligiousTraditions. *International Jownal of Contemporary Hospitality Management* 11(2–3): 91–95.

HADDEN, J. K (1987). Towards Descralizing Secularization Theory. *Social Forces* 65(3): 587–611.

HEINTZMAN, R. (2011). *Rediscovering Reverence: The Meaning of Faith in a Secular World.* Montreal and Kingston: Me-Gill-Queen's University Press.

JORDAN, M. (2007). Seeking Answers with Field Trips in Faith. *Washington Post.* Available at http:// www.washingtonpost.com/wp-dyn/content/article/2007/06/24/AR2007062401422.html- Accessed on 15 May 2013.

KAELBER, L. (2002). The Sociology of Medieval Pilgrimage: Contested Views and Shifting Boundaries. In Swatos Jr, W. H. and Tomasi, L. (Eds) *From Medieval Pilgrimage* to *Religious Tourism.* London. Praeger: 51–74.

KAELBER, L. (2006). Paradigms of Travel: From Medieval Pilgrimage to the Postmodem Virtual Tour. In Timothy, D. J. and Olsen, D. H. (Eds) *Tourism, Religion and Spiritual Journeys.* London and New York. Routledge: 49–63.

KAUFMANN, E., GOUJON, A. and SKIRBEKK, V. (2012). The End of Secularization in Europe? A Socio-Demographic Perspective. *Sociology of Religion* 73(1): 69–91.

LUCKMANN, T. (1963). The *Invisible Religion:* The *Problems of Religion in MDdern Society.* New York. Macmillan.

MACCANNELL, D. (1976). The *Tourist: A New* Theory *of the leiswe Class.* New York. Schocken.

MARTIN, D. (2005). *On Secularization: Towards a Revised General* Theory. Aldershot. Ashgate.

MARTIN, D. (2011). The *Futwe of Christianity: Reflections on Violence and Democracy, Religion and Secularization.* Farnham, UK Ashgate.

MAOZ, D. and BEKERMAN, Z. (2010). Searching for Jewish Answers in Indian Resorts. *Annals of Tourism Research* 37(2): 423–439.

MOSCARDO, G., MURPHY, L. and BENCKENDORFF, P. (2012). Tourism Types and Labels. In Proceedings of the 22"'i Annual Conference *CAUTHE 2012:* The *New Golden Age ofTowism and Hospitality: Book* 2. Melbourne, Australia. La Trobe University: 424–429.

NORMAN, A. and JOHNSON, M. (2011). World Youth Day: The Creation of a Modem Pilgrimage Event for Evangelical Intent. *Jownal of Contemporary Religion* 26(3): 371–385.

OLSEN, D. H. (2008). Contesting Identity, Space and Sacred Site Management at Temple Square in Salt Lake City, Utah. PhD Dissertation, University of Waterloo.

OLSEN, D. H. (2010). Pilgrims, Tourists and Max Weber's "Ideal Types". *Annals of Tourism Research* 39(3): 848–851. OSTROWSKI, M. (2000). Pilgrimages or Religious Tourism. In Jackowski, A. (Ed.) *Peregrinus Cracoviensis.* Cracow,

Poland: Publishing Unit, Institute of Geography, Jagiellonian University: 53–61.

PFAFFENBERGER, B. (1983). Serious Pilgrims and Frivolous Tourists: The Chimera of Tourism in the Pilgrimage of Sri Lanka *Annals of Tourism Research* 10: 57–74.

PICKEL, G. (2011). Contextual Secularization: Theoretical Thoughts and Empirical Implications. *Religion and Society in Central and Eastern Europe* 4(1): 3–20.

RYMARZ, R (2007). Who Goes to World Youth Day? Some Data on Under-18 Australian Participants. *Journal of Beliefs & Values: Studies in Religion & Education* 28(1): 33–43.

SALTZMAN, D. (2010). Faith-Based, Religious Travel Strong and Growing. *Travel Market Report.* Available at http://www.travelmarketreportcom/leisure?articleiD=4545&LP=1- Accessed on 15 May 2013. SELWYN, T. (1994). Tourism as Religion. *Contours* 6(6): 28–30.

SHARPLEY, R. (2009). Tourism, Religion, and Spirituality. In Jamal, T. and Robinson, M. (Eds) The *SAGE Handbook of Tourism Studies.* Thousand Oaks, CA. SAGE Publications Ltd: 237–253.

SHARPLEY, R. and SUNDARAM, P. (2005).Tourism: ASacred Journey? The Case of Ashram Tourism, India *International Journal of Tourism Research* 7: 161–171.

STARK, R. (1999). Secularization, RI.P. Sociology *of Religion* 60(3): 249–273.

STARK, R. and BAINBRIDGE, W. S. (1985). The *Future of Religion.* Berkeley. University of California Press.

SMITH, V. L. (1992). Introduction: The Quest in Guest *Annals of Tourism Research* 19: 1–17.

SWARBROOKE, J. and HORNER, S. (1999). *Consumer Behaviour in Tourism.* Oxford. Butterworths Heinemann.

SWATOS Jr., W. H. and CHRISTIANO, K J. (1999).Secularization Theory: The Course of a Concept. Sociology *of Religion* 60(3): 209–228.

TILLEY, C. (2006). Identity, Place, Landscape and Heritage. *Journal of Material Culture* 11(1/2): 7–32.

TOWNER, J. (1995). What is Tourism's History? *Tourism Management* 16(5): 339–343.

TURNER, V. and TURNER, E. (1978). *Image and Pflgrimage in Christian Culture.* New York. Columbia University Press.

URIELY, N. (2005). The Tourist Experience: Conceptual Developments. *Annals of Tourism Research* 32(1): 199–216.

URIELY, N. (2009). Deconstructing Tourist Typologies: The Case of Backpacking. *International Journal of Culture, Tourism and Hospitality Research* 3(4): 306–312.

VUKONIC, B. (2002). Religion, Tourism and Economics: A Convenient Symbiosis. *Tourism Recreation Research* 27(2): 59–64.

WALLACE, F. C. (1966). *Religion: An Anthropologiazl View.* New York. Random House.

WICKENS, E. (2002). The Sacred and the Profane: A Tourist Typology. *Annals of Tourism Research* 29(3): 834–851.

WILFORD, J. (2010). Sacred Archipelagos: Geographies of Secularization. *Progress in Human Geography* 34(3): 328–348.

Submitted: June 1, 2013
Accepted: June 4, 2014

Continued ....

To Be or Not to Be a Tourist: The Role of Concept-Metaphors in Tourism Studies

Noel B. Salazar

A Conceptual Introductory Note

Tourists have been labelled, metaphorically, in multiple ways (Dann 2002). This includes descriptions of tourists as (secular) pilgrims in a quest of authenticity (MacCannell1976) but also as travellers on a sacred journey (Grabum 1978, 1983, 2001). In contrast, the stereotypical image that tourists are hedonists is related particularly (but not exclusively) with sun, sand and sex, and is associated with, sometimes unbridled, consumerism (cf. Salazar 2010a). This research probe deals with the question whether a tourist is a 'secular pilgrim' or a 'hedonist in search of pleasure'. Both descriptors refer to a debate among tourism scholars that started back in the 1970s. It is important to put the question and possible answers to it within the historical context of this discussion.

First, I would like to state that, in my humble opinion, this probe is not about tourist typologies. Knox and Hannam suggest otherwise: 'any time we set out to explore a dualism we go through a predictable process of arriving at a conclusion whereby we might reject the notion of the categories as mutually exclusive or discover instead an additional set of categories. Such is the stuff that tourist typologies are made of. As taxonomies do not make assertions, they cannot be judged true or false. Like tools, they may be found more or less useful for a particular purpose (e.g., tourism marketing and management strategies). As I will discuss below, the scholars who started this debate had something very different in mind from trying to classify tourists. The descriptions of 'ideal type' tourists as (secular) pilgrims or hedonists, however, are related to broader visions of contemporary society. They are best conceived as a concept-metaphor. Henrietta Moore defines concept-metaphors as 'a kind of conceptual shorthand... They are domain terms that orient us towards areas of shared exchange, which is sometimes academically based... Their exact meanings can never be specified in advance-although they can be defined in practice and in context-and there is a part of them that remains outside or exceeds representation' (Moore 2004:73).

Concept-metaphors act as framing devices, and as such, they are perspectival The advantage of using them is that they facilitate comparison. The problem with concept-metaphors such as 'secular pilgrim', however, is that by their nature they continue to have a shifting and unspecified tie to physical relationships in the world. As Moore (2004) argues:

> If *concept-metaphors are to be relevant in a disciplinary context then they must connect to the construction of composite theories. Composite theories are those that contain ontological, epistemological and empirical claims. Concept-metaphors that merely act as a descriptive gloss or posit causal forces that remain unexamined are essentiaUy suffering - at the very least -from under-theorization(p. 80)*

In order to answer the research probe properly, we thus need to assess the composite theories underlying the concept-metaphors and the ontological, epistemological and empirical claims they make.

The Tourist as Concept-Metaphor

> *Of all the metaphors used to capture the postmodern condition, none has perhaps been employed more frequently than that of 'the tourist'. Just as modernity had its metaphor of 'the traveller', seeking the rational goal of educational improvement, the moral path of spiritual renewal, the scientific and imperialistic exploration of unknown te"itories, so too did post modernity seize upon the tourist as connotative of a dilettante life of fun in the sun and hedonism ad libitum in placeless destinations where the 'other' was cheerfully ignored in favour of the unbridled pursuit of individualism sans frontieres. (Dann 2002: 6)*

Although Dean MacCannell (1976) did not coin the term 'secular pilgrim', he did provide the theoretical context to conceptualize the tourist as a secular pilgrim in quest of authenticity. Comparing tourism with pilgrimage, he writes:

> The *motive behind a pilgrimage is similar to that behind a tour: bothare quests forauthentic experiences. Pilgrims attempt to visit a place where an eventofreligious*

importanre aduaUy occu"ed. Tourists present themselves at places of social, historical and cultural importance. (MacCannell 1973: 593)

Importantly, MacCannell begins his seminal monograph, The Tourist: A New Theory of the Leisure Class, by specifying that he means two different things by tourist: (1) an 'actual person' and (2) a 'model for modem-man-in-general' (1976:1). The notion of 'secular pilgrim' is related more closely to the second meaning. In other words, the description of tourist as (secular) pilgrim functions as a concept-metaphor. It is not so much about what a tourist does and is, as tourist, but how the figure of the tourist is emblematic for the times in which we live. This important distinction seems lost in the piece of Knox & Hannam. They argue that 'contemporary tourism is now so diverse that the original terms of the opposition set up between secular pilgrims and hedonists arguably no longer apply. There is little difference between pilgrims and tourists and many pilgrimages are hedonistic'. That was, of course, not the point of the original argument.

Inspired by Emile Durkheim's study of primitive religion and Levi-Strauss's structural anthropology, MacCannell sees tourism as an icon of the rootlessness and alienation of modem life. The search for meaning in (secularized) modem societies encourages pilgrimage to the sites of differentiation created by modernity and a search for the 'primitive' and pre-modem cultures it has displaced-a quest for 'authenticity' that is ultimately doomed as it is hindered by locals and tourism service providers that stage reality and 'real live' as mere attractions. MacCannell 's work is thus an 'ethnography of modernity' (1976:2) in which tourism functions as a modern surrogate religion in connection with mass leisure.

MacCannell's theory is partially a reaction to the earlier work of Boorstin (1964), for whom tourism is essentially an aberration, a trivial, frivolous, superficial, and vicarious activity. Boorstin bemoans the disappearance of the travelling of yesteryear, which was an individual, painstaking and educational experience. He believes that modem (mass) tourists are no more than sheer hedonists, unable to experience reality directly, thriving on and finding pleasure only in the inauthentic and, therefore, taking pleasure in contrived experiences, attractions and 'pseudo-events' created by tourism service providers and the media. Early on in the debate, Cohen (1979), points out that it is inaccurate to assume that all tourists are either dopes or secular pilgrims. According to him, most are simply out to have a good time-'travelling for pleasure'. Only a few, of the non-institutionalized variety (also known as 'independent travellers'), look for meaning in their lives by touring the world inhabited by the 'Other'.

The use of the tourist as a metaphor for larger societal developments has also been popular outside of tourism studies. According to Appadurai, for example, tourists metaphorically represent 'the shifting world in which we live' (1996:33). Bauman (1996) takes the pilgrim as emblematic of modernity, the tourist of post-modernity: 'in the same way as the pilgrim was the most fitting metaphor for the modern life strategy preoccupied with the daunting task of identity-building, the stroller, the vagabond, the tourist and the player offer jointly the metaphor for the postmodem strategy moved by the horror of being bound and fixed' (Bauman 1996:26). For Bauman, the main difference is situated in the kind of mobility that characterizes tourism:

The tourist moves on purpose (or so he thinks). His movementsarefirstofall'inorder to',and only secondarily (if at all) 'because of. The *purpose is* new *experience; the tourist is a conscious and systematic seeker of experience, of a* new *and different experience, of the experience* of *difference and novelty- as the joys* of *the fomi1iar wear off and cease to allure. The tourists want to immerse themselves in a strange and bizarre element (a pleasant feeling,a tickling and rejuvenating feeling, like letting oneself be buffeted by sea waves) -on condition,though, that it wi1l not stick to the skin and thus can be shaken off whenever they wish (Bauman 1996:29).*

The Tourist as Pilgrim

One needs to move beyond the Turnerian structural approach in an appreciation of the poststructuralist developments that have influenced sociology and anthropology during the 1980s. Nonessentialistacrountsoftourism need to be explored, making use of recent pilgrimage studies focusing on the deconstruction of such unitary categories as pi1grim and tourist. Behind the superficial analogies between pt1grimage and tourism, there lies a more complex world of dissonance, ambiguity, and conflict that one is now beginning to explore through the analysis of official attempts to organize people's activities and beliefs and unofficial resistance to the power of those who claim to know what is both right and good (Eade 1992: 31).

Many scholars have characterized tourism metaphorically as 'pilgrimages' (something scholars of religion have repeatedly criticized). Some seem to suggest that tourism evolved out of pilgrimage. While tourist, as a concept, only appeared at the beginning of the nineteenth century in English and French, one can identify people most of us would call tourists, as well as their actions, long before that. The term itself, derived from the Greek tornos (a carpenter's tool for drawing a circle), refers to an individual who makes a circuitous journey (usually for pleasure) and returns to the starting point. Pilgrim, on the other hand stems from the Latin peregrinus, which originally meant foreigner, wanderer, exile, and traveller, as well as newcomer and s ran er. The meaning of pilgrimage histoncally developed to become 'a journey claimed to be undertaken for reasons involving religious sacrifice' (Palmer et al. 2012: 71). Others use a much broader definition, making the difference with tourism become much smaller: 'a journey undertaken by a person in quest of a place or a state that he or she believes to embody a valued ideal' (Morinis 1992: 4).

The metaphorical conceptualization of tourists as pilgrims, like many other aspects of tourism (Salazar 2013), has largely been influenced by the work of anthropologists (Badone and Roseman 2004). Victor and Edith Turner famously wrote that 'a tourist is half a pilgrim, if a pilgrim is half a tourist' (1978:20). According to the Turners, pilgrimage, like tourism, is organized, bureaucratized and uses the same infrastructure mediated by travel agencies. They see evident links between tourism and pilgrimage in terms of both the journey and the experience of communitas, although they distinguish between the obligatory nature of many traditional rituals and the voluntary nature of tourism. Victor Turner himself notes that the simultaneous rise of the anthropology of tourism along with that of pilgrimage is no accident, since both areas of study have become metaphors for a world on the move, 'where rapid transportation and the mass media are moving millions literally or mentally out of the stasis of localization' (1992: viii).

In a Durkheimian tradition, Grabum (1978, 1983) maps tourism to Victor Turner's tripartite structure of rites of passage, situating the tourist's quest as a pursuit of the 'sacred' (non-ordinary), as separate from the 'profane' (ordinary). Importantly, tourists are being metaphorically compared to pilgrims as being on a 'sacred journey'. Tourism is seen as a secular and universal equivalent of religion operating in non-ordinary time:'even when the role[s] of tourist and pilgrim are combined, they are necessarily different but form a continuum of inseparable elements' (Graburn 1983: 16). Even history-distorting theme parks such as Disney World have been viewed as contemporary secular equivalents to traditional centres of faith, where the icons of civil religion are ritually worshipped and consumed (Moore 1980). Pfaffenberger (1983) sees present-day tourism arising out of a long tradition of religious pilgrimage. At the same time, he opposes serious (pious) pilgrims against frivolous (hedonistic) tourists. According to Colin Turnbull, too, tourists travel for 'hedonistic purposes'. Pilgrims, on the other hand gain 'a sense of belonging to a religious or spiritual heritage rather than a cultural one' (1981: 14).

In a special issue of Annals of Tourism Research on the relationships between pilgrimage and tourism, Valene Smith argues that 'tourist encounters can be just as compelling [as pilgrimage] and almost spiritual in personal meaning' (1992: 2). She sees social approval as the most important factor differentiating tourists from pilgrims. Her observation, similar to the Turners (Turner and Turner 1978), infers that from a tourism perspective there is (superficially) no difference between tourists and pilgrims: they share leisure time, income, and social sanctions for travel and, in most instances, the same infrastructure. Pilgrimage and tourism may be conceptualized as 'two parallel, interchangeable lanes' (Smith 1992: 15). People can 'travel either lane, or switch between them, depending on personal need or motivation, and as appropriate to time, place, and cultural circumstances' (Smith 1992: 15). Smith relies on Durkheim to link pilgrimage to the sacred and tourism to the secular and, in between these two, religious tourism, which is related to the profane. In that same special issue, John Eade (1992) calls to reconstruct the terms tourist and pilgrim in order to capture the lack of harmony that exists between them.

Interestingly, Knox and Hannam never fully define how they understand the concepts of pilgrim and hedonist. This allows them to 'play' around with the terms (e.g., discussing the hedonistic aspects of pilgrimage). Although they recognize the 'metaphor of the pilgrim', they seem stuck in a logic of tourism typologies. That is why Knox and Hannam argue that the'categories' of tourist and pilgrim 'have in some cases grown towards and overlap each other so that we can identify elements of the pilgrimage in hedonistic forms of tourism and hedonistic behaviour among supposed pilgrims'. Such statements suggest wrongly that there was a time in which 'pure' (and, thus, 'authentic') types of pilgrims and tourists existed. In addition, pilgrimage is imagined, in an evolutionary fashion, as 'ancestral' to tourism. On the one hand, Knox and Hannam seem to suggest that (cultural or creative) tourism has replaced pilgrimage (or, at least, the 'original' form of the practice). On the other hand, the 'mystical and spectacular behaviour of the faithful' the pilgrims who are still around - has become part of the attraction for tourists visiting religious sites.

Knox and Hannam write that 'some tourists very clearly fall into the camp of being hedonists in search of pleasure'. This should come as no surprise. For most people, tourism involves more hedonism and conspicuous consumption than learning or understanding. However, with tourism becoming a phenomenon of the masses, the label of 'tourist' has received increasingly negative connotations. Middle and higher social classes try to 'distinguish' (Bourdieu 1984) themselves by engaging in tourism activities that are esteemed to be of higher (moral) value (cf.Munt1994). This strategy does not always seem to be successful. As Knox and Hannam note, 'cultural tourism is no longer clearly separate from mass forms of tourism'. Importantly, hedonists and pilgrims are not two 'types'of tourists,but concept-metaphors that reflect the societal role that scholars attribute to tourism. Again, Knox and Hannam think otherwise: 'That hedonists are in pursuit of pleasure suggests that the lot of the pilgrim, whether secular or otherwise, is somewhat more ascetic as they self-flagellate themselves from museum to gallery to heritage attraction'.

Conclusion

Knox and Hannam end their piece by arguing that'while the twin concepts of pilgrim and hedonist continue to have some utility they are not sufficiently nuanced or all-encompassing to describe, let alone explain, the totality of contemporary tourist practice'. Of course not, because this would imply an essentialist stance according to which tourists should be classified either as (secular) pilgrims or as hedonists. As Cohen noted long ago, 'tourism spans the range of motivations between the desire for mere pleasure characteristic of the sphere of "leisure" and the quest for meaning and authenticity, characteristic of the sphere of "religion"' (1979: 193). Tourism overlaps with pilgrimage, but also with business, migration and other phenomena (Salazar 2010b; Salazar and Zhang 2013). The two descriptors of this probe capture the complexity neither of tourists nor, by extension, the society they are supposed to represent.

As I made it clear above, this is not a discussion about tourist typologies or, worse, tourism market segments. Instead of asking whether tourists are secular pilgrims or hedonists, we should be asking what we gain, analytically, by using these concept metaphors. For whom does this matter and how does tourism (or at least the study of it) benefit by the use of these terms? The fruitfulness of the original theoretical discussion is that it opened up an in-depth reflection on the nature of tourism in relation to the wider socio-economic context in which it develops. The role of concept-metaphors such as 'secular pilgrims'is to'open up spaces for future thinking' (Moore 2004: 74). Their 'purpose is to maintain ambiguity and a productive tension between universal claims and specific historical contexts' (Moore 2004: 71). As the text by Knox and Hannam illustrates, however, the concepts have started to lead their own life and have almost become 'tourism imaginaries' (Salazar 2012; Salazar and Graburn 2014) in their own right - socially transmitted representational assemblages that are used as meaning making and world-shaping devices in tourism and beyond.

Despite my criticism and reservations, there is one point on which I wholeheartedly agree with Knox and Hannam, namely, that 'this is a highly situated, contingent and Eurocentric debate that fails to take account of tourist practices outside of the Western World'. Indeed, this conceptual discussion has been dominated since the very start by Western scholars and this research probe is a failed opportunity to open up the debate more globally. A similar argument could be made when it comes to gender. These criticisms seriously weaken the universal validity of the arguments made. Or, as Graham Dann argues, 'because the iconicity of metaphor depends on cultural codes, and cultures themselves vary, there can be no universal metaphors' (2002: 1).

References

APPADURAI, A. (1996). *Modernity at Large: Cultural Dimensions of Globalization*. Minneapolis. University of Minnesota Press.

BADONE, E. and ROSEMAN, S. R. (Eds) (2004). *Intersecting Journeys: The Anthropology of Pilgrimage and Tourism*. Urbana University of Illinois Press.

BAUMAN, Z. (1996). From Pilgrim to Tourist- or A Short History of Identity. In Hall, S. and Gay, P. D. (Eds) *Questions of Cultural Identity*. London. Sage: 18–36.

BOORSTIN, D. (1964) The *Image: A Guide* to *Pseudo-events in American Society*. New York. Atheneum.

BOURDIEU, P. (1984). *Distinction: A Social Oitique of the Judgement of Taste* (translated by R. Nice) Cambridge. Harvard University Press.

COHEN, E. (1979). A Phenomenology of Tourist Experiences. *Sociology* 13(2):179–201.

DANN, G. M.S. (Ed) (2002). The *Tourist as a Metaphor of the Social World*. Wallingford CABI.

EADE, J. (1992). Pilgrimage and Tourism at Lourdes, France. *Annals of Tourism Research* 19(1):18–32.

GRABURN, N. H. H. (1978) Tourism: The sacred Journey. In Smith, V. L. (Ed) *Hosts and Guests:* The *Anthropology of Tourism*. Oxford. Blackwell: 17–31.

GRABURN, N. H. H. (1983). The Anthropology of Tourism. *Annals of Tourism Research* 10(1): 9–33.

GRABURN, N. H. H. (2001). Secular Ritual: A General Theory of Tourism. InSmith, V. L. and Brent, M. A. (Eds) *Hosts and Guests Revisited: Tourism Issues of the 21st Century*. New York. Cognizant Communication Corporation: 42–50.

MACCANNELL, D. (1973). Staged Authenticity: Arrangements of Social Space in Tourist Settings. *American Journal of Sociology* 79(3): 589–603.

MACCANNELL, D. (1976). The *Tourist A New* Theory *of the Leisure Class*. New York. Schocken Books.

MOORE, A. (1980). Walt Disney World: Bounded Ritual Space and the Playful Pilgrimage Center. *Anthropological Quarterly* 53(4): 207–218.
MOORE, H.L. (2004). Global Anxieties: Concept-metaphors and Pre-theoretical Commitments in Anthropology. *Anthropologiazl Theory* 4(1): 71–88.
MORINIS, E. A. (Ed) (1992). *Sacred Journeys: The Anthropology* of *Pilgrimage.* Westport. Greenwood Press.
MUNT, I. (1994). Eco-tourism or Ego-tourism? *Race and Class* 36(1): 49–59.
PALMER,C. T., BEGLEY,R 0.and COE,K (2012). In Defence of Differentiating Pilgrimage from Tourism. *International Journal of Tourism Anthropology* 2(1): 71–85.
PFAFFENBERGER, B. (1983). Serious Pilgrims and Frivolous Tourists: The Chimera of Tourism in the Pilgrimages of Sri Lanka. *Annals* of *Tourism Research* 10(1): 57–74.
SALAZAR, N. B. (2010a). *Envisioning Eden: Mobilizing Imaginaries in Tourism and Beyond.*Oxford. Berghahn
SALAZAR, N. B. (2010b). Towards an Anthropology of Cultural Mobilities. *Crossings: Journal* of *Migration and Culture* 1(1): 53–68.
SALAZAR, N. B. (2012). Tourism Imaginaries: A Conceptual Approach. *Annals of Tourism Research* 39(2): 863–882.
SALAZAR, N. B. (2013). Imagineering Otherness: Anthropological Legacies inContemporary Tourism. *Anthropological Quarterly* 86(3): 669–696.
SALAZAR, N. B. and GRABURN, N.H. H. (Eds) (2014).*Tourism Imaginaries: Anthropological Approaches.*Oxford Berghahn.
SALAZAR, N.B. and ZHANG, Y. (2013). Seasonal Lifestyle Tourism: The Case of Chinese Elites. *Annals of Tourism Research* 43(4): 81–99.
SMITH, V. L. (1992). Introduction: The Quest in Guest *Annals* of *Tourism Research* 19(1): 1–17.
TURNBULL, C. (1981). A Pilgrimage in India. *Natwal History* 90(7): 14–20.
TURNER, V. (1992). Foreword. In Morinis, E. A. (Ed) *Sacred Journeys: TheAnthropology of Pilgrimage.*Westport. Greenwood Press: vii–viii.
TURNER, V. and TURNER, E. (1978). *Image and Pilgrimage in Christian Culture: Anthropological Perspectives.* New York. Columbia University Press.

Submitted: July 1, 2013
Accepted: May 25, 2014

Concluding Remarks

When first considering the question as to whether the "tourist is a secular pilgrim or a hedonist in search of pleasure" we were inclined towards offering a one word answer: No. We now recognize that this is a very complicated question that cuts to the very core of contemporary tourism studies. Margry asserts dissatisfaction with the term'secular pilgrimage' with which we would concur but which fails to engage with the popular currency of the metaphor. Salazar's account of both the tourist and tourism as pilgrimage as concept-metaphors illustrates that 'secular pilgrims' have become 'tourism imaginaries'. This is an important point since the intentions of the original scholarly arguments have generally been lost in the later adoption and application of those ideas by others whether in the scholarly community or the travel and tourism industry.

For Salazar, however, the debate is not about typologies or market segmentation. Tourism as pilgrimage and the tourist as pilgrim have taken on the status of ready-made ideas in tourist studies as well as in popular accounts of tourism. Our strategy here was to extend the metaphor of the pilgrim far beyond its usefulness in understanding contemporary tourism. Olsen questions whether or not pilgrims (tourists) really self-flagellate when visiting museums, galleries and heritage attractions. The answer to this question depends on whether or not you wish to remain wedded to the pilgrimage metaphor, extend it (playfully or otherwise) or wholly abandon it. If one practice is equated with another metaphorically then surely all component parts of that practice are potentially semantically available for play? This is the problem of metaphors in general in that they are based upon a limited number of points of comparison but that such connections or similarities may not bear further analysis.

A post-secular age is one in which attitudes to religion are characterised by indifference which is only possible following a period of securlarisation (Kyrlezhev 2008). Within secular thinking, religion becomes simply another identity choice open to the individual and was not given prominence over civil institutions. Ironically, this has created a circumstance within which the belief systems relating to post Enlightenment culture and art have taken on some of the characteristics of religion in the sense of requiring belief, commitment and adherence to a moral code. Durkheim ([1915] 2008) identified the sacred I profane distinction as a central characteristic of religion and the decline in the importance in this distinction is not so much concerning the dichotomy itself but in relation to the dichotomy in the terms of dominant organized religious structures (i.e. Christianity). In societies such as those of Europe and North America the formal influence of organized religions has tended to wane and more individualistic pursuits of a spiritual nature have come to prominence. In these circumstances, sacred I profane distinctions are evident within cultural fields and artistic movements in the sense of inside I outside distinctions.

It was never our intention to suggest that Christian pilgrimage was anything other than "alive and welf', nor to denigrate the important spiritual implications it retains for pilgrims. However, we feel that for Olsen to argue that equating hedonistic forms of tourism with religious experiences is potentially offensive and further to question whether "authenticity" and the "emotional and spiritual pleasures" of such tourism are "the same as those sought for and experienced by pilgrims or even cultural tourists" is to continue to make the same mistakes in terms of elevating some cultural practices above others. Similarly, Margry's statement concerning youth tourism that "they merely engage in a (yearly) holiday within the reach of their budget and social peers, in which all the experiences of a vacation must take place" adds little in the way of critical insight.

To specifically respond to Margry's concerns about our own moral tone, while we make no claim for having conducted a neutral analysis, it is possible that a certain degree of irony has been overlooked. The concern over who defines certain tourist motivations and performances as morally superior or inferior was central to our opening article and remains a key concern (see also Mostafanezhad and Hannam 2014 for further analysis of moral encounters in tourism). The answer to this problem in the short term is not to develop further typologies but to develop research agendas that ask different questions. Finally, we should note that all of the contributing authors were disappointed that that the participants in this research probe are Western scholars when other voices might usefully have been incorporated.

References

DURKHEIM, E. (1915). The *Elementary Forms* of *Religious Ufe.* Oxford: Oxford World's Classics.

MOSTAFANEZHAD, M. and HANNAM, K. (Eds) (2014). *Moral Encounters in Tourism.* London. Ashgate.

KYRLEZHEV, A. (2008). The Postsecular Age: Religion and Culture Today. *Religion, State* and Society 36: 21–31.

Self-expression and Play: Can Religious Tourism Be Hedonistic?

Sean Lochrie, Ian W. F. Baxter, Elaine Collinson, Ross Curran, Martin Joseph Gannon, Babak Taheri, Jamie Thompson and Ozge Yalinay

ABSTRACT

Using data collected from 538 Iranian tourists undertaking the religious pilgrimage of Umrah (i.e. voluntary travel to the holy city of Mecca at any time throughout the year), this study investigates the concept of *play* and its relationship with self-expression and hedonism in an Islamic tourism context. By testing a theoretically derived structural model, the findings suggest that self-expression strongly influences tourists' sense of play. Here, *play* is realised when tourists feel that they can express themselves freely and augment their self-image while travelling, irrespective of the religious context. Nonetheless, the pilgrimage environment *is* characterised by the religious congregation – with tourists engaging in communal experiences that reflect themselves and their faith. Therefore, the ideal experience, where tourists feel a sense of escapism from the pressures of everyday life, is a product of their ability to project, develop, and ratify their self-concept. Finally, the study suggests that pilgrimage managers and marketers should focus on the importance of play, enabling their destinations to heighten the intensity of the 'enjoyable' elements of religious travel (e.g. group camaraderie, escapism and positive emotional reactions) alongside their inherent religious benefits.

Introduction

Islamic tourism represents the unusual union of religion and leisure and serves as one notable example of the contemporary interpretation of *pilgrimage* (Jafari & Scott, 2014). Unlike its mass-market counterparts, Islamic tourism is not typically considered to embody the hedonic characteristics of travel (Henderson, 2009). Instead, the emphasis is placed on its inherent religious and spiritual importance, underpinned by opportunities for self-reflection and communal worship (Sonmez, 2001). Nonetheless, given its fundamental association with the movement of people across national and international borders, it is unsurprising that interest in 'pilgrimage' continues to flourish within the domain of tourism research (Aukland, 2017; Gannon et al., 2017).

Similarly, many religious sites are internationally recognised as holding significant cultural importance, with visitor numbers to match (Fleischer, 2000; Jutla, 2002). For example, two million visitors travelled to the holy city of Mecca while undertaking *Hajj* in 2017 (Smith, 2017). While studies debate the differences between 'tourist' and 'pilgrim', and their motivations for visiting sacred destinations (Sharpley & Jepson, 2011; Hellman, 2017), pilgrims nonetheless exhibit many of the same characteristics, behaviours, and expectations as general tourists (Gannon et al., 2017). As such, while the spiritual disposition of pilgrims has traditionally been given priority over the materialistic or secular outcomes emerging from such travel (Nyaupane, Timothy, & Poudel, 2015); recent studies question this approach (Taheri, 2015). Thus, this study aims to further explore this disparity by considering some of the less-spiritual characteristics of pilgrimage. In order to do so, we examine whether *'self-expression' and 'play' stimulate perceived hedonic value within the context of Islamic tourism.*

To this end, religious tourists undertaking pilgrimage do not eschew everything we have come to know about their secular counterparts. Such pilgrims still seek high-quality experiences, opportunities to socialise, appropriate accommodation, and the prospect of experiencing and ingratiating themselves with the local culture (Gannon et al., 2017). As such, despite its haughty and edifying characteristics, pilgrimage may offer opportunities for Islamic tourists to engage in 'play'. Play is an important emotional state or activity and is defined by Barnett (2000, p. 442) as a 'behavioural disposition, characterised by pleasure, enjoyment, freedom and spontaneity, which elicits engagement by participants and which is manifest in a variety of different forms'.

Given its focus on sensuality and euphoria, it is understandable that play has been afforded considerable attention within extant tourism discourse (Gyimóthy & Mykletun, 2004), with some contextual focus on the importance of sacred rituals and pilgrimage (Belk, Wallendorf, & Sherry, 1989). However, other significant elements of the tourist experience relate to self-expression and hedonism. Self-expression is associated with involvement and consumption. Simply, participation, whether by obtaining a specific product or actively *being* in a particular situation, goes some way to demonstrating to others *who* the individual is and acts as a medium through which tourists can augment their self-image (Prayag & Hosany, 2014; Whiting & Hannam, 2015). Further, within the consumption experience hedonism relates to intangible, multi-sensual emotions felt by consumers (Hirschman & Holbrook, 1982) and is regularly considered when assessing tourists' travel evaluations.

Both self-expression and hedonism are intricately associated with the notion of play (McGinnis, Gentry, & Gao, 2008; Wearing, McDonald, & Wearing, 2013). Nonetheless, despite their ubiquity in tourism research, there remains little investigation into neither the relationships between self-expression, play and hedonic value nor their importance specific to the Islamic tourism context (Taheri, 2015). Therefore, this study is underpinned by a desire to examine whether self-expression and play stimulate perceived hedonic value within this context. By addressing this gap, it contributes to the growing body of knowledge on Islamic tourism by enhancing nascent understanding of the effect of self-expression on *play,* and the subsequent impact of both on the hedonic value derived from this under-researched form of tourism (Olsen & Timothy, 2006). This is underpinned by the notion that when tourists express themselves and openly engage with consumption experiences, their sense of 'play' and the perceived hedonic value derived from such travel increases.

The paper continues by first discussing literature focused on the concepts of play (underpinned by *communitas*, *flow* and *ecstasy*), self-expression, and hedonism. Next, hypotheses and a newly developed conceptual model are presented, with the structural model subsequently tested using data collected from Iranian tourists travelling to Mecca, Saudi Arabia, on the voluntary religious pilgrimage of *Umrah*. While important in terms of conveying religious devotion, Umrah can nonetheless 'be undertaken at any point throughout the year and is not considered compulsory' when compared to Hajj, which is obligatory and considered 'one of the five pillars of Islam' (Gannon et al., 2017, pp. 448–449). In doing so, this study also advances the application of partial least squares structural equation modelling (PLS-SEM) in tourism research. The paper concludes by discussing the theoretical and managerial implications of the findings, with suggestions for future research also presented.

Literature review

Academic interest in the psychological antecedents of religious tourism is long established, with emphasis often placed on the inherent religiosity, spirituality, and sociality of such travel (Delener, 1994; Mathras, Cohen, Mandel, & Mick, 2016). Having shaped much of human history, religiosity nonetheless remains an important contemporary cultural phenomenon capable of influencing individuals' attitudes, values and behaviours (Eid & El-Gohary, 2015; Spilka, Hood, Hunsberger, & Gorsuch, 2003; Zamani-Farahani & Musa, 2012). Schmidt et al. (1999, p. 10) define religion as 'systems of meaning embodied in a pattern of life, a community of faith, and a worldview that articulates a view of the sacred and of what ultimately matters'. From a psychological perspective and by adapting Schmidt et al.'s (1999) definition, Mathras et al. (2016) and Saroglou (2011) stress that religion influences consumer behaviour via four dimensions: beliefs, rituals, values, and community. In marketing studies, the concept of religion is largely concerned with segmentation, brand relationships, and compensatory consumption (Spilka et al., 2003). However, Islamic marketing literature also highlights the influence of religion on consumer behaviour and purchasing decisions (El-Bassiouny, 2014; Taheri, 2015; Fam, Waller, & Erdogan, 2004). As such, religious tourism can be considered a subset of general tourism, imbued with the same concerns surrounding consumption and behaviour, nonetheless differentiated by its emphasis on visiting and engaging with religious destinations (Albayrak et al., 2018; Terzidou, Scarles, & Saunders, 2018; Jafari & Scott, 2014).

To this end, Muslim tourists evaluate destinations not only through their value as pilgrimage sites (i.e. travel underpinned by the importance of Islamic values) but also in terms of their leisure offerings (i.e. experientially engaging in the cultural consumption process and interacting with others) (Albayrak et al., 2018; Jafari & Scott, 2014). Psychological constructs such as cosmopolitanism, self-identity, socialisation, engagement, emotional connection, and involvement (Albayrak et al., 2018; Gannon et al., 2017; Jafari & Scott, 2014; Taheri, Gori, O'Gorman, Hogg, & Farrington, 2016; Terzidou et al., 2018), have been studied within the context of religious tourism. However, extant research in the area is not exhaustive, and our overall understanding of the psychological constructs that influence Islamic tourism remains incomplete. As such, by considering self-expression, play, and hedonism within this distinct context, this study may shed further light on the assertion that Islamic tourism is underpinned by the two-way interaction between the tourist and the destinations that they visit.

Play and pilgrimage

Definitions of *'play'* often emerge from the notion of self-determined freedom without restrictions (Turner, 1969). Across research exploring consumption, play is a prevalent construct (McGinnis, Gentry, & Gao, 2012), and McGinnis and Gentry (2004, p. 407) argue that when individuals experience 'play' in the context of undertakings perceived as sacred, it makes 'the meaning of [the] activity more salient and more memorable'. They continue by asserting that, 'play in the context of a sacred ritual experience can be defined in terms of communitas, flow, and ecstasy' (McGinnis & Gentry, 2004, p. 407). As such, in order to investigate this assertion further, we turn to Islamic pilgrimage and Islamic tourism as a less immediately obvious vessel for *play*, nonetheless endowed with the requisite mechanisms to stimulate feelings of communitas, flow, and ecstasy in those undertaking such travel.

Pilgrimage represents the 'journey resulting from religious causes, externally to a holy site, and internally for spiritual purposes and internal understanding' (Barber, 1993, p. 1) and, given its emphasis on both the journey and destination, it has long served to interest tourism scholars (Collins-Kreiner, 2010; Moufahim & Lichrou, 2019). Here, the focus has been placed on exploring and understanding the characteristics, motivations, and experiences of those undertaking religious travel (Collins-Kreiner & Kliot, 2000; Devereux & Carnegie, 2006). To this end, extant research recognises the role of pilgrims as consumers, albeit with an emphasis on experiencing, interacting with, and engaging with objects and artefacts of established spiritual importance (Scott & Maclaren, 2013; Moufahim & Lichrou, 2019). Therefore, it is important to recognise that while *pilgrimage* is underpinned by spirituality and religiosity, this is not necessarily religious travellers' sole concern *in practice* (Gannon et al., 2017). Indeed, this study is underpinned by the assumption that pilgrims share many of the same characteristics and motivations as non-religious tourists (Collins-Kreiner, 2010), moving beyond the functional and reductive notion that pilgrimage should be considered exclusively as 'journeying to a sacred place or shrine as a devotee' (Griffin, 2007, p. 16). Indeed, extant research suggests that religious travel, including pilgrimage, can be imbued with a range of hedonic aspects from which pleasure is derived (e.g. souvenir purchases, general recreation, and opportunities for relaxation) (Olsen & Timothy, 2006); social considerations, such as sharing the pilgrimage experience with family and friends, or of feeling a part in a wider collective of likeminded individuals (Gannon et al., 2017); and hedonic feelings of immersion, involvement, complicity and participation (Hellman, 2017) – which together may reinforce notions of play. Therefore, given the aforementioned potential for religious tourism to stimulate feelings of communitas, flow, and ecstasy, irrespective of the Islamic tourism context, we respect Taheri et al.'s (2016) call for conceptualising play as a second-order multi-dimensional construct comprised of three sub-scales: (1) communitas, (2) flow, and (3) ecstasy. Here, we thus echo Vukonic (1996), who suggests that pilgrims undertake religious travel to satisfy both their spiritual and material needs, further blurring the distinction between traditional pilgrim and contemporary religious tourist.

Play: communitas, flow and ecstasy

As mentioned prior, play is conceptualised by Taheri et al. (2016) as a second-order multi-dimensional construct underpinned by three first-order dimensions: communitas, flow, and ecstasy. Here, communitas is described as a situation where the social order is 'an undifferentiated, homogeneous whole, in which individuals confront one another integrally, and not 'segmentalised' into statuses and roles' (Turner, 1969, p. 177). Communitas therefore represents an exceptional experience shared with others. When a sense of communitas is manifest, individuals' existing social, professional and economic characteristics are secondary – with the shared experience and emotions attached forthwith considered paramount (Sharpe, 2005; Wichmann, 2017). As a result, there is a sense of belonging; a harmony with those sharing the experience and a sense of mutuality, offering the opportunity to gain a greater understanding of the place, beliefs, and experiences converging to produce this communal spirit of togetherness (Wichmann, 2017).

Communitas serves as the catalyst for developing connections to like-minded individuals, with emphasis often placed on those holding similar interests with regards to religion, sport, or music and the sense of belonging that can emerge from such interactions. With this in mind, some tourists are motivated by travel that provides them with the opportunity to meet new people (Cordina, Gannon, & Croall, 2018), whereas others consider it a fortunate by-product of tourism.

The importance of communitas is central to discourse on collective and experiential consumption activities, such as festival tourism (Crompton & McKay, 1997), white-water rafting (Arnould & Price, 1993), skydiving (Celsi, Rose, & Leigh, 1993) and the World Gymnaestrada (Wichmann, 2017). Communitas has also been investigated with regards to tourists' consumption of the 'sacred' (McGinnis et al., 2012). Further, the 'need' to experience communitas, and the desire to revel in the spirit of community, has an impact on tourists' perceptions of the sanctity of certain pursuits, and it can thus solidify and develop consumption practices and subsequently act as 'vehicles for experiencing the sacred' (Belk et al., 1989, p. 1).

There are a number of ways in which communitas is engendered through shared experiences in places of religious importance. Pilgrimage occurs in all major religions (Collins-Kreiner & Kliot, 2000; Olsen & Timothy, 2006; Jutla, 2002) and is generally considered as serving a dual purpose; internally to reinforce faith and religiosity, and externally as the functional travel to sites of established spiritual importance (Barber, 1993). However, this reductive view of pilgrimage is being superseded by acknowledgement of the importance of the social elements of pilgrimage, which extends beyond the religious emphasis on friendship and family (Hyde & Harman, 2011).

Therefore, the second dimension underpinning play is *flow* (Taheri et al., 2016), which is described as 'the holistic experience people feel when they act with total involvement' (Csikszentmihalyi, 1975, p. 36). When experiencing *flow,* individuals lose track of their self-consciousness and time, making such experiences both unforgettable and effortless (Vitterso, Vorkinn, & Vistad, 2001). As with communitas, the links between flow and pilgrimage have been acknowledged (Turner, 1969). Here, those who experience flow may believe that the domain in which it occurred is sacred (McGinnis et al., 2008). As a result, tourists often remember those places that they have visited – where they have felt a sense of flow – with admiration and pleasure (McGinnis et al., 2008). Research has suggested that flow is core to the leisure experience (Heo, Lee, McCormick, & Pedersen, 2010), due to the acknowledgment that contemporary leisure activities (and indeed tourists) are underpinned by a focus on providing/gaining specific experiences (Vespestad & Lindberg, 2011).

As such, discourse suggests that destination marketers should focus on encouraging feelings of escapism, adventure, and participative involvement among tourists (Mannell & Iso-Ahola, 1987). Tourism stimulates the ideal circumstances for experiencing intensified states of being (i.e. flow) (Heo et al., 2010) and, due to the exceptional nature of many travel experiences; flow is considered enchanting and addictive (Celsi et al., 1993). Therefore, it consequently affects the tourists' satisfaction levels and intentions to revisit or recommend destinations (Gannon et al., 2017; Thompson et al., 2018). To this end, numerous studies have identified the relationship between flow and hedonic value (McGinnis et al., 2008) and the subsequent derivation of positive emotions and enjoyment (Wu & Liang, 2011).

Finally, ecstasy and its related emotional concepts (i.e. overwhelming joy and pleasure) are associated with an individual's attendance or consumption of events, activities, and/or experiences (Belk et al., 1989). Such consumption experiences are considered positive if feelings of ecstasy are stimulated in the consumer (Fillis & Mackay, 2014), and this intertwines with the other components of play (McGinnis et al., 2008). Similar to communitas and flow, ecstasy emerges from activities that require immersion in the process (Taheri et al., 2016). Across tourism studies, play also affects consumers' participation in interactive environments (Taheri et al., 2016). For example, exploring adventure tourism in the Artic, Gyimóthy and Mykletun (2004) found that play was associated with the adventurers' pursuit of 'high arousal'. Ecstasy has been linked to attendance at religious sites, events, and pilgrimage (McGinnis & Gentry, 2004; Turner & Turner, 1978; Taylor & Taylor, 1997).

There is a natural spiritual connection associated with attendance at religious events, which can reassure one's faith, result in pleasure through a sense of belonging and togetherness, or simply emerge through a sense of awe and wonder inspired by the sanctity of the place and/or journey (Mazumdar & Mazumdar, 2004; Taylor & Taylor, 1997). This uplifting sense of spirituality and ratification of faith can result in elation, which transcends that of normal, everyday religious experiences (Belk et al., 1989; Grappi & Montanari, 2011; Jafari & Scott, 2014). As such, as vehicles for play, religious tourism and the pilgrimage environment may stimulate feelings and outcomes that are not necessarily considered 'religious', with emphasis on the non-spiritual, intrinsic, and hedonic value derived by the tourist in such circumstances. Indeed, Eid and El-Gohary (2015) highlight three analogous constructs which reinforce the importance of play in stimulating satisfaction within an Islamic tourism context: the traditional *social* and *emotional* value derived from such travel, alongside value underpinned by the 'Islamic non-physical attributes'. As such, given the ubiquitous importance placed upon shared worship (i.e. social value) and its inherent ability emotionally stimulate tourists (Henderson, 2009), the pilgrimage context may provide the ideal circumstances to foster the sense of communitas and ecstasy crucial to ensuring tourists experience 'play' (Taheri et al., 2016).

Self-expression and hedonism

Self-expression is 'the degree to which participants express their self-concept or individuality through the situation or object of study' (Gross & Brown, 2008, p. 1141). Self-expression is linked to the concept of conspicuous consumption, implying that individuals derive pleasure from activities which demonstrate their stature, knowledge, tastes, and social-standing (Schau, Gilly, & Wolfinbarger, 2009). Stemming from the suggestion that behaviour is influenced by a desire to confirm self-image, one of the core facets of conspicuous consumption is that individuals use products to convey their identity to others (Kleine, Kleine, & Kernan, 1993). In the tourism context, self-expression is important (Prayag & Hosany, 2014; Whiting & Hannam, 2015), as it serves as the symbolic meaning people give to specific leisure activities (Gross & Brown, 2006). To this end, destination-choice and participation in activities therein serve as a means to express individuality and contribute to crafting the impression tourists hope to convey to others (Kyle & Chick, 2002).

Self-image has a role in influencing the destinations that individuals elect to visit (Moore, Cushman, & Simmons, 1995), with many studies considering the role of self-expression in exploring elements of the tourist experience. For example, research contends that, if an individual ascribes robust personal meaning to a leisure activity, they are more likely to experience feelings of ecstasy or hedonic value (Jang, Lee, Park, & Stokowski, 2000; Prayag & Hosany, 2014; Wearing et al., 2013). Others highlight the positive relationship between self-expression and elements of 'play' (Hopkinson & Pujari, 1999; McGinnis et al., 2008). Here, when exploring enduring involvement with extended service encounters, McGinnis et al. (2008) found positive relationships between flow and communitas on self-expression. This is supported by research identifying the links between self-expression and communitas (McGinnis & Gentry, 2004). These studies highlight that *meaning* in consumption often ascends from self-expression, communitas, and a close connection to others (Hopkinson & Pujari, 1999).

As such, extant research highlights the associations between a tourist's ability to express their individuality through consumption experiences imbued with elements of play – communitas, flow and ecstasy (Hopkinson & Pujari, 1999; McGinnis & Gentry, 2004). An individual has the potential to experience play if they are in an environment, or have access to a specific artefact, which helps to convey and amplify their self-image (McGinnis et al., 2008). Indeed, within the Islamic tourism context this may hold greater emotional value (Moufahim & Lichrou, 2019), where the very act of undertaking pilgrimage can help to strengthen an individual's religious identity, with travel and any associated consumption on-site serving as the physical manifestation of this religious self-expression and the sense of flow and communitas experienced therein (Eickelman & Piscatori, 2013). It is thus likely that self-expression positively influences play and, within the context of those travelling for Umrah, this study therefore hypothesises:

H1. Self-expression is positively related to play.

Further, within the consumption experience, hedonism relates to the intangible, multi-sensual emotions felt by consumers (Hirschman & Holbrook, 1982). Through consumption, hedonism stimulates feelings of arousal, fantasy, enjoyment and pleasure (Babin, Darden, & Griffin, 1994; Grappi & Montanari, 2011). In the post-consumption evaluation process, hedonism features significantly (Babin et al., 1994). It is important to tourism (Hightower, Brady, & & Baker, 2002), where the purchase of tourism products and services are often inherently hedonic in nature (Curran et al., 2018). Hedonism is also positively associated with high levels of satisfaction and is linked to feelings of excitement and ecstasy (Dunham, Freeman, & Liedtka, 2006). Therefore, hedonism is a fundamental element of experiential consumption (e.g. tourism) and lives-long within an individual's memory (Mannell & Kleiber, 1997).

Hedonism is also associated with communitas (Arnould & Price, 1993), with Hopkinson and Pujari (1999, p. 274) highlighting that when '... the individual enters a transcending community of camaraderie the individual also experiences ecstasy and flow'. Therefore, hedonism, ecstasy and flow can be regarded as closely associated (Hopkinson & Pujari, 1999; McGinnis et al., 2008). However, this is not necessarily solely a contemporary concern. Historically, the purpose of undertaking a pilgrimage was partially based on the hedonic value derived from the journey, process, and experience. This is based on the notion that pilgrimage offers travellers 'an adventure, an escape, or a chance to experience worldly pleasures denied them at home' (Hyde & Harman, 2011, p. 1345). However, as time has elapsed, and the religious importance of tourism and heritage sites has continued to develop, the personal pleasure derived from visiting such sites has been considered secondary (Gannon et al., 2017). Nonetheless, extant research does provide evidence to support the hedonic value of the journey and experience manifest post-pilgrimage, particularly in an Islamic context (Gannon et al., 2017; Taheri, 2015).

Hedonism is also important in establishing tourists' satisfaction and memories, as it has the ability to influence their future behaviours (Wearing et al., 2013). For example, the hedonic feelings of patronage are positively associated with the intention to recommend an experience to others (Grappi & Montanari, 2011; Hosany & Gilbert, 2010). Service providers must, therefore, ensure that their offering fosters hedonistic feelings (Arnold & Reynolds, 2012; Mason & Paggiaro, 2012). In doing so, they may take advantage of the emotional value experienced while undertaking pilgrimage, where visitors derive pleasure from the positive emotions engender through visiting such sites (Gannon et al., 2017). Just as leisure tourists derive pleasure from positive travel experiences (Mason & Paggiaro, 2012); pilgrims may derive hedonic value when their visit exceeds their expectations. For those tourists already familiar with the significance of a pilgrimage site, their interest can be solidified through visitation and the experience itself. Indeed, such travel is encouraged in Islam, where believers are encouraged to experience the 'Greatness of God' through travel (Eid & El-Gohary, 2015; Jafari & Scott, 2014; Taheri, 2015). This can be self-affirming in so much as their belief can be reinforced, or their decision to travel what is often significant distance to a site can be justified to themselves on the basis that the experience was enjoyable, informative, and reassuring – with hedonic value derived from this experiential faith affirmation.

Nonetheless, some scholars suggest that hedonism and religiosity occupy different ends of the psychological spectrum (Roccas, 2005; O'Shaughnessy & Jackson-O'Shaughnessy, 2002). Yet, pilgrimage represents a vessel for symbolic consumption, 'where the sacred and the profane overlap and where exploration of the relationship between religion, spirituality, and consumption can be undertaken' (Moufahim & Lichrou, 2019, p. 322). While the hedonic value of symbolic consumption is long established (Allen, 2001), the sanctity of Islamic tourism represents an interesting, under-explored manifestation of this couched within the confines of religion (Moufahim & Lichrou, 2019). To this end, and within the context of Islamic tourism, Eid and El-Gohary (2015) suggest that the hedonic value derived from consumption contributes to the level of satisfaction experienced by Muslim tourists.

As such, both play and self-expression are expected to have a positive influence on hedonism (Arnould & Price, 1993; Hopkinson & Pujari, 1999; McGinnis et al., 2008). Given that play is related to feelings of heightened enjoyment, the loss of one's self in the moment, ability to escape everyday reality, and joy derived from a connection with similar groups of people (Wu & Liang, 2011), the relationship to the stimulation of hedonic feelings is expected to be positive. The ability for an individual to project their self-concept and express themselves through an experience is also suggested to link positively to hedonism (Prayag & Hosany, 2014; Schau et al., 2009). Therefore, as per Figure 1, this study hypothesises:

H2. Play is positively related to hedonism.

H3. Self-expression is positively related to hedonism.

Finally, while tourism literature hypothesises the direct effect of self-expression on hedonism (Eid & El-Gohary, 2015; Gannon et al., 2017), and the direct effect of play on hedonism (McGinnis et al., 2012; Taheri, Farrington, Gori, Hogg, & O'Gorman, 2017), few studies evaluate the indirect effects of play on the relationship between self-expression and hedonism. Hence, this study proposes the following hypothesis centred on examining the mediating role of play:

H4: Play mediates the relationship between self-expression and hedonism.

Research methodology

Data collection

The focus of this study is Iranian tourists to the holy city of Mecca (Saudi Arabia) for the purpose of Umrah. Jafari and Scott (2014) describe Umrah as a smaller pilgrimage experience and a voluntary trip when compared to the compulsory Hajj. Hajj is therefore considered a more serious religious pilgrimage experience, as it is obligatory for Muslims to undertake it at-least once in their lifetime, during a specific period, whereas Umrah can be undertaken at any time throughout the year (Gannon et al., 2017). The data was collected via a self-reported survey of Iranian tourists who had undertaken Umrah on organised tours with travel agencies, and who had subsequently returned to Iran. The agents used paper copies of the survey to collect data, and one member of the research team remained present to answer any questions arising from participants. Overall, 650 surveys were distributed, with 538 usable responses obtained. Here, Hair, Sarstedt, Ringle, and Gudergan (2017) suggest that the sample size must be ten times larger than the highest number of items used to capture a single construct. In this study, the highest number of items belongs to the ecstasy, communitas and flow constructs (yielding a minimum sample size of **70**). The G*Power demonstrates that the sample size for this study is appropriate as, to generate a power of 0.95 for the conceptual model, the minimum sample required would be 138 respondents (Faul, Erdfelder, Buchner, & Lang, 2009).

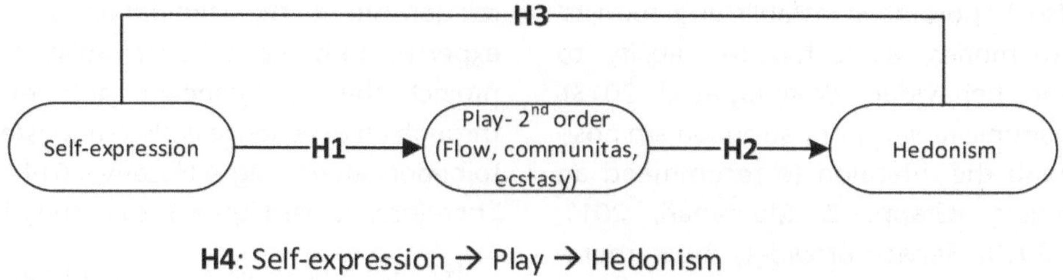

Figure 1. Conceptual model.

A back-translation approach was used to convert the survey into Farsi, the native language of Iran, in order to ensure an accurate translation. A pilot test was undertaken, which was vital in confirming that all statements within the survey were clear to participants. Overall, the sample was comprised of 47.4% male and 52.6% female. 10.5% of participants were between 18 and 35 years old, 14.5% were between 36 and 45 years old, 29% were between 46 and 55 years, and 46.1% of the sample was 56 years old or older. Non-response bias was also tested. Early and late versions of the questionnaire were compared for any systematic differences in socio-demographic attributes, with none identified between these groups (Armstrong & Overton, 1977).

With regards to the quality of the data collected, the possibility of the violation of common method bias was also tested for. As the dependent and independent variables were collected from the same destination in a single contextual setting, common method bias could have become a problem for this study (Podsakoff, MacKenzie, Lee, & Podsakoff, 2003). In response to this, the survey also included several additional construct items in order to dilute the probability of participants guessing the potential relationships outlined in the conceptual model. Further, the dependent and independent variables were allocated to different parts of the questionnaire. Two statistical tests were also used to detect possible threats: (1) Harman's single-factor test was used to test for common method variance (CMV). The results of the exploratory factor analysis (EFA) exposed six factors with Eigenvalues above 1, and the highest portion of variance described by one factor was 40.641% (Podsakoff et al., 2003). (2) Following Liang et al.'s (2007) procedure, a common method factor was presented to the PLS model. The common method factor embraces all indicators of the model. While the constructs of the model explain 0.66 on average, the common method factor only explains 0.01 on average, yielding an acceptable ratio of 66:1. Hence, this suggests that CMV is not a concern in this study.

To ensure content validity, the statements of all constructs were adapted from scales established in the extant literature. All items were measured with a 7-point Likert scale (from 1 = 'strongly disagree' to 7 = 'strongly agree'). Self-expression has six items and was adapted from Gross and Brown (2006, 2008). Further, respondents indicated their level of agreement regarding general involvement and feelings towards tourism activities. Play, the second-order construct, was formed reflectively by three first-order dimensions, adapted from Taheri et al. (2016) and McGinnis et al. (2012): (1) Communitas (6 items), (2) Flow (7 items), and (3) Ecstasy (7 items). Thus, the second-order play construct is comprised of 20 items in total. For each, the respondents indicated their level of agreement during their Umrah experience. Hedonism was measured with 4 items borrowed from Grappi and Montanari (2011). Again, the respondents indicated their level of agreement regarding their travel experience while undertaking Umrah.

Analytical technique

Partial least squares structural equation modelling (PLS-SEM) was employed to analyse the research model for a variety of reasons. First, PLS-SEM does not require the data to be normally distributed (Hair et al., 2017; Smith & Ram, 2017). For this study, a multivariate normality examination (by calculating z-scores for kurtosis and skewness for all items) was used. The results showed that some items held skewness and kurtosis above the mandatory cut-off point of either −3 or + 3 (Wells, Taheri, Gregory-Smith, & Manika, 2016). Second, PLS-SEM can deal with complex models compared to covariance-based SEM (Wells et al., 2016). Third, it can be used in reflective, formative and higher-order modes (Hair et al., 2017; Lee, Hallak, & Sardeshmukh, 2016; Taheri, Jafari, & O'Gorman, 2014). In this study, both higher-order and reflective constructs were used, and the measurement and structural models were investigated within SmartPLS 3.0 (Ringle, Wende, & Becker, 2014).

Results

Analysis of reflective measurement models

The reliability and validity of the reflective constructs were tested using composite reliability (CR), Cronbach's

Table 1. Assessment of the measurement model.

Construct item	Loading	CR	AVE	α
Play – Ecstasy		0.733	0.524	0.733
I often lose control when I undertake religious experiences due to extreme excitement	0.773			
When I undertake religious experiences, I often become spiritually uplifted	0.744			
When I undertake religious experiences, I often lose my mind in rapture	0.760			
When I undertake religious experiences, I get caught in a state of euphoria	0.752			
When I undertake religious experiences, I easily become ecstatic	0.746			
When I experience religion, I often experience a joy that is beyond measure	0.713			
When I undertake religious experiences, I often become elated	0.787			
Play – Communitas		0.928	0.682	0.906
When I undertake religious experiences, I feel a sense of camaraderie	0.863			
When I undertake religious experiences, I feel a bond with my fellow visitors that I could not experience away from the experience	0.756			
When I undertake religious experiences, I feel a sense of belonging with others	0.781			
When I undertake religious experiences, I feel a sense of harmony with the others	0.907			
When I undertake religious experiences, I feel a sense of sharing feelings with the people there	0.844			
This religious experience really allows me to get to know my fellow visitors	0.796			
Play – Flow		0.907	0.585	0.880
This religious experience is an enjoyable release from the everyday grind	0.745			
This religious experience is the best way to relieve my stress	0.720			
When involved in religious experiences, it receives my total concentration	0.806			
When involved in religious experiences, I am surprised to find I have lost track of time	0.734			
When involved in religious experiences, I can become totally involved in what I am doing	0.835			
When involved in religious experiences, time seems to rush by quickly	0.812			
I find a religious experience to be very liberating	0.785			
Self-expression		0.891	0.581	0.852
When I engage in tourism experiences I can really be myself	0.804			
When I engage in tourism experiences, others see me the way I want them to see me	0.760			
My choice of tourism experiences says a lot about who I am	0.786			
Where I engage in tourism experiences gives a glimpse of the type of person I am	0.734			
You can tell a lot about a person by whether or not they engage in tourism experiences	0.824			
My tourism experiences are a reflection of my lifestyle	0.836			
Hedonism		0.851	0.613	0.751
This experience was truly enjoyable	0.901			
I truly felt it was an escape	0.843			
I enjoyed the experience for its own sake	0.908			
I truly felt delighted	0.725			

Alpha (α), factor loadings, and average variance extracted (AVE). The factor loadings, CR, and α all showed values above the obligatory threshold of 0.7 (Hair, Black, Babin, & Anderson, 2010). The AVE surpassed the threshold of 0.5 for all constructs (Hair et al., 2010) (Table 1). Finally, discriminant validity was tested following Fornell and Larcker's (1981) criterion, which requires a construct's AVE to be larger than the square of its largest correlation with any other construct (Table 2). Further, Henseler, Ringle, and Sarstedt (2015) highlight that the heterotrait-monotrait ratio of correlations (HTMT) approach indicates superior performance (by means of a Monte Carlo simulation) when compared to the Fornell-Larcker criterion. In this study, all HTMT values were below the threshold (0.85), ranging from 0.172 to 0.675. $HTMT_{inference}$ criterion (using complete bootstrapping) was calculated; all HTMT values differed significantly from 1 (ranging from 0.358–0.721) (Wells et al., 2016). Thus, discriminant validity was established.

Analysis of the play higher-order model

In order to establish 'play' as a higher-order construct with a second-order factor characterised reflectively by three first-order dimensions, an EFA test with an Oblique Rotation was performed. Here, the results demonstrated that the second-order 'play' construct was represented by 3 underlying dimensions. All items' loadings were above the minimum threshold (≥0.5) (Hair et al., 2010). As indicated in Table 1, the three first-order constructs (*communitas, flow* and *ecstasy*) showed values above the recommended suggestion for CR, Cronbach's Alpha, and AVE. Finally, the repeated measures approach for estimation of the hierarchal component models (HCMs) in PLS was employed (Becker, Klein, & Wetzels, 2012; Lee et al., 2016). Following the HCMs procedure, all items for the three first-order constructs were assigned to their respective constructs reflectively. All 20 items underpinning 'play' were connected reflectively to the second-order play construct (Figure 2) and the three first-order constructs were reflectively linked to the second-order construct. The results indicated that the relationship between the three first-order constructs and the second-order construct were strong and significant ($p < .05$) with R^2 values higher than 0.5. Further, the play construct explained more than 50% of the variance in its respective single-order constructs (Figure 2). As such, the results supported the

Table 2. Correlations between constructs.

	(1)	(2)	(3)	(4)	(5)	(6)	Mean	SD
(1) Communitas-play	0.826						4.880	1.390
(2) Ecstasy-play	0.435**	0.723					3.927	0.947
(3) Flow-play	0.513***	0.547***	0.765				4.466	1.1313
(4) Hedonism	0.332	0.555***	0.605**	0.783			4.787	1.314
(5) Play	0.540**	0.595**	0.454	0.507*	n/a		n/a	n/a
(6) Self-expression	0.572**	0.645***	0.484	0.558***	0.576*	0.762	4.874	1.416

Note: Significant at *$t > 1.96$ at $p < .05$; **$t > 2.57$ at $p < .01$; ***$t > 3.29$ at $p < .001$.

assertion that play serves as a second-order construct, represented reflectively by the three dimensions of communitas, flow, and ecstasy.

Structural model analysis

After confirming the measurement model, the hypothesised relationships among constructs was tested through PLS to determine: (1) path coefficients, (2) Effect sizes (f^2), and (3) the R^2 values of the endogenous variables (Figure 2) and Standardised Root Mean Square Residual (SRMR) fit index (Hair, Hult, Ringle, & Sarstedt, 2014; Henseler, Hubona, & Ray, 2016). A nonparametric bootstrapping procedure was used to test the structural model (Hair et al., 2017). As shown in Figure 2, self-expression has a strong positive and significant influence on play (**H1**: $\beta = 0.776$, $p < .001$; $f^2 = 0.514$). Play has a positive and significant impact on hedonism (**H2**: $\beta = 0.30$, $p < .001$; $f^2 = 0.126$). Finally, self-expression has a positive and significant influence on hedonism (**H3**: $\beta = 0.525$, $p < .001$; $f^2 = 0.281$). The f^2 values for the three hypothesised paths were all above 0.02, suggesting satisfactory effects for the endogenous latent constructs. The SRMR value was 0.046; less than the recommended value of 0.08 (Henseler et al., 2016).

Following Williams and MacKinnon (2008) and Lee et al.'s (2016) recommendation, mediation analysis within the bootstrapping method was used. This technique designates that if the direct effect between two constructs is significant, the results demonstrate partial mediation. However, if the direct effect between two constructs is not significant, the findings indicate full mediation. The results indicated that self-expression indirectly impacts on hedonism through play (CI: 0.143–0.308). As the direct influence was significant, the findings showed that play partially mediates the impact of self-expression on hedonism, confirming **H4** (Table 3).

Table 3 . Estimates of indirect paths.

Path	Path coefficient	t-values	Low CI	High CI
Self-expression→ Play→Hedonism	0.233	5.371	0.143	0.308

Note: 95% confidence intervals (CI) via bootstrapping and 5000 resamplings.

Discussion and implications

Using data collected from Iranian tourists who travelled to Mecca and experienced Umrah, this study aimed to gain a greater understanding of the concept of play and its relationship to self-expression and hedonism in a novel and under-explored context. A theoretically derived structural model was thus developed to explore the relationships between play, self-expression and hedonism within the domain of Islamic tourism. The findings highlight significant relationships among the constructs, with these relationships and theoretical implications now discussed. First, the analysis of the conceptual model reveals the potency of play as a second-order construct underpinned by communitas, flow and ecstasy. Therefore, the analysis supports the assertion that these three dimensions serve as one aggregated construct – play (McGinnis & Gentry, 2004). This reinforces previous studies which emphasise the interdependent nature of communitas, flow and ecstasy, and how together each serves as vital components of 'play' (Taheri et al., 2016). As such, this study also posits the relevance of this second-order construct in another, less obvious setting. Extant research has investigated play in nightclubs and commercial hospitality (Taheri et al., 2016), service encounters (McGinnis et al., 2008), adventure tourism (Gyimóthy & Mykletun, 2004), and sporting activities (McGinnis et al., 2012). Instead, the relevance of this construct is now established in the context of religious pilgrimage.

This paper contributes to the investigation of the self-expression construct in relation to play (McGinnis et al., 2008). Indeed, the findings suggest that self-expression has a strong positive and significant influence on play. Play is realised if (and when) an individual can express themselves and augment their self-image through a specific experience, situation or product (Gross & Brown, 2008). The findings, therefore, correspond with existing research, where individuals engaging in activities and consuming products which allow them to express their self-concept, are likely to experience feelings of exhilaration or ecstasy (Prayag & Hosany, 2014). Tourists undertaking Umrah regard it as a mechanism to expose their true selves, subsequently experiencing euphoria. Given the religious nature such travel, these

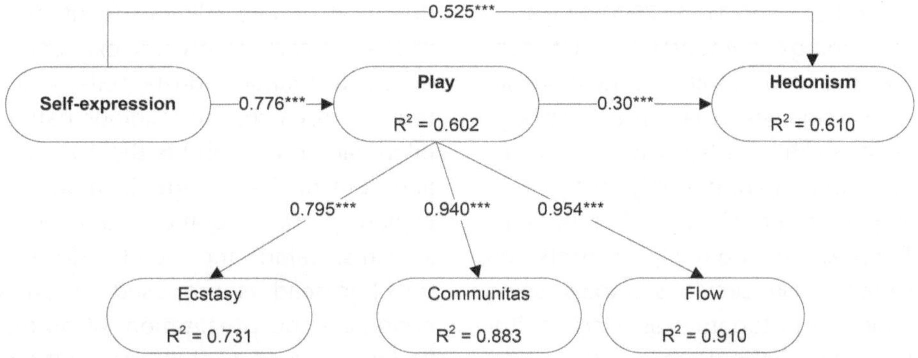

Figure 2. Results of structural model.

religious tourists are surrounded by fellow worshippers in an environment where the event expresses and reinforces one aspect of their identity (Henderson, 2009). Therefore, feelings of ecstasy are likely to ensue from this situation that reflects their self-image.

Similarly, the association between self-expression and play found in this study is also supported by prior research (e.g. McGinnis et al., 2008; Hopkinson & Pujari, 1999). Indeed the Umrah and pilgrimage environment is one characterised by a religious congregation where individuals engage in an experience enabling self-reflection. The ideal experience or flow, where individuals lose a sense of time while escaping from the pressures of everyday life, is a product of the tourist's ability to project their own self-concept in a particular social group. This reinforces research suggesting that the potency of sacred places as consumption environments stimulate feelings of flow (McGinnis et al., 2008). Further, the findings support the association between self-expression and communitas (Hopkinson & Pujari, 1999; McGinnis & Gentry, 2004). Feelings of togetherness, *belonging* and collective harmony among those experiencing Umrah and the ability to express their true identity fit well within the pilgrimage context. Studies highlight that communitas occurs during shared experiences (Gannon et al., 2017), and those travelling for Umrah experience feelings of camaraderie and shared experiences as they regard such tourism as reflecting themselves and exposing their self-concept, manifest through sharing the pilgrimage with fellow worshippers (Jafari & Scott, 2014). Given that self-expression is a significant element of the Umrah experience, and pilgrimage is a setting which brings travellers of similar beliefs together for the same purpose, the motivation to undertake shared experiences and get to know fellow travellers is expected.

The hedonic nature of consumption experiences, especially in tourism, is established (Hightower et al., 2002). However, this study sheds further light on the relationship between play and hedonism. The findings highlight that *play* has a positive and significant impact on hedonism. The pleasure gained from '*sacred*' experiences has been recognised in previous studies (Hyde & Harman, 2011; Mason & Paggiaro, 2012). The collective nature of pilgrimage, and voluntary nature of Umrah, therefore reinforces the value of Islamic tourism by providing an environment where shared experiences and a sense of belonging are typical (Eickelman & Piscatori, 2013). The findings, therefore, suggest that through a sense of amity with a situation saturated with notions of worship and social cohesion, feelings of enjoyment (with regards to both the spiritual and non-spiritual value derived from such travel) are cultivated. Therefore, the findings support previous studies by highlighting the relationship between play and hedonism (Arnould & Price, 1993; Hopkinson & Pujari, 1999), while extending Islamic tourism and consumption research by reinforcing the interconnected nature of the 'sacred' and the 'selfish' within the pilgrimage environment (Eid & El-Gohary, 2015; Moufahim & Lichrou, 2019)

The opportunity for individuals to escape their daily lives and engage in unforgettable activities transcending time is indicative of travel and tourism, and this is reflected in our findings. Specifically, the voluntary nature of Umrah and feelings of flow which result from tourists' experiences naturally stimulate delight and escapism. Similarly, feelings of ecstasy are associated with hedonism (Dunham et al., 2006). Indeed, the enjoyment of a sacred experience has been inherently associated with euphoria (Grappi & Montanari, 2011). Especially, the pilgrimage environment engenders heightened levels of ecstasy, beyond that of everyday religious experiences. Therefore, this positive relationship, particularly with regards to ecstasy, is somewhat anticipated. Nevertheless, the positive relationship between play and hedonism highlighted within this study is theoretically significant, adding to the growing body of knowledge on the non-religious, rather than spiritually traditional, outcomes of the pilgrimage experience.

Supporting other studies (Jang et al., 2000; Prayag & Hosany, 2014), the findings demonstrate that self-expression has a positive and significant relationship with hedonism. This signifies that the symbolic meaning of the activity, in this context Umrah, enables tourists to express themselves comfortably, and generates feelings of enjoyment and pleasure. The findings indicate that self-expression indirectly impacts on hedonism through play. Unlike previous studies, using self-expression and hedonism together as a progenitor for enduring involvement (McGinnis et al., 2008), this study provides different perspectives of the relationship of self-expression in relation to play and hedonism. Here, it suggests that self-expression is positively associated with play, resulting in feelings of enjoyment or hedonism; self-expression is, therefore, a significant precursor for playful and pleasurable experiences, even in the religious pilgrimage travel context.

As such, by prioritising the importance of play, destination managers and marketers involved in developing the experiential elements of Umrah can increase tourists' enjoyment of the non-religious aspects of pilgrimage. While the spiritual nature of Islamic travel remains dominant (Damari & Mansfeld, 2016), the hedonistic nature of such travel is an important element for those undertaking voluntary pilgrimage manifest as Umrah. Managers should, therefore, strive to create environments that catalyse play – where group camaraderie, escapism and positive emotional reactions are also considered sacrosanct – adding to the experiential nature of the pilgrimage in their own unique way. This could be accomplished by providing opportunities for organising group activities, e.g. day trips and shared eating spaces – enabling the interaction and shared experiences that even religious tourists crave (Gannon et al., 2017; Yalinay et al., 2018).

Further, those organising and managing Umrah should not shy away from promoting the 'playful' and self-gratifying elements of this voluntary pilgrimage (Eid & El-Gohary, 2015). Far from focussing solely on the religious importance of Umrah, managers should instead embrace the fact that it is not obligatory; instead focussing on the social and hedonic benefit that can be derived from enjoyable travel which simultaneously has the power to affirm religious faith. Indeed, as a result of its *voluntary* nature, Umrah may, in fact, represent an as-yet untapped vessel for self-expression and play in the Islamic tourism context, where religious tourists *choose* to undertake pilgrimage for a wider, more rounded and self-fulfilling purpose than its innate spiritual value. As such, understanding the importance of the interplay between self-expression, play and hedonism outlined in the findings is also useful for destination marketers. Those marketing Umrah, despite its innate religious focus, should consider using images of companionship, excitement and escapism to appeal to Islamic tourists' self-image. While not new in tourism research, the religious nature and sensitivity of pilgrimage often subdue the more pleasurable elements and 'lure of the voyage' in marketing activities. When exploring the hedonic nature of 'mass tourism' for Muslims, Jafari and Scott (2014) state that 'Islamic travel instead is proposed as an alternative to this hedonic conceptualisation of tourism'. However, our findings suggest that destination marketers must remain cognisant of both the hedonistic and spiritual nature of religious travel in order to reflect the subtlety of tourists' expectations, behaviours, and motivations inherent within the Islamic pilgrimage context.

Limitations and future research

Despite offering embryonic insight into the non-spiritual characteristics of religious tourism, the limitations of this study must be acknowledged. First, questionnaires were used to collect the data. While this approach is established and valid, it is not without weakness. A qualitative approach may provide greater insight into different perspectives in deeper descriptive depth. Therefore, future studies may wish to explore the conceptual model presented in this study using a qualitative approach. This could further inform the model and highlight interesting avenues arising tangentially from the findings of this study. Second, the specific nature of the context of this research – Iranian tourists visiting Mecca on the voluntary holy pilgrimage of Umrah – limits the generalisability of the model and findings due to its religious and geographical specificity. Future research should explore the model proposed in this study in different contexts – both religious and non-religious – in order to extend its application and test it further. Finally, as mentioned prior, the sample was restricted to Iranian visitors to Umrah. Given the inherent differences in cultures that exist across nationalities, future studies should investigate the conceptual model using a sample of visitors from other nationalities to determine its validity across various sub-sections of the Islamic tourism community.

Disclosure statement

No potential conflict of interest was reported by the authors.

References

Albayrak, T., Herstein, R., Caber, M., Drori, N., Bideci, M., & Berger, R. (2018). Exploring religious tourist experiences in Jerusalem: The intersection of Abrahamic religions. *Tourism Management, 69*, 285–296.

Allen, M. W. (2001). A practical method for uncovering the direct and indirect relationships between human values and consumer purchases. *Journal of Consumer Marketing, 18*(2), 102–120.

Armstrong, J. S., & Overton, T. S. (1977). Estimating nonresponse bias in mail surveys. *Journal of Marketing Research, 14*(3), 396–402.

Arnold, M. J., & Reynolds, K. E. (2012). Approach and avoidance motivation: Investigating hedonic consumption in a retail setting. *Journal of Retailing, 88*(3), 399–411.

Arnould, E. J., & Price, L. L. (1993). River magic: Extraordinary experience and the extended service encounter. *Journal of Consumer Research, 20*(1), 24–45.

Aukland, K. (2017). Pilgrimage expansion through tourism in contemporary India: The development and promotion of a Hindu pilgrim circuit. *Journal of Contemporary Religion, 32*(2), 283–298.

Babin, B. J., Darden, W. R., & Griffin, M. (1994). Work and/or fun: Measuring hedonic and utilitarian shopping value. *Journal of Consumer Research, 20*(4), 644–656.

Barber, R. W. (1993). *Pilgrimages*. London: Boydell Press.

Barnett, L. (2000). Play. In. J. Jafari (Ed.), *Encyclopedia of tourism* (pp. 442–444). London: Routledge.

Becker, J. M., Klein, K., & Wetzels, M. (2012). Hierarchical latent variable models in PLS-SEM: Guidelines for using reflective-formative type models. *Long Range Planning, 45*, 359–394.

Belk, R. W., Wallendorf, M., & Sherry, J. F. (1989). The sacred and the profane in consumer behavior: Theodicy on the odyssey. *Journal of Consumer Research, 16*(1), 1–38.

Celsi, R. L., Rose, R. L., & Leigh, T. W. (1993). An exploration of high-risk leisure consumption through skydiving. *Journal of Consumer Research, 20*(1), 1–23.

Collins-Kreiner, N. (2010). Researching pilgrimage: Continuity and transformations. *Annals of Tourism Research, 37*(2), 440–456.

Collins-Kreiner, N., & Kliot, N. (2000). Pilgrimage tourism in the holy land: The behavioural characteristics of Christian Pilgrims. *GeoJournal, 50*(1), 55–67.

Cordina, R., Gannon, M. J., & Croall, R. (2018). Over and over: Local fans and spectator sport tourist engagement. *The Service Industries Journal*. doi:10.1080/02642069.2018.1534962

Crompton, J. L., & McKay, S.L. (1997). Motives of visitors attending festival events. *Annals of Tourism Research, 24*(2), 425–439.

Csikszentmihalyi, M. (1975). *Beyond boredom and anxiety*. San Francisco, CA: Jossey-Bass.

Curran, R., Baxter, I. W., Collinson, E., Gannon, M. J., Lochrie, S., Taheri, B., ... Yalinay, O. (2018). The traditional marketplace: Serious leisure and recommending authentic travel. *The Service Industries Journal, 38*(15–16), 1116–1132.

Damari, C., & Mansfeld, Y. (2016). Reflections on pilgrims' identity, role and interplay with the pilgrimage environment. *Current Issues in Tourism, 19*(3), 199–222.

Delener, N. (1994). Religious contrasts in consumer decision behaviour patterns: Their dimensions and marketing implications. *European Journal of Marketing, 28*(5), 36–53.

Devereux, C., & Carnegie, E. (2006). Pilgrimage: Journeying beyond self. *Tourism Recreation Research, 31*(1), 47–56.

Dunham, L., Freeman, R. E., & Liedtka, J. (2006). Enhancing stakeholder practice: A particularized exploration of community. *Business Ethics Quarterly, 16(1)*, 23–42.

Eickelman, D. F., & Piscatori, J. (2013). *Muslim travellers: Pilgrimage, migration and the religious imagination*. London: Routledge.

Eid, R., & El-Gohary, H. (2015). The role of Islamic religiosity on the relationship between perceived value and tourist satisfaction. *Tourism Management*, *46*, 477–488.

El-Bassiouny, N. (2014). The one-billion-plus marginalization: Toward a scholarly understanding of Islamic consumers. *Journal of Business Research*, *67*, 42–49.

Fam, K. S., Waller, D. S., & Erdogan, B. Z. (2004). The influence of religion on attitudes towards the advertising of controversial products. *European Journal of Marketing*, *38*(5/6), 537–555.

Faul, F., Erdfelder, E., Buchner, A. & Lang, A. G. (2009) Statistical power analyses using G*power 3.1: Tests for correlation and regression analyses. *Behavior Research Methods*, *41*(4), 1149–1160.

Fillis, I., & Mackay, C. (2014). Moving beyond fan typologies: The impact of social integration on team loyalty in football. *Journal of Marketing Management*, *30*(3–4), 334–363.

Fleischer, A. (2000). The tourist behind the pilgrim in the holy land. *International Journal of Hospitality Management*, *19*(3), 311–326.

Fornell, C., & Larcker, D. F. (1981). Structural equation models with unobservable variables and measurement error: Algebra and statistics. *Journal of Marketing Research*, *18*(3), 382–388.

Gannon, M. J., Baxter, I. W., Collinson, E., Curran, R, Farrington, T., Glasgow, S., … Yalinay, Ozge (2017). Travelling for umrah: Destination attributes, destination image, and post-travel intentions. *The Service Industries Journal*, *37*(7–8), 448–465.

Grappi, S., & Montanari, F. (2011). The role of social identification and hedonism in affecting tourist re-patronizing behaviours: The case of an Italian festival. *Tourism Management*, *32*(5), 1128–1140.

Griffin, K. A. (2007). The globalization of pilgrimage tourism? Some thoughts from Ireland. In R. Raj & N. D. Morpeth (Eds.), *Religious tourism and pilgrimage festival management: An international perspective* (pp. 15–34). Wallingford: CAB International.

Gross, M. J., & Brown, G. (2006). Tourism experiences in a lifestyle destination setting: The roles of involvement and place attachment. *Journal of Business Research*, *59*(6), 696–700.

Gross, M. J., & Brown, G. (2008). An empirical structural model of tourists and places: Progressing involvement and place attachment into tourism. *Tourism Management*, *29*(6), 1141–1151.

Gyimóthy, S., & Mykletun, R. J. (2004). Play in adventure tourism: The case of Arctic trekking. *Annals of Tourism Research*, *31*(4), 855–878.

Hair, J. F. J., Black, W. C., Babin, B. J., & Anderson, R. E. (2010). *Multivariate data analysis: A global perspective*. Upper Saddle River, NJ: Pearson.

Hair, J. F. J., Hult, G. T. M., Ringle, C. M., & Sarstedt, M. (2014). *A primer on partial least squares structural equation modeling (PLS-SEM)*. Thousand Oaks, CA: Sage.

Hair, J. F., Sarstedt, M., Ringle, C. M., & Gudergan, S. P. (2017). *Advanced issues in partial least squares structural equation modeling*. Thousand Oaks, CA: SAGE Publications.

Hellman, J. (2017). Pilgrim guides and pilgrims in productive complicity: Making the invisible visible in west Java. *Tourist Studies*. doi:10.1177/1468797617723765.

Henderson, J. C. (2009). Islamic tourism reviewed. *Tourism Recreation Research*, *34*(2), 207–211.

Henseler, J., Hubona, G., & Ray, P. A. (2016), Using PLS path modelling in new technology research: Updated guidelines. *Industrial Management and Data Systems*, *116*(1), 2–120.

Henseler, J., Ringle, C. M., & Sarstedt, M. (2015). A new criterion for assessing discriminant validity in variance-based structural equation modeling. *Journal of the Academy of Marketing Science*, *43*(1), 115–135.

Heo, J., Lee, Y., McCormick, B. P., & Pedersen, P. M. (2010). Daily experience of serious leisure, flow and subjective well-being of older adults. *Leisure Studies*, *29*(2), 207–225.

Hightower Jr, R., Brady, M. K., & Baker, T. L. (2002). Investigating the role of the physical environment in hedonic service consumption: An exploratory study of sporting events. *Journal of Business Research*, *55*(9), 697–707.

Hirschman, E. C., & Holbrook, M. B. (1982). Hedonic consumption: Emerging concepts, methods and propositions. *The Journal of Marketing*, *46*(3), 92–101.

Hopkinson, G. C., & Pujari, D. (1999). A factor analytic study of the sources of meaning in hedonic consumption. *European Journal of Marketing*, *33*(3/4), 273–294.

Hosany, S., & Gilbert, D. (2010). Measuring tourists' emotional experiences toward hedonic holiday destinations. *Journal of Travel Research*, *49*(4), 513–526.

Hyde, K. F., & Harman, S. (2011). Motives for a secular pilgrimage to the Gallipoli battlefields. *Tourism Management*, *32*(6), 1343–1351.

Jafari, J., & Scott, N. (2014). Muslim world and its tourisms. *Annals of Tourism Research*, *44*, 1–19.

Jang, H. C., Lee, B., Park, M., & Stokowski, P. A. (2000). Measuring underlying meanings of gambling from the perspective of enduring involvement. *Journal of Travel Research*, *38*(3), 230–238.

Jutla, R. S. (2002). Understanding Sikh pilgrimage. *Tourism Recreation Research*, *27*(2), 65–72.

Kleine III, R. E., Kleine, S. S., & Kernan, J. B. (1993). Mundane consumption and the self: A social-identity perspective. *Journal of Consumer Psychology*, *2*(3), 209–235.

Kyle, G., & Chick, G. (2002). The social nature of leisure involvement. *Journal of Leisure Research*, *34*(4), 426–448.

Lee, C., Hallak, R., & Sardeshmukh, S. R. (2016). Innovation, entrepreneurship, and restaurant performance: A higher-order structural model. *Tourism Management*, *53*, 215–228.

Liang, H., Saraf, N., Hu, Q., & Xue, Y. (2007). Assimilation of enterprise systems: The effect of institutional pressures and the mediating role of top management. *MIS Quarterly*, *31*(1), 59–87.

Mannell, R. C., & Iso-Ahola, S. E. (1987). Psychological nature of leisure and tourism experience. *Annals of Tourism Research*, *14*(3), 314–331.

Mannell, R. C., & Kleiber, D. A. (1997). *A social psychology of leisure*. State College, PA: Venture Publishing Inc.

Mason, M. C., & Paggiaro, A. (2012). Investigating the role of festivalscape in culinary tourism: The case of food and wine events. *Tourism Management*, *33*(6), 1329–1336.

Mathras, D., Cohen, A. B., Mandel, N., & Mick, D. G. (2016). The effects of religion on consumer behavior: A conceptual framework and research agenda. *Journal of Consumer Psychology*, *26*(2), 298–311.

Mazumdar, S., & Mazumdar, S. (2004). Religion and place attachment: A study of sacred places. *Journal of Environmental Psychology*, *24*(3), 385–397.

McGinnis, L. P., & Gentry, J. W. (2004). Examining the mediating relationship of aplay@ on ritual enduring involvement. *Advances in Consumer Research*, 31, 405–411.

McGinnis, L. P., Gentry, J. W., & Gao, T. (2008). The impact of flow and communitas on enduring involvement in extended service encounters. *Journal of Service Research, 11*(1), 74–90.

McGinnis, L.P., Gentry, J.W., & Gao, T. (2012). Antecedents to consumer perceptions of sacredness in extended service experiences: The case of golf. *Journal of Service Research, 15(4)*, 474-488.

Moore, K., Cushman, G., & Simmons, D. (1995). Behavioral conceptualization of tourism and leisure. *Annals of Tourism Research, 22*(1), 67–85.

Moufahim, M., & Lichrou, M. (2019). Pilgrimage, consumption and rituals: Spiritual authenticity in a Shia Muslim pilgrimage. *Tourism Management, 70*, 322–332.

Nyaupane, G. P., Timothy, D. J., & Poudel, S. (2015). Understanding tourists in religious destinations: A social distance perspective. *Tourism Management, 48*, 343–353.

Olsen, D. H. & Timothy D. J. (2006). Tourism and religious journeys. In D. J. Timothy and D. H. Olsen (Eds.), *Tourism, religion and spiritual journeys* (pp. 1–22). Abingdon: Routledge.

O'Shaughnessy, J., & Jackson-O'Shaughnessy, N. (2002). Marketing, the consumer society and hedonism. *European Journal of Marketing, 36*(5/6), 524–547.

Podsakoff, P. M., MacKenzie, S. M., Lee, J., & Podsakoff, N. P. (2003). Common method variance in behavioral research: A critical review of the literature and recommended remedies. *Journal of Applied Psychology, 88*(5), 879–903.

Prayag, G., & Hosany, S. (2014). When Middle East meets West: Understanding the motives and perceptions of young tourists from United Arab Emirates. *Tourism Management, 40*, 35–45.

Ringle, C. M., Wende, S., & Becker, J. M. (2014). *SmartPLS 3.0*. http://www.smartpls.com/

Roccas, S. (2005). Religion and value systems. *Journal of Social Issues, 61*(4), 747–759.

Saroglou, V. (2011). Believing, bonding, behaving, and belonging the big four religious dimensions and cultural variation. *Journal of Cross-Cultural Psychology, 42*(8), 1320–1340.

Schau, H. J., Gilly, M. C., & Wolfinbarger, M. (2009). Consumer identity renaissance: The resurgence of identity-inspired consumption in retirement. *Journal of Consumer Research, 36*(2), 255–276.

Schmidt, R., Sager, G. C., Carney, G., Jackson, J. J., Zanca, K., Muller, A., & Jackson, J. (1999). *Patterns of religion*. Belmont, CA: Wadsworth Publishing.

Scott, L. M., & Maclaren, P. (2013). Consuming the mists and myths of Avalon. In D. Rinallo, L. M. Scott, & P. Maclaren (Eds.) *Consumption and spirituality*. (Vol. 16, pp. 195–210, p. 16). New York, NY: Routledge.

Sharpe, E. K. (2005). Delivering communitas: Wilderness adventure and the making of community. *Journal of Leisure Research*, 37(3), 255–280.

Sharpley, R., & Jepson, D. (2011). Rural tourism: A spiritual experience? *Annals of Tourism Research, 38*(1), 52–71.

Smith, L. (2017). Hajj 2017: Two million Muslim pilgrims from all over the world head to Mecca. [online]. Retrieved from: https://www.independent.co.uk/news/world/middle-east/hajj-2017-muslim-pilgrimage-mecca-islam-pilgrims-saudi-arabia-a7917851.html

Smith, M., & Ram, Y. (2017). Tourism, landscapes and cultural ecosystem services: A new research tool. *Tourism Recreation Research, 42*(1), 113–119.

Sonmez, S. (2001). Tourism behind the veil of tourism: Women and development in the Middle East. In Y. Apostolopoulos, S. Sonmez and D. J. Timothy (Eds.), *Women as producers and consumers of tourism in developing regions* (pp. 113–142). Westport, CT: Praeger.

Spilka, B., Hood, R., Hunsberger, B., & Gorsuch, R. (2003). *The psychology of religion: An empirical approach*. New York, NY: The Guilford Press.

Taheri, B. (2015). Emotional connection, materialism, and religiosity: An Islamic tourism experience. *Journal of Travel and Tourism Marketing, 33*(7), 1011–1027.

Taheri, B., Farrington, T., Gori, K., Hogg, G., O'Gorman, K. (2017). Escape, entitlement, and experience: Liminoid motivators within commercial hospitality. *International Journal of Contemporary Hospitality Management, 29*(4), 1148–1166.

Taheri, B., Gori, K., O'Gorman, K., Hogg, G., & Farrington, T. (2016). Experiential liminoid consumption: The case of nightclubbing. *Journal of Marketing Management, 32*(1–2), 19–43.

Taheri, B., Jafari, A., & O'Gorman, K. (2014). Keeping your audience: Presenting a visitor engagement scale. *Tourism Management, 42*, 321–329.

Taylor, M., & Taylor, R. (1997). Something for the weekend, sir? Leisure, ecstasy and identity in football and contemporary religion. *Leisure Studies, 16*(1), 37–49.

Terzidou, M., Scarles, C., & Saunders, M. N. K. (2018). The complexities of religious tourism motivations: Sacred places, vows and visions. *Annals of Tourism Research, 70*, 54–65.

Thompson, J., Baxter, I. W., Curran, R., Gannon, M. J., Lochrie, S., Taheri, B., & Yalinay, O. (2018). Negotiation, bargaining, and discounts: Generating WoM and local tourism development at the Tabriz bazaar, Iran. *Current Issues in Tourism, 21*(11), 1207–1214.

Turner, V. (1969). *The ritual process: Structures and antistructure*. Chicago, IL: Aldine Publishing Company.

Turner, V., & Turner, E. (1978). *Image and pilgrimage in Christian culture*. New York, NY: Columbia University Press.

Vespestad, M. K., & Lindberg, F. (2011). Understanding nature-based tourist experiences: An ontological analysis. *Current Issues in Tourism, 14*(6), 563–580.

Vitterso, J., Vorkinn, M., & Vistad, O. I. (2001). Congruence between recreational mode and actual behavior – a prerequisite for optimal experiences? *Journal of Leisure Research, 32*(3), 137–159.

Vukonic, B. (1996). *Tourism and religion*. Oxford: Pergamon.

Wearing, S. L., McDonald, M., & Wearing, M. (2013). Consumer culture, the mobilisation of the narcissistic self and adolescent deviant leisure. *Leisure Studies, 32*(4), 367–381.

Wells, V. K., Taheri, B., Gregory-Smith, D., & Manika, D. (2016). The role of generativity and attitudes on employees home and workplace water and energy saving behaviours. *Tourism Management, 56*, 63–74.

Whiting, J., & Hannam, K. (2015). Creativity, self-expression and leisure. *Leisure Studies, 34*(3), 372–384.

Wichmann, A. (2017). Participating in the world gymnaestrada: An expression and experience of community. *Leisure Studies, 36*(1), 21–38.

Williams, A., & MacKinnon, D. P. (2008). Resampling and distribution of the product methods for testing indirect effects in complex models. *Structural Equation Modeling, 15*(1), 23–51.

Wu, C. H. J., & Liang, R. D. (2011). The relationship between white-water rafting experience formation and customer reaction: A flow theory perspective. *Tourism Management, 32*(2), 317–325.

Yalinay, O., Baxter, I. W., Collinson, E., Curran, R., Gannon, M. R., Lochrie, S., … Thompson, J. (2018). Servicescape and shopping value: The role of negotiation intention, social orientation, and recreational identity at the Istanbul Grand Bazaar, Turkey. *Journal of Travel & Tourism Marketing*, *35*(9), 1132–1144.

Zamani-Farahani, H. & Musa, G. (2012). The relationship between Islamic religiosity and residents' perceptions of socio-cultural impacts of tourism in Iran: Case studies of Sare'in and Masooleh. *Tourism Management*, *33*(4), 802–814.

Pilgrimage: Journeying Beyond Self

Chris Devereux and Elizabeth Carnegie

ABSTRACT

This paper explores how the experience of pilgrimage can contribute towards the subsequent sustained transformation of individual and community well-being. In so doing it examines four areas. The first is to draw a conceptual map that emphasizes the linkages between pilgrimage and wellness tourism and explores the larger meaning of the words 'wellness' and 'spirituality'. The intention is to form a working definition and context in which to examine the well being aspects of pilgrimage and the experiences it provides. The second area focuses briefly on what spirituality and pilgrimage mean. In so doing, it goes beyond the idea of pilgrimage as a journey to a sacred place, and concentrates more on the journeying itself, the importance of community and the space the journey affords for reflection on its physical, emotional and spiritual aspects. The third area considers two case studies from the pilgrimage experience. The first of these draws on the experiences of those who have travelled overland to Santiago de Compostela in Spain, while the second considers the experiences of people who have undertaken various charity treks as another form of pilgrimage. The individual experiences, drawn from those with religious affiliations and those with none, indicate how a challenging physical and emotional journey often - but not always - results in not only an enhanced physical well being, but also a better understanding of self and others, a chance for renewal, and a learning experience that can be carried forward into daily life. The fourth area discusses how wellness tourism might react to those seeking more meaning to their lives through the journeying experience rather than the arrival.

Introduction

The process of pilgrimage, and journeyssuch as charity treks, can offer experiences that extend the individuals beyond the norms of their daily lives. These experiences can be seen to offer opportunities for individuals to develop personally in ways that promote sustained wellness.

Such journeys can be viewed as part of a developing body of alternative tourism determined by Wearing 'as forms of tourism that set out to be consistent with natural, social and community values which allows both hosts and guests to enjoy positive and worthwhile interactions and experiences' (Wearing 2001: 32). He goes on to argue that increased urbanization has led to a growth in the use of rural landscapes as 'a means of escape from the pressures of modern living' and cites Welton (1978) who contends that, 'The only way to counteract the unnaturalness of urban life is to periodically renew oneself in the wilderness' (Wearing 2001: 90). This paper argues that as more and more people feel isolated within an increasingly urbanized and secularized society, there is a growing need for leisure activities, which reinforce their sense of self and community. Indeed Bauman (2001) argues that it is the decline of community within urbanized environments that has lead to insecurity and feelings of loss. Where communities have to be voluntarily formed and shaped. '.... all homogeneity must be 'hand-picked' from a tangled mass of variety through selection, separation and exclusion; all unity needs to be made...' (Bauman: 2001: 14)

In considering pilgrimage and wellness this paper examines how pilgrimage, in two different contexts, has increasing relevance in a postmodern and fragmented world where people are seeking meaning and sense within their own lives. To understand this claim, and before offering a theory of how pilgrimage and wellness tourism interact, it is worth considering briefly how recent changes in postmodernism impact on these two areas.

Postmodernity exists traditionally as a generalized body of thinking that is concerned with the 'now', where the past is included with the present. And this 'presentness' is found in its approach to culture and capitalism, science and technology, wherein it challenges the classical notions of identity, objectivity, truth, and explanation. Connor (2004) identified four phases to postmodernism, namely: - accumulation, synthesis, autonomy and dissipation.

It is this last stage, that of dissipation - emerging throughout the 90s which has begun to challenge 'old' postmodernism, one associated with an ephemeral western societal culture which is 'ungrounded, diverse, unstable...and sceptical about...the objectivity of truth, history and norms' (Eagleton 1996: vii). It is now being replaced by one in which ethics, religion and justice are tempering an absolutist view. As Philippa Berry suggests, 'the putative cultural triumph of secular reason is haunted or shadowed by its presumed opposite' (Berry 2004: 170). So as we move towards what we might call 'hypermodernity', we find that key postmodernists such as Derrida have outlined the futility of trying to sustain absolute distinctions between such aspects as belief and doubt (Derrida 1998). Derrida refers to the return of religions, and the joining together of materialism and 'spirit', but these religions are not the traditional ones, nor are they new fundamentalist ones. Rather, he sees them as signs of spirituality that are 'post-religious' or 'quasi-religious' and are outside formal structures. Berry quotes Levinas (1979) in defining this new area (to postmodernism) as being one in which the individual 'is required to 'open oneself as space' -or, put another way, 'to create a space for some direct or indirect form of communication.' (Berry 2004: 175).

These important developmentsseem to suggest that in the context oftoday's hypermodern world, secularity and spirituality will, in future, begin to coexist more easily as boundaries become increasingly blurred. This means that in exploring the idea of pilgrimage and its relationship to wellness tourism, one's thinking needs to operate at a meta level, which holds simultaneously the secular and the spiritual, the 'belief and doubt' of Derrida, and the acceptance that pilgrimage - in the context of this paper - is not about visiting a sacred site, but about what happens on the physical and spiritual journey.

Methodology: Grounded Theory in the Study of Pilgrimage and Wellness Tourism

Figure 1 is a conceptual map of the dimensions of pilgrimage, in relation to wellness and tourism. It is developed from grounded theory, which is considered to be the most appropriate way of researching interrelated complexities. Data for case study one was drawn from the narratives of seven individuals who either cycled or walked to Santiago de Compostela during the summer of 2002. Case study two took a similar narrative approach with 6 individuals who took part in charity treks in 1999–2002. In a field where one is attempting to derive meaning from the narratives of individuals- based on their own understanding of their journey- traditional methodologies are inappropriate.

Grounded theory does not begin with a preconceived theory in mind. 'Rather, the researcher begins with an area of study and allows the theory to emerge from the data.' (Strauss and Corbin 1998). Just as Lave and Wenger (1991) argue the necessity of understanding the personal and social world of the learner before designing teaching interventions, so grounded theory needs to gain overarching themes which emerge from individual narratives which, in turn, inform ways of approaching pilgrimage and its meaning in relation to wellness. In bringing together the often intimate narratives of pilgrims and some of the existing data on well ness, the object was to categorize the generic themes in the varied data to build a picture of what research on pilgrimage and well ness might use as a working conceptual model.

As a result, categories were identified through the process of open coding, which consists of 'the analytic process through which concepts are identified and their properties and dimensions are discovered in data.' (Strauss and Corbin 1998: 101). This requires going 'beyond static analysis to multiple layers of meaning' (Gray 2004: 330). These multiple layers are depicted in Figure 1 in relation to pilgrimage. On the right side of the diagram there are the three main concepts of Community, Journey, and Spirituality which emerged from the data. On the left the literature on wellness, and tourism are disaggregated and rejoined again as Wellness Tourism.

The figure as a whole is intended to represent the dimensions of the data that need to be borne in mind when representing the nature of pilgrimage as revealed by the voices of pilgrims. So the suggestion that pilgrimage is more to do with journey than arrival is one borne out by their own narratives. The potential conflict between route-based tourism and pilgrimage is recognized by Murray and Graham (1997) whilst other authors, for example, Gonzales and Medina (2003) and Santos (2002) centre on the importance of particular arrival sites in pilgrimage which are sometimes at odds with the ideas presented here.

The concepts emerging from grounded theory attempt to map the minds of both pilgrims and writers. This was facilitated by initially the six concepts identified through open coding. However, a form of axial coding helps to recognize relationships between certain categories, For example, 4.4 Kindness of Strangers, 6.2 Fulfilment through others and 1.3 Friendships all have an affinity, whilst Community itself is recognized in 4 and in 6.2, 1.4, 2.3 and 3.1. Although Strauss and Corbin (1998) remain the classic authors on grounded theory, other aspects of qualitative research which bear on the way data have been used in this paper can be found in

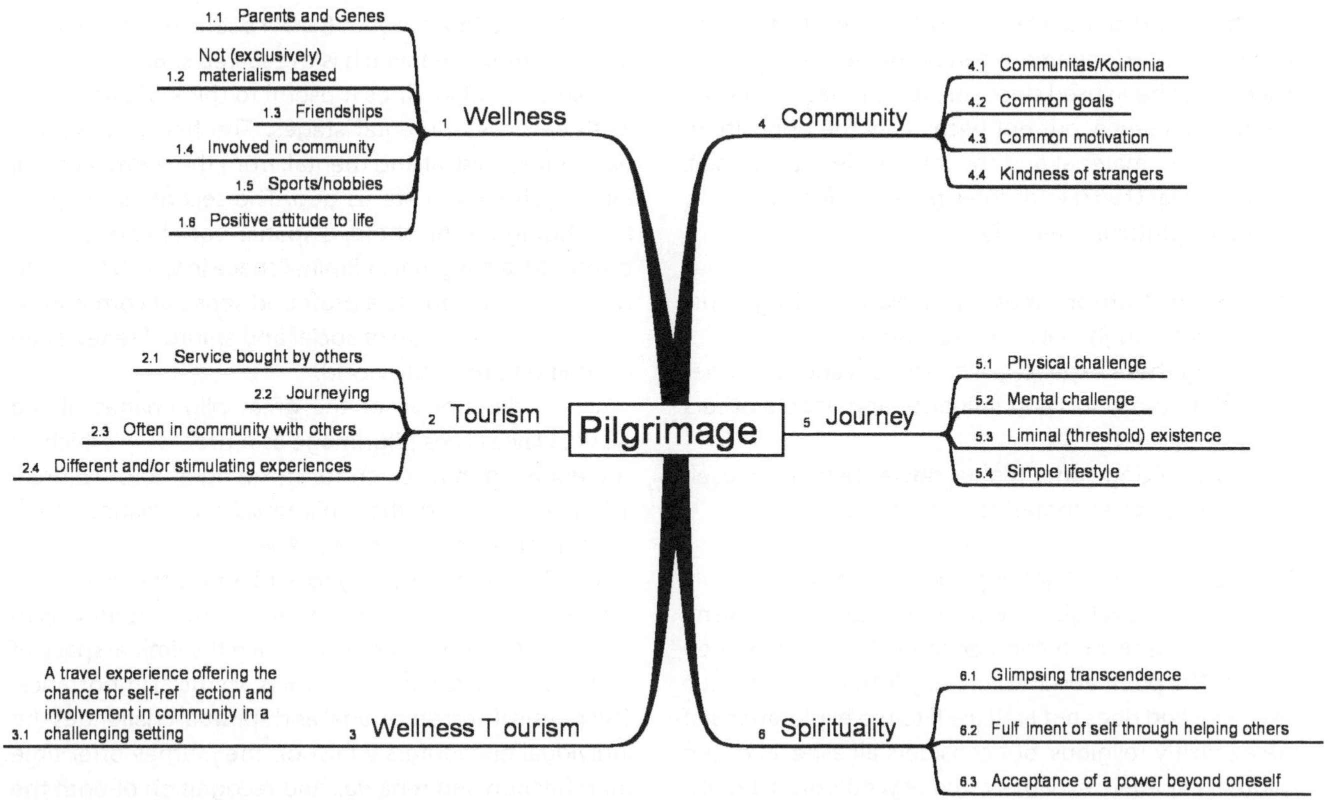

Figure 1. Dimensions of Pilgrimage.

Gray (2004), Charmaz (1995) Sandelowski (1995), Clough (2002) and Phillimore and Goodson (2004) amongst others.

There is also a considerable literature on the nature of wellness which informs this paper. Shah and Marks (2004), in their study, relate to well-being, 'as more than just happiness. As well as feeling satisfied and happy, well-being means developing as a person, being fulfilled, and making a contribution to the community' (Shah and Marks 2004: 2), all of which are found in pilgrimage.

The authors draw from a wide range of research literature in both the economic, health, social, psychological and leisure fields. The Wellness category in Figure 1 summarizes their views as well as those of others in the field of well-being at work and leisure, e.g., Keyes (1998), Csikszentmihalyi (1997), Gardner (1993), Diener and Seligman (2002), and Sarafino (2002). The overview suggests that well being, though initially dependent on genes and upbringing, is also positively influenced by community involvement, sports and hobbies and a positive attitude towards life. It is the intentional activities that people undertake which bring happiness and once basic needs have been met, money makes no difference to well-being. In fact, 'people who are materialistic are less happy than those who value other things.' (Shah and Marks 2004: 6, quoting Kasser 2002).

Category 4, the Community branch of Figure 1, considers the community aspect of pilgrimage, incorporating the themes of common goals, motivation and reciprocal generosity. This was the original meaning of the Latin and Greek for 'community'. The way Community is represented in the conceptual map presented here has parallells with Turner's ideas. Victor Turner's concept of liminality and communitas built on the work of Arnold van Gennep who described the 'liminal phase of rites of passage' as a three phase process of, 'separation, margin (or Limen, signifying threshold in Latin) and aggregation'. (Turner 2002: 359), Pilgrimages, he argues, 'exhibit in their social relations the act of communitas; and this quality of communitas in long established pilgrimages becomes articulated in some measure with the environing social structure through their social organisation'. (Turner 1974: 167). The social organization of pilgrimage requires a relinquishing of the usual social mores, structures and values of everyday life and new communities are formed based on shared goals, mutual needs and the reliance on strangers for support.

What is Spirituality?

Spirituality is used in this paper in a number of ways, embracing both the secular and multi-faith ideas of how people experience it as a sense of individual well-being within society. The word 'spirituality' comes from the Latin 'spiritualitas', an abstract word, related to the Greek word 'pneuma' meaning breath, the essence of life. It has been used since very early times to emphasize the difference, but not necessarily the opposition, between the material and the inner or immanent self. Von Balthasar (1967) finds three different dimensions to the word spirituality, namely:

- Seeking truth and meaning of existence by glimpsing it through real lived experience;
- Using these experiences to express and fulfil oneself in the world and influence and enable others as well; and
- Recognizing some higher power beyond oneself to guide one's thoughts and actions.

The use of the word 'higher power' may not necessarily be a God in the religious sense. In the case of Marxism, it is society itself, or in another context it could be seen as Gaia. The point about spirituality is that it is a holistic discipline and does not limit itself to the 'explorations of the explicitly religious' but considers all aspects of the spiritual experience, namely, 'the psychological, bodily, historical, political, aesthetic, intellectual and other dimensions of the human subject of spiritual experience' (Schneiders 1989: 693).

Consequently, in tackling the experience of pilgrimage, an understanding of the spiritual (in its widest sense) as well as the temporal needs of the pilgrim are crucial. Davie (who contributed to the European Values Study of 1981 (Barker *et al.* 1992), defines the growth in interest in the spiritual as 'believing without belonging' (Davie 1994). Brown (2001) argues that: 'What emerges *(in contemporary societY)* is a story not merely of church decline but of the end of Christianity as a means by which, men and women as individuals, construct their identities and their sense of self (Brown 2001: 2). However, Davie argues that although the marked decline of participation in church attendance is widespread throughout Europe, belief persists but becomes 'increasingly personal, detached and heterogeneous particularly among young people' (Davie 2002: 8).

What is Pilgrimage?

Pilgrimage is a physical journey, which symbolizes and reflects the life journey of the individual: 'the need to set in physical search for a spiritual goal is one common to all mankind' (Barber 1991: 6). Pilgrimageisgenerallyseenasa journey to a sacred place or shrine, which is the ultimate reason for the journey. In the two examples which follow, the journey to Santiago de Compostela and the journeying involved in charity treks- both are considered pilgrimages in Barber's sense of the word- it is the journey itself which is the sacred space.

Carrasco (1996) finds it useful to think of pilgrimage in three developmental stages. The first is a separation, both physical and mental, from the normal spatial and psychological status quo. The second is the journey through a threshold, a spatial, social and spiritual threshold, a marginal or liminal space in which bonding with others generates a profound sense of community. The third is the stage of social and spiritual renewal on returning to the daily world.

In his description of the great pilgrimages of the world, Davies sees pilgrimage as a process in which all 'are engaged in a search for meaning and for spiritual advancement and their pilgrimage dramatizes their quest for the divine' (Davies 1998: 1).

In selecting the journey to Santiago as the first casestudy and charity treks as the second, the intention is to see pilgrimage as ajourney in which the liminal space of that journey and not the arrival is of crucial importance. The physical and emotional and spiritual challenges the individual encounters whilst on the journey offer time for reflection and renewal, and recognition of both the immanent and the transcendentin the course of the journey. Post (1996: 4-6) dwells heavily on the past and its tradition in pilgrimage as providing an anchor in an ever-changing restless world. However, the assumption in the present paper is to acknowledge the past, but also to look to the future, whilst still in the 'presentness' of the journey. This process could be called a 'remembered future', namely, one in which the past is not looked back on with nostalgia as some form of escapism, but is a reflection of one's own life experience to date and gives insights into what one might yet be.

Case Study 1: Santiago de Compostela

Recent years have shown a rapid increase in those who choose to walk, or ride by cycle or horse to Santiago de Compostela, a medieval pilgrimage route from central France to the northwest tip of Spain.

In 1987, 2,905 pilgrims walked, cycled or travelled by horse to Compostela. In 1997, this number had risen to 25,179. In 2003, this had nearly tripled to 74,614. Of this number 36.4 per cent were between 16 and 30 years of age. 69.9 per cent of all pilgrims saw the main motivation for their journey as being a spiritual one. (Registro de la Oficina de Acogida de Peregrinos 2004). These data support the findings of Davie and the European Values Study (Davie 1994, 2002; Barker *et al.* 1992) outlined earlier, namely, that in spite of increased secularism, young people especially feel a need to seek a spiritual well-being in ways that do not reside in orthodox religions.

The journey to Santiago de Compostela, is generally called the Camino de Santiago (or the Chemin de StJacques in France) and many pilgrims who walk and cycle the route keep their own diaries or recordings and there are also a number of published accounts including Internet sites. Post (1996), examined a number of pilgrim accounts of their journey to Santiago de Compostela and classifies them into three clusters, namely:

- ritual, the journey, the essence of being a pilgrim
- meeting and relations
- the past (Post 1996: 4)

The themes of the Journey, the Community and Spirituality outlined in Figure 1 are expanded in relation to Santiago in the form of:

- the daily route itself and simple living
- the community of the Camino and its effects
- arrival and the return to daily life

The Journey

The road to Santiago is not an easy one. From Le Puy en-Velay to Santiago de Compostela is just over 1000 miles. Walkers who start from StJean de Pied de Port on the French side of the Pyrenees still have 500 miles to walk before they reach Santiago. Many accounts talk of the worry and the excitement of a physical challenge and the preparation for it. As the following quotes drawn from the narratives of pilgrims, people feared the thought of the journey and its likely hardships.

> *'I couldn't believe the steepness of the hill from Le Puy. I thought, 'How am I going to make it if this first hill is so bad?'*
>
> *'I worried about how bad my blisters would get.'*

Early on most pilgrims experienced times when they felt the journey would defeat them. But very soon, many individuals ceased to be concerned with strict schedules and began to come to terms with what they had taken on.

> *'After a few days I realized that every door would open. There would be a place to sleep, somewhere to buy food, and my bicycle would eventually get me there. I ceased to worry.'*
>
> *'I slowed down after two days. I was missing things that I had come to see. Mentally, I was still at work trying to meet deadlines. When I slowed down, time somehow expanded so that I still got to roughly the place I intended and I spoke to far more people. I began to relax and let the road carry me rather than me trying to fight it. '*

This sense of relaxing into the journey and becoming separated from the spatial and psychological status quo to which Carrasco (1996) refers, brings with it a heightened experience of the present.

> *'You don't have to worry about anything. All you have to do effectively is walk and eat and sleep. It puts into context what you worry about normally, and how unnecessary much of it is.'*

This sense of the present within the rural, or even wildernesssetting where there is a freedom, albeit temporarily, from urban spaces and urban values also bears outWearing's (2001) earlier arguments.

> *'I had this sense of living in the present, the present moment, aware of how my blisters hurt right now, of not knowing, or minding, what would happen next. We never booked anything, or thought about distance travelled. After a time I noticed the towns we passed through were very straining and made me feel tired. '*
>
> *'I decided to avoid Pamplona and took a different route over the mountains to Estella. I couldn't face the noise and stress of a big city. '*

Community

New communities become formed on the route both with other pilgrims and with people who live on the route or offer support or hospitality. The community exists within liminal space as pilgrims become suspended in a paradoxical timeless present whilst moving onwards, (the word 'Ultreia!' meaning 'Onwards!' is the greeting pilgrims give to one another).

It is this closely bonded community, all with the same goal of reaching Santiago, all experiencing physical and emotional challenges and all separated in a timeless present that makes the Camino so special. It goes beyond just physical well-being, towards an inner well-being engendered by being part of a pilgrim community with common goals and common motivation. In pilgrim accounts and in interviews, the daily small acts of kindness given and received, and the ability of total strangers to be more open with each other than they would normally be, appears to be one of the key 'presents' -as one pilgrim put it- that one receives from the journey.

> *'People are more open than in normal life. We had substantial conversations with all sorts.'*
>
> *'The level of hospitality was amazing. You walked in, used the facilities people offered, and then moved on. It was if you were part of their family. And these people did this to thousands of pilgrims each year and were still smiling. 'That's why we're here.' they said. How could I not appreciate it?'*
>
> *'I knocked at a nearby door to see if I could get my pilgrim record stamped, but she said the office next door was closed and they wouldn't be back till tomorrow. As I was*

going she said; 'It is mystical, seeing the constant flow of people like you, passing day after day. It is an example to us all. One day I want to make the journey. Please remember me when you get to Santiago'. I was touched by her words and I did remember her when I got to Santiago.'

Pilgrims report that the simplicity of life on the Camino and the constant meeting andre-meeting of others on the road engenders a relationship in which those acts of kindness from total strangers are generously given and gratefully received. The expensive walking poles donated to someone with a bad leg, the sharing of food and drink, the cyclist taking a message from one walker to another, some way ahead, are the daily lived experiences that take people out of themselves for brief moments and give meaning to the practical nature of a lived spirituality. They are looking beyond self both in their acts of kindness to others and also in the generosity shown to them. Many report events that transcend normal explanation and understanding. One such is the account of a couple meeting with a man living under polythene sheets in the woods in Galicia. He asked them to read aloud the cards people had sent him, as he himself could not read. He, in turn, recognized that the companion of the reader was limping. Declaring he was a healer, he manipulated the knee and the pain and the limp went away.

'I couldn't explain it really. Perhaps he was an ace physiotherapist, one who couldn't read, and lived under a few polythene sheets? But it did the trick. I just don't know.'

Arrival and Return

The highs and lows of the journeying and the experience of weeks or months on the road, ill prepares pilgrims for the noise and bustle of what might be called the religious tourism of Santiago. There is the feeling of corning down from a high to renew one's relations with consumerism. 'All the emphasis is on the pilgrimage, the journey itself. In some cases the arrival is something of an anticlimax. "Is that all...?'(Post 1996: 3). Immediate comments are:

'After an hour in Santiago I wanted to be back on the road again'
or
'It's a nice medieval town, but I felt a bit empty and lost in Santiago.'

But the overall pilgrimage experience often builds an inner confidence and understanding of self, which becomes more apparent after the pilgrimage.

'I learnt the meaning of forgiveness'
'It reinforced the stuff I already believed'
'The Camino is not a path. it is a way of life'

Common to all is a sense of achievement and purpose.
For example:

'I can live from a bag that weighs 9 kilos for over a month'
or
'my own Camino back here now begins,'

confirms the confidence in self that often accompanies the return from Santiago.

In some cases, the effect of the pilgrimage brings about real life changes. One pilgrim told how he gave up his job as a commercial photographer and became a teacher. Another walked into a job he had always wanted after having been unemployed for a lengthy period. This is not to say that participants would not have done these things in any case, but the effect of pilgrimage does lead them, from their own point of view, to take more control over their lives and experience a sense of emotional and physical well-being.

Returning to Figure 1, it can be seen that the journey and community aspects coalesce to enhance wellness, but also contribute towards the fulfilment of self by attending to others. (see branch 6.2 in Figure 1).

Case Study 2: Charity Treks

Charity treks are adventure holidays or expeditions in which people earn the right to participate by raising a sum of money over and above the real cost of the trip so that a donation can be made to a charity concerned. This is the crucial way that charity trips vary from similar or identical trips, which are sold directly to the consumer as a holiday. Although many charities still choose to operate their own treks utilizing the services of a tour operator, Charity Challenge is the main organizing body, which provides a 'one stop shop' for people who generally chose their trip first and then decide on the charity they wish to support. Charity Challenge supports 100 challenges each year for around 2,000 participants, resulting in £15 million being raised for over 550 charities (http://www.charitychallenge.com/about. php).Numbers for each event tend to be small, generally between 15 and 30 participants, although numbers can be greater (up to 70) for bike rides. Charity Challenge actively promotes responsible tourism as members of the Centre for Environmentally Responsible Tourism.

The trips included in this case study are (a) a ten day bike trip through the Grand Canyon organized on behalf of the charity Scope in 1999; and (b) a shorter weekend trip to climb Mulhacen, Spain's highest mountain in 2002 undertaken on behalf ofBarnardo's. Quotes used in this case study are drawn from a pilot project involving unstructured life history interviews with participants and carried out after the individuals had returned from their trips.

The journey

Charity treks, embody many of the qualities of pilgrimage in that they have the common themes of physical and mental challenge and the sense, like Santiago, of traveling through a liminal space, as outlined in Figure 1. But more than that, the true charity trekker brings to their journey that sense of 'doing things for others' which can be missing from those who go on charity treks as independent travellers with little or no interest in the charity they are 'supporting'.

For an idea of the difference between the two profiles, the conceptual map of Figure 1 shows that the three interlocking themes of Community, Journey and Spirituality are necessary for pilgrimage and in particular, branch 6.2 (fulfillment of self through helping others) is an essential part of the pilgrimage.

Whether they are cycle rides in America or Africa, whitewater rafting in Nepal, treks to Jordan or Peru, dog sledding in Lapland, or horse riding in Mongolia, these adventures mirror that of pilgrims, in that although they are journeys of self-discovery, they are within a community. As in pilgrimage, treks are all challenging and test emotional and physical stamina and require to some degree, a simple lifestyle, involving basic subsistence level camping and foods. Participant Gil, who undertook the cycle trip through the Grand Canyon in 1999, emphasized that the fund raising element was the main goal that led her to undertake the trip.

Those who hold religious beliefs told of occasions when they linked their experiences with their faith community back home. One woman who climbed Spain's Mount Mulhacen experienced the trip as both arduous and spiritual.

> *'I am a Christian and I do have a strong faith and one of the things that really brings it home is when you are climbing and you can really see the beauty of the world ... It was about the same time when our church would be on so I was thinking, Oh put up a prayer for me, just on that last bit. I thought I can't do this.'*

But all such experiences need to look outward as well as inward. Charity treks act as pilgrimage when they give insights into self, but do so through daily lived experience, which involves interaction with others in a way those others can appreciate. The pursuit of happiness and well-being is likely to be chimerical if it does not involve others. Just as spirituality and community involve generosity to others, so does well-being. Shah and Marks note that evidence shows that happy people are, amongst other things, more:

- Altruistic
- Generous
- Tolerant and
- Sociable.

(Shah and Marks 2004: 5).

In short, for pilgrimage to be a reality in charity treks, it also needs to involve spiritual and community elements.

Community

The trek conforms to the idea of the journey suspended continually in a threshold or liminal space where people divest themselves of worldly goods and symbols of rank and status and form new communities, which have common goals and common motivation. The tour organizers provide everything; food, shelter, bikes, ponies or any equipment needed for subsistence living and is usually utilitarian, at odds with what people might have used in training at home. Participants are encouraged to bring only the minimum requirements. In this way, the threshold existence and the simple lifestyle issimilar to pilgrims on their way to Santiago.

Within these new communities different hierarchies can develop, and in the case of the bike ride, fitness determined the groupings. Although pilgrims may or may not train for their ordeal, trekkers are encouraged to undergo a programme of fitness over some months to ensure they are able to undertake the trip. Trips are also graded according to how arduous they are.

> *'There were some really fit people from Aberdeen who wanted to go ahead · cos everybody was too slow for them and keeping them back. We actually came in last to the whole bike ride but we were all just concerned with keeping together'.*

The difference between the charity trekker and the holidaymaker is one of motivation and of something outside the immediate experience. The charity trekker has in mind those for whom they are really doing the trek for (i.e. not themselves). The distinction is that the true charity trekker joins a community that has a goal beyond the immediate task of reaching a destination, namely, to help others less fortunate than themselves. In so doing they commit to a task that is beyond the immediate activity in a way that a holidaymaker does not. This quote from an interviewee sums up the tension between the willingness of the charity trekker and the expectations of the holidaymaker.

'If you'd paid for this you would say I'm not doing this. You wouldn't tolerate the terrible conditions, the lack of sleep, the really bad kind of organization and the fact that it rained. You would look on it in a different way. You put up with the gruelling-ness of it because it is for charity'.

Charity trekkers will accept some hardships. Holidaymakers require order, structure and service in the form of local guides who support the trip and carry bags, food and bedding from camp to camp. In terms of the community aspect of the journey, Shah and Marks' (2004) observations that altruism, generosity and tolerance are characteristics of well-being seem to suitably describe the necessary attributes for the charity trekker community.

Arrival and Return

As in the previous case study, on arrival, participants interviewed experienced a sense of anti-climax or did not want the trip to end despite the hardships.

'My lowest moment was the end. I remember thinking about two days before that I didn't want it to end, as hard as it was, because we were all going to part. I expected the gala dinner to be like a graduation ball. Off course it was nothing like that. It was like some sort of roadside cafe/ restaurant in America. We all dressed up for it.'

In real terms the benefits of increased physical fitness or having 'time out' from the stresses of everyday life can promote wellness;- witness branches 1.5 Sports/ hobbies and 5.1 Glimpsing transcendence, 5.3 Threshold existence and 5.4 Simple lifestyle in the conceptual map. It could even be argued that physical fitness creates a 'new spirituality' facilitated through leaving the urban setting to commune with self, nature and others through experiencing a contrived level of physical hardship. For this to be true, and link charity trips to pilgrimage and wellness, there needs to be something that looks beyond self rather than just into self and elements of spirituality need to be seen as an integral part of the role of charity trekkers.

Conclusion

The search for something beyond self, something more than consumerism, is recognized in the latest stages of postmodernism and in work such as the European Values Study (Barker *et* al. 1992). Pilgrimage to Santiago and the generosity required by the charity trek gives some idea of how people are prepared to look beyond the immediate in their lives and explore ideas and possibilities in community with others.

How can wellness tourism respond to the needs of such pilgrims? The conceptual map of Figure 1 attempts to define wellness tourism as:

'A travel experience offering the chance of self-reflection and involvement in community in a challenging setting.'

This definition incorporates the community and the sports/hobbies aspect of wellness, links this with the journeying aspect of tourism/pilgrimage, and places it in a setting in which the chance for self-reflection is offered to those who wish to take it. So wellness tourism- in relation to pilgrimage - needs to allow individuals to become reacquainted with themselves and their community at all levels, and with their own physical, spiritual and emotional needs.

In this context, wellness tourism needs to go beyond a service 'offer' to pilgrims or trekkers and instead work towards a more collaborative approach in which things are done 'with' pilgrims rather than 'to' them in ways they themselves appreciate. Schneiders' (1989) view is that it is the practical, day-to-day lived experience and the context in which it is perceived that is important. And each individual is different.

Wellness tourism has several important aspects to offer to those searching for meaning in their lives. Two particular aspects are offered here. They are:

- 'Flow' and the nature of challenges
- The building of social well-being

Flow

The journeying aspect of pilgrimage entails physical and mental challenges. The process of experiencing challenging activities and having the abilities to meet those challenges Csikszentmihalyi (1997) has termed as 'flow'. 'Flow' occurs when we are completely absorbed in an activity, whether work or play, and though difficult it, does not provoke excessive stress.

Wellness tourism can extend this concept of 'flow', by concentrating more on the journey aspect of pilgrimage and on different types of pilgrims, such as those who may not wish to walk or cycle, but still want to engage in the three themes of pilgrimage set out in Figure 1. Such people may not have the ability to walk extensively, or may have limited time. The skill needed by wellness-tourism providers is to find ways on the journey, however long or short it may be, to challenge people of different abilities, but not excessively so. Everyone needs to gain 'flow'.

Social well-being

Social well-being is defined by Keyes as 'the appraisal of one's circumstance and functioning in society' (Keyes 1998: 122) and is similar to Wearing's 'social value', which is 'the mutual process of learning and personal growth' (Wearing 2001: 172). Both of these are based on perceptions of how people feel about their own sense of belonging in society and how much they contribute. Social well-being differs from the more ubiquitous social capital in that the latter is more to do with systems approaches to norms, network structures and ways of connecting in society.

Wellness tourism can make significant contributions in fostering individual and community well-being in the way they adopt approaches that interpret and influence the journeying aspect as a vehicle for developing the wellness of not just individuals but the community they travel in. In short, it is the 'scaffolding' that wellness tourism providers build around the process that will enable individuals and groups to learn more, not just about themselves, but also about the communities they engage with.

Pilgrimage is complex because it is about developing the self and one's place within the community. To tackle this area with any success wellness tourism is in for the long haul. Crucial to its success in the medium term is an understanding not only of the psychology ofwellness and of 'flow', but also the ritual and logistics of journeying. It also requires an enabling sensitivity towards those trying to give meaning to their lives that lies somewhere beyond themselves.

References

BARKER, D., HALMAN, L. and VLOET, A. (1992). *The European Values Study 1981–1990:* Summary *Report.* London. Gordon Cook Foundation on behalf of the European Values Group.

BAUMAN, Z. (2001). *Community: Seeking Safety in an Insecure World.* London. Polity.

BERRY, P. (2004). Postmodernism and Post religion. In Connor, S. (ed) *The Cambridge Companion to Postmodernism.* Cambridge. Cambridge University Press.

BROWN, C. (2001). *The Death of Christian Britain.* London. Routledge.

CARRASCO, D. (1996). *Those Who Go on a Sacred Journey: The Shapes and Diversity of Pilgrimages.* London. SCM Press Ltd.

CHARMAZ, K. (1995). Grounded theory. In Smith, J. A., Harre, R. and Langenhove, L. V. (eds) *Rethinking Methods in Psychology.* London. Sage.

CLOUGH, P. (2002). *Narratives and Fictions in Educational Research.* Buckingham. Open University Press.

CONNOR, S. (2004). *The Cambridge Companion to Postmodernism.* Cambridge. Cambridge University Press.

CSIKSZENTMIHALYI, M. (1997). *Finding Flow, the Psychology of Engagement in Everyday Life.* New York. Perseus Books.

DAVIE, G. (1994). *Religion in Britain since 1945: Believing without Belonging.* Oxford. Blackwell.

DAVIE, G. (2002). *Europe: The Exceptional Case - Parameters of Faith in the Modern World.* London. Darton. Longman and Todd.

DAVIES, J. G. (1998). *Pilgrimage: Yesterday and Today: Why? Where? How?* London. SCM Press Ltd.

DERRIDA, J. (1998). Faith and Knowledge: The Two Sources of 'Religion' at the Limits of Reason Alone. In Derrida, J. V., Gianni, V. (eds) *Religion.* Cambridge. Polity.

DIENER, E. and SELIGMAN, M. (2002). Very Happy People. *Psychological Science* 13: 81-84. EAGLETON, T. (1996). *The Illusions of Postmodernism.* Oxford. Blackwell Publishing.

GARDNER, H. (1993). *Frames of Mind: The Theory of Multiple Intelligences.* New York. Basic Books.

GONZALES, R. and MEDINA. J. (2003). Cultural Tourism and Urban Management in Northwestern Spain. *Tourism Geographies* 5: 446-460. GRAY, D. E. (2004). *Doing Research in the Real World.* London. Sage.

KASSER, T. (2002). *The High Price of Materialism.* Cambridge, MA. MIT Press. KEYES, C. (1998). Social Well-being. *Social Psychology Quarterly* 61: 121–140.

LAVE, J. and WENGER, E. (1991). *Situated Learning: Legitimate Peripheral Participation.* Cambridge. Cambridge University Press.

LEVINAS, E. (1979). *Otherwise than Being,* or *Beyond Essence, trans. A Lingis.* The Hague. Martinus Nijhoff.

MURRAY, M. and GRAHAM, B. (1997). Exploring the Dialectics of Route-based Tourism: The Camino de Santiago. *Tourism Management* 18: 513–524.

PHILLIMORE, J. and GOODSON, L. (eds) (2004). *Qualitative Research in Tourism: Ontologies, Epistemologies and Methodologies.* London. Routledge.

POST, P. (1996). The Modern Pilgrim: A Christian Ritual Between Tradition and Postmodernity. *Consilium* 4: 1–9.

REGISTRO DE LA OFICINA DE ACOGIDA DE PEREGRINOS (2004). Oficina de Sociologia del Arzobispado de Santiago de Compostela. Santiago de Compostela, Oficina de Sociologiadel Arzobispado de Santiago de Compostela.

SANDELOWSKI, M. (1995). Aesthetics of Qualitative Research. *Image* 27: 205–209.

SANTOS, X. M. (2002). Pilgrimage and Tourism at Santiago de Compostela. *Tourism Recreation Research* 27(2): 41–50.

SARAFINO, E. (2002). *Health Psychology.* New York. Wiley.

SCHNEIDERS, S. (1989). Spirituality in the Academy. *Theological* Studies 50: 676–697.

SHAH, H. and MARKS, N. (2004). *A Well-being* Manifesto for *a Flourishing* Society. London. New Economics Foundation.

STRAUSS, A. and CORBIN, J. (1998). *Basics of Qualitative Research: Techniques and* Procedures for *Developing Grounded Theory.* Thousand Oaks. Sage.

TURNER, V. (1974). *Fields and Metaphors: Symbolic* Action in *Human Society.* London. Cornell University Press.

TURNER, V. (2002). Liminality and Communitas. In Lambeck, M. (ed) *A* Reader in *Anthropology of* Religion. London. Blackwell.

VON BALTHASAR, H. U. (1967). Das Evangelium als Norm und Kritik aller Spiritualitat. Spiritus Creator: Skizzenzur *Theologie:* 247-267. WELTON, C. E. (1978). Natural Freedom and Wilderness Survival. *Journal of Physical Education and Recreation* 4: 1–29.

Websites Reference

WEARING, S. (2001). *Volunteer Tourism: Experiences that Make a Difference.* New York. CABI. Websites Reference http://www.charitychallenge.com/about.php. Accessed on April 10 2005.

Submitted: April 4, 2005
Accepted: December 24, 2005

Writing on the Graceland Wall: On the Importance of Authorship in Pilgrimage Landscapes

Derek H. Alderman

ABSTRACT
The journey to Graceland-the final residence and resting place of singer and entertainer Elvis Presley-has become a major pilgrimage within modem popular culture and tourism. As a cultural religious landscape, Graceland plays an important role in both the official and vernacular construction of Elvis as a sacred figure. This paper examines the vernacular aspects of Graceland, specifically the practice of visitors inscribing messages on the fieldstone wall outside the estate. By focusing on the inscription process, the paper emphasizes the idea of Graceland visitors as 'authors' of Elvis' memory and religiosity. Elvis remains an important part of the American imagination because his meaning is not fixed but in a constant state of redefinition by fans and others. A reading of the Graceland wall reveals the existence of several discourses or ways of representing the image of Elvis. This study aims to advance not only our understanding of Graceland as a place of cultural religious heritage but also the importance of authorship and agency in pilgrimage landscapes.

Perhaps no other figure is more identified with modem popular music in English than singer Elvis Presley. Since his death in 1977, Elvis has transcended the boundaries of rock n'roll celebrity to become a ubiquitous image and icon in contemporary culture. According to Rodman (1996:1), Elvis' image appears in "songs, movies, television shows, advertisements, newspapers, magazines, comic strips, comic books, greeting cards, trading cards, T-shirts, poems, plays, short stories, novels, children's books, academic journals, university courses, art exhibits, home computer software,cookbooks, political campaigns, postage stamps, and innumerable other comers of the cultural terrain". Presley is an omnipresent icon for many reasons, some of which will never be understood. Elvis' fans are an important but often under-analyzed factor behind his continued popularity. They play an active role in "preserving and controlling Elvis' history" and"revising and redeeming his historical memory" (Doss 1999: 64). Devotion to the King runs so deep that public remembrance of Elvis Presley has taken on religious-like meaning to certain social groups inside and outside the United States.

Fans connect with Elvis through a variety of cultural practices, from impersonation to the collection of images and objects related to him. Arguably, the best known and most public of these practices is the pilgrimage to Graceland-Elvis' last residence and final resting place in Memphis, Tennessee. King (1993) has gone so far as to compare the Graceland estate to the shrines of medieval European saints, noting similarities in the sense of validation and devotion felt by pilgrims. Approximately 700,000 visitors tour the mansion each year, with daily admission reaching as much as 5,000 people in the busy summer months (Los Angeles Times 1997). As Coffey (1997: 289) contended, "Graceland may be the second most famous house in America, next to the White House." When faced with the choice, 18 per cent of registered voters surveyed by Fox News in 1997 actually said they would rather visit Graceland than the White House (Fox News 1997).

While now a popular destination for avid fans and curious observers across the country and the globe, Graceland originated out of Presley's need for sanctuary. Arguably, there would have never been a Graceland if not for the first home bought by Elvis on 1034 Audubon Drive. The house in Audubon Park, a quiet, upper middle-class neighbourhood in Memphis, became a centre of fan commotion with the arrival of the Presley family, a fact that tried the patience of Elvis' mother Gladys and horrified neighbours. According to Memphis historians Hazen and Freeman (1997:89), these neighbours "banded together and offered to buy the Presley home until they discovered that Elvis was the only homeowner without a mortgage." In March 1957, less than a year after buying the house on Audubon Drive, the Presleys purchased the more spacious and private Graceland mansion. After moving in, Elvis built a fieldstone wall around the estate to shield himself and his family from the traffic of admirers and observers. Vernon Chadwick, a noted Elvis scholar, described the important symbolic role that this wall has come to play in the Elvis commemorative movement and the journey to Graceland.

> *Ironically, the wall and gates that were meant to secure Elvis' privacy soon became a magnet for the 24-hour vigil maintained by legions of fans, some of whose scribblings, like initials of ancient tourists etched in the pyramids of Gezer, now can be read by all. While Elvis was alive, but increasingly after his death, visitors to Graceland found the wall a tempting space for a variety of communiques: personal greetings, jokes, prayers, poems, love letters, votive offerings.What normally would be considered acts of vandalism were in fact acts of devotion by some of the politest and best-intentioned fans in the world (quoted in Wright 1996: 7).*

This paper has two purposes. First, it offers a conceptual framework for interpreting Graceland and the specific cultural practice of visitors making inscriptions on the wall outside the mansion. Graceland is a pilgrimage landscape devoted to"cultural religious heritage."It plays an important role in both the official and vernacular construction of Elvis as a sacred figure in popular culture. While the estate's management attempts to fix and control the official public image of Presley, writing on the Graceland wall represents vernacular and more diverse ways of sanctifying the singer. By focusing on the inscription process, this study emphasizes the idea of Graceland visitors as 'authors' or creators of Elvis' memory and religiosity.

Second, the study suggests a way of analyzing the writings found on the Graceland wall and uses a published collection of these inscriptions to document some of the ways in which visitors interpret and construct Elvis' image. This approach is grounded in the belief that pilgrimage landscapes are the visible articulation of multiple, sometimes competing discourses about religious heritage. Several discourses flow through inscriptions on the Graceland wall, thus revealing the versatile way in which Elvis is made socially important and meaningful to visitors. The paper is an attempt to advance not only our understanding of Graceland as a place of religious heritage but also the importance of authorship and agency in pilgrimage landscapes.

Graceland as a Pilgrimage Landscape: Visitors as Authors

Campo (1998) noted the popularity and diversity of pilgrimage landscapes in contemporary America. He introduced a typology for categorizing pilgrimages and, specifically, the journey to Graceland. First, there are pilgrimages associated with organized religions, such as the Catholic pilgrimage to The National Shrine of St. Jude in Chicago or the Mormon pilgrimage to Temple Square in Salt Lake City. Second, there are pilgrimages associated with American civil religion or the "interconnection of God and country, commemoration of heroes...and martyrs, and the attribution of patriotic significance to the natural landscape" (Campo 1998: 48). Such patriotic journeys take people to the Gettysburg National Military Park, Mount Rushmore, the United States Holocaust Museum, and the Martin Luther King, Jr. National Historic Site.

Third and most important to the subject at hand, are pilgrimage landscapes associated with cultural religion, the sanctification of figures and themes from popular culture and mass media.These pilgrimage landscapes are often sites for the production of nostalgia and the commodification of culture. According to Campo (1998), Disney theme parks and Graceland are the two best examples of pilgrimage landscapes associated with American cultural religion. Campo's designation of Graceland as a cultural pilgrimage landscape recognizes the sacred-like status of Presley while avoiding the obvious pitfalls of viewing his memorialization in the same way as conventional faiths. Indeed, from the perspective of many fans, the commemoration of Elvis does not necessarily replace or conflict with other forms of religious heritage. Visitors to Graceland often celebrate Elvis' memory by referring to and embracing the entertainer's dedication to the organized religion of Christianity as well as his commitment, through military service, to the'civil religion' of American patriotism (Olson and Crase 1990).

Campo (1998) used the term pilgrimage landscape, as this paper does, to stress the interrelationships between people and place. No place is intrinsically sacred. Pilgrimages, and their attendant landscapes, are social constructions-a perspective that recognizes the role played by "humans in their creation, appropriation, organization, and representation" (Campo 1998: 42). Pilgrimage landscapes do not simply emerge but undergo what Seaton (1999) calls sacralization-a sequential process by which tourism attractions are 'marked' as meaningful, quasi religious shrines. Graceland is not a cultural religious landscape simply because it was once the home of Elvis Presley. Rather, its holiness stems from the behaviors, attitudes, and actions displayed by visitors (Rodman 1996: 117). The cultural significance of Graceland, according to Rodman (1996:103), is that it serves as"a physical point of articulation, where a global community of Elvis' fans regularly congregate and acquire a true sense of themselves as self-defined community." As Rodman also pointed out, however, Graceland represents a point of articulation for non-fans as well, where visitors can recognize and respond, sometimes negatively, to the canonization of Elvis and the existence of a community of fans. Graceland functions as a site for controlling, redefining, and perhaps even challenging the deification of Elvis. This paper adopts such a perspective as it attempts to analyze how visitors use the wall outside of Graceland as a text for marking and interpreting the cultural religious importance of Elvis and the journey to Memphis.

As a pilgrimage landscape, Graceland is a site for constructing both official and vernacular images of Elvis. This distinction between official and vernacular cultural expression is drawn from the work of Bodnar (1992), known for his analysis of public commemoration in America, and Jackson (1984), known for his analysis of America's vernacular landscapes. Both authors suggested that official expressions of culture are intended to maintain social order, present a unified vision of reality, and remind citizens of their obligations and history. In contrast, vernacular constructions of culture result from the changing expressions, values, and decisions of individuals at the grassroots level. While less stable than their official counterparts, vernacular landscapes provide a mode of expression to social actors and groups not normally heard from or seen.

As the headquarters of Elvis Presley Enterprises, Graceland is deeply involved in constructing and marketing an official image of the King through the selling of copyrighted memorabilia in gift shops and the leading of tours through the estate. Visitors on these tours wear headsets playing recorded narration. There are opportunities for visitors to tour and internalize at their own pace and sequence. From this perspective, however, visiting Graceland is very much about being a consumer and allowing the estate's management to script and regulate much of the sacralization process. The consumer metaphor is perhaps a good word choice given that 60 per cent of visitors to Graceland buy something at the gift shops before leaving (Arthur 1989: 49). As Bodnar (1992) suggested, the goal of official commemoration is to construct and present a unified narrative about the past. Graceland actively participates in not only sacralizing the image and life story of Elvis but also 'sanitizing' it. For example, the estate's official line of tours, museum displays, and video documentaries make no mention of the racial and sexual controversies that surrounded Elvis' rise to stardom, the prescription drug abuse that led to his death, or even the fact that the king of rock 'n' roll died in a bathroom (Rodman1996: 120).

The process of sacralizing a tourist attraction such as Graceland is not just controlled by the sponsors of the site. The changing motives and perspectives of visitors and pilgrims also shape it (Seaton 1999). As Jackson (1984) and Bodnar (1992) argued, landscape change can be vernacular, or reflect the diverse interests and views of ordinary people. Two vernacular memorial practices stand out at Graceland: the leaving of personal items and offerings at Elvis' grave site, and more important to the subject at hand, the inscribing of visitors' names, hometowns, and messages on the wall in front of the estate. The line between official and vernacular is perhaps not as distinct as Jackson and Bodnar suggest. Writing on the Graceland wall is not entirely unofficial. Graceland's management has built a drive-up lane adjacent to the wall to make writing on the wall safer and more accessible. They also pressure wash the wall when profane or irrelevant inscriptions appear (Rodman 19%: 119; Wright 1996: 39). Moreover, many of the devoted fans who visit Graceland and write on the wall are members of fan clubs that carry on an official relationship with Elvis Presley Enterprises, although not all of these clubs are on friendly relations with the estate (Doss 1999).

Although not independent of official organizers, inscriptions on the Graceland wall do represent a departure from the single, unified historical narrative presented by Elvis Presley Enterprises. They reveal the plurality of ways in which visitors remember and meaningfully connect with Elvis as part of their cultural religious heritage. Visitors frequently have their photographs taken near the wall, signifying the importance of writing and reading about Elvis as a ritual performance at Graceland. From this perspective, making the pilgrimage to Graceland is about being an author of Elvis' memory rather than simply a consumer of it. Viewing Graceland visitors as authors of Elvis' memory is consistent with recent research by Doss (1999). She stressed the agency and power of fans in sustaining and shaping the popularity of Elvis' image. On this point, Doss (1999: 31) wrote: "Fans do not simply derive meaning from Elvis' image, but actually 'make' Elvis in dramatic and deeply emotional ways." Of course, not everyone who inscribes a message on the Graceland wall is a devoted fan of Elvis Presley. In fact, some inscriptions are critical of Elvis' legacy. This fact does not lessen the importance of the entertainer as a cultural icon. It illustrates, as Doss (1999:16) contended, that "Elvis' image is a model of hybridity: different images of Elvis, or in other words, different Elvises hold different meanings to different viewers." On a more general level, it signifies that the authorship of pilgrimage landscapes is not restricted to one category of visitor or pilgrim.

Toward a Discourse Analysis of Pilgrimage Landscapes

If the Graceland wall is a public text for writing about and making Elvis' image, then how should we read or analyse it and other landscapes like it? Schein (1997) has provided a useful framework for analysing cultural landscapes, although he has not addressed the specific issues of religious heritage and pilgrimage. He suggested that landscapes are the product of countless individual decisions, each of which is embedded within a discourse. A discourse is a way of talking about, writing, or otherwise representing the world to make it intelligible and meaningful. Discourses are more than simply ideas and language. They are shared,common sense assumptions that shape our realities and hence the landscapes we construct (Barnes and Duncan1992). Schein (1997: 663) characterized a landscape as the materialization of several discourses and pointed to "the possibility that the cultural landscape can itself capture different, even competing, sets of meaning, or independent, thematic networks of knowledge."

This paper suggests that pilgrimage landscapes can be read or interpreted as the tangible product of several discourses about the meaning of religious heritage. Galbraith (2000) advocated a similar idea when examining the experiences of young Poles on a walking pilgrimage to the Black Madonna of Czestochowa. She viewed pilgrimages as occasions for both constituting and contesting religious communities. In common with this paper, Galbraith (2000) stressed the vernacular experiences and actions of individual pilgrims over official and institutional organizers.She characterized pilgrimage as an "arena for competing religious and secular discourses," in which pilgrims express a wide range of motivations for making the journey and, in some instances, actually challenge the official goals and interpretations of the pilgrimage (Galbraith 2000: 65). In this paper, Graceland's wall is interpreted as the product of several discourses about Elvis as a culturally important figure, some of which run counter to the official tourism narrative maintained and presented by Elvis Presley Enterprises.

As suggested by scholars such as Rodman (1994), Chadwick (1997), and Doss (1999), Elvis' cultural importance is rooted in the very fact that his image is multifaceted.Elvis continues to have an active posthumous career, according to Rodman (1994), because his life story dovetails with a series of important cultural mythologies or narratives about race, gender, and class. Images of Elvis can be contradictory, such as seeing Elvis as a musical/ racial integrator versus seeing him as an appropriator of black music and heritage. The imagery of Elvis has found its way into "struggles between high culture and low culture, youthfulness and adulthood, the country and the city, rebellion and conformity" (Rodman 1994: 458). References to Elvis have surfaced in the most unlikely of discussions, such as those about the Gulf War, the fall of Communism, and abortion (Rodman1994).Doss (1999) also noted the versatility of Elvis' image and how he can exist at many different levels to many different people depending on their personal and social identity. She further contended that the popularity of Elvis' image would die if left solely in the hands of his official sponsors. "But if Elvis remains elastic, so that fans and others can continue to look at him and talk about him and remake him...then Elvis may well survive into the twenty-first century as America's premiere icon" (Doss 1999: 259). According to Chadwick (1997), the Elvis text, that is, how Elvis' image is made part of the language of modern culture, is the product of several discourses.

Analyzing Inscriptions on the Graceland Wall

What sort of Elvis discourses can we find flowing through the inscriptions found on the Graceland wall? To answer this question, this study drew upon the work of Wright (1996), who transcribed and published hundreds of messages from the wall into a popular book entitled Dear Elvis: Graffiti from Graceland.While offering little in the way of analysis, Wright's work presents the researcher with a useful database of cultural expressions. Wright does not disclose how he chose which inscriptions to publish, other than to say he selected "the best." However, the emphasis here is not on obtaining a representative sample of writings, if this is even possible. As pointed out by Berg and Kearns (1996), discourse analysis is characterized by an absence of: (1) a single set of codified procedures; (2) a single way of reading texts; and (3) a single truth to be discovered. However, this type of analysis requires close and skillful reading and interpretation.

Wright's book Dear Elvis contains approximately 320 messages collected from the Graceland wall in the early 1990s. This paper presents a small sample of these inscriptions with the hope of illustrating the potential value of conducting a much larger and more empirically-based study, one that would combine collecting wall messages with observing and interviewing Graceland visitors. Based on an analysis of wall inscriptions collected by Wright (1996), visitors employ and emotionally connect with several different discourses about Elvis-Elvis as the American Dream, Elvis as Food and Consumption, Elvis as Sex and Romance and Elvis as

Family.These are not the only themes evident but represent some of the most dominant and interesting ones that Wright found in his study of the Graceland Wall.The inscriptions examined here illustrate how the sacralization or marking of Graceland as a meaningful shrine is not just carried out by the estate's management but by ordinary, vernacular interests.

Elvis as the American Dream

As Doss (1999) and Rodman (1994) found in their studies, Elvis is often represented as the embodiment of the American Dream of social mobility. Fans embrace him as someone who made tremendous gains in wealth and status without forgetting his working class roots. While this discourse does not dominate the messages published in Wright's collection, some visitors to Graceland do define Elvis'image in terms of how far he had come economically and socially. For instance, one inscription read: "You came to us, a poor boy from Tupelo, and you gave us all you had! You rocked the world! You will always be the King!" Some fans dwell less on Elvis' humble beginnings and more on the conspicuous wealth and success exhibited at Graceland. For instance, a visitor named Ben left this inscription: "Now that I've been to Graceland and seen it, I have the utmost respect for you." In his message on the wall,Jeff wrote: "Elvis, I loved that Jungle Room!" In a direct reference to Graceland's plush decor, another inscription read, "E. Thanks for carpeting the ceilings of our hearts!" One brave visitor to the wall requested,"Elvis, lend us some money." Not all visitors are incomplete awe ofGraceland, however. Connie left this observation behind: "You had a weird decorator. Thank God he didn't write music!" Elvis' penchant for buying automobiles was on the mind of Mary Lou as she contemplated his premature death and wrote: "You wouldn't have liked the way Caddies look today, anyway." One visitor expressed a critical perspective on the usually euphoric American Dream image of Elvis' "Too much, too soon, Elvis." This inscription is useful in illustrating how the Graceland wall is not simply a medium for fans to express their undying devotion to Elvis. Rather, it is also a place for some visitors to conduct a sober dialogue about the ultimate meaning of the singer's life and image.

Elvis as Food and Consumption

Elvis' connection with food and eating is another discourse found flowing through the wall messages collected by Wright. The official Graceland tour openly discusses the King's love of certain types of food such as fried peanut butter and banana sandwiches. This official narrative, however, never discusses gluttony or the weight problems Elvis experienced in the last years of his life. In fact, his weight gain is blamed on chronic back problems. Fans and other visitors to the wall do not always present such a sanitized image. One inscription read, for instance, "Turkey loves Elvis and vice-versa I bet!" Other messages proclaimed,"In the name of Elvis, I eat cheese," and "I saw Elvis at Denny's." "Elvis: How about a tennis lesson?" was perhaps a biting reference to the entertainer's exercise regiment (or lack thereof). Most devoted fans would take issue with these characterizations, arguing that Elvis lived a clean, athletic lifestyle through almost all of his years. Nevertheless, the discourse found within these inscriptions presents a more varied and richer understanding of how Graceland visitors view, identify with, and comment on Elvis' image. Wright did find inscriptions that made a more positive equation of Elvis with food. Cheryl wrote, for instance, "Have a peanut butter sandwich on me." Another inscription read, "Elvis, you're all that and a bag of chips." Tany had this food for thought: "Nothing better than a cheese steak, a pint, and Elvis at Cauley's." In a more puzzling and perhaps sensual food representation of Elvis, an unidentified visitor to Graceland left these words: "Elvis, you were in my dream and you were eating hot buttered popcorn." The identification of Elvis with food, while often the fuel for jokes and ridicule, deals with the fundamental issues of consumption and how one represents the body of Elvis as a cultural fact. The next section presents a very different discursive construction of Elvis' body.

Elvis as Sex and Romance

Many wall messages focus on the romantic and sexual aspects of Elvis' image. As mentioned earlier, this is a discourse not frequently found in the official narratives presented by Graceland's management. Gina wrote, for instance, "Elvis, I wish I was your belt buckle for a day!" Jenny issued this invitation:"Elvis-You are the sexiest man I have ever seen. Meet me at K-Mart (in lingerie)!" Rachel posed a question perhaps on the minds of many Elvis fans: "Hey Elvis, Are you lonesome tonight?" A visitor named Linda appears to be lonesome herself. She wrote this to the King: "I'm still waiting for my one night with you! Love you forever!" Linda will have to make room for Valerie, who inscribed "Lay your head on my pillow. Lay your warm and tender body next to mine." Anna described Elvis as "one handsome hound dog" while Cammi proclaimed, "Elvis, I lust you hard." Of course, the sexualization of Elvis' image may not appeal to all who visit Graceland. Indeed, one message read: "Elvis, Elvis, Let me be. Keep your pelvis far from me!"

Connecting to the memory of Elvis through the discourse of romance and sex is not limited to women visitors, however. Wright transcribed the following bizarre message from the Graceland wall: "Elvis, I'm bearing your ghost child, and I'm confused because I'm a man!" Upon visiting Graceland, Dave issued this compliment:"You had great taste in women. Priscilla is a babe." According to one visitor, Elvis' romantic skills are believed to carry on into the afterlife: "Please show Jackie 0. a good time!" Romance was on the mind of Chris when he asked: "Yo Elvis! How about getting me and Muriel together?" As this message from Chris shows, fans connect with the memory of Elvis at different levels and through different discourses depending on their personal identity and needs.

Elvis as Family

The discourse of family frequently surfaces as visitors to Graceland emotionally connect with the memory of Elvis. The entertainer is often remembered as part of a loving family which he supported and showered with gifts. Embracing this idea, a visitor addressed the following message to the King's parents: "Dear Gladys & Vernon You done good!" Elvis is represented as a family man even in death. "There is only one King, and I'm sure he is having fun with his mother and father in Heaven," wrote one fan. Some fans, however, have used the wall to comment on the surviving members of the Presley family. For example, Mary pleaded: "Hey Elvis, Lisa Marie made a big mistake! Come back and straighten her out!" Stephanie added: "Elvis, How 'bout that son-in-law?" These messages, of course, refer to the short-lived marriage between Michael Jackson and Elvis' daughter Lisa Marie Presley.

Visitors also invoke the discourse of family as they recount what Elvis meant to their own loved ones. For example, Michelle wrote: "Elvis, I have grown to love you because of my Mom." Tracy showed her appreciation by leaving this message: "Thank you for making my Mom as happy as Roger has made me!" Unfortunately, Rima could not say the same: "My momma doesn't like your music, but I'm kosher." While representing two different relationships between Elvis and mothers, both messages affirmed the importance of family in contextualizing the experience of Graceland.Insome cases, visitors use the wall as a medium for honouring not only Elvis but also a lost family member.Debbie wrote the following message to her father: "Dad, you are now in heaven with the King of rock n-roll! You and Elvis will stay in my heart and mind forever." Another visitor requested that Elvis "Say Hi to Dad for me. You're both in my heart." Flo provided one of the saddest family-related inscriptions,"Elvis, my husband Mac wanted to come with me, but now he's in heaven with you." Fans identity with Elvis' devotion to family and claim him as part of their own families and family histories.This was perhaps no more evident than in a message left by Kathy: "Elvis, we love you so much that we named our little boy Chadwik after you in Blue Hawaii.We miss you."

Concluding Remarks

Studying the journey to Graceland can be a difficult project for two major reasons. First, there is uncertainty about who exactly travels to Graceland, an uncertainty perpetuated by the proprietary manner in which Elvis Presley Enterprises protects visitor information for marketing purposes. Second and more importantly, scholars and the lay public have typically viewed the memorialization of Elvis as a strange, almost laughable part of popular culture. Doing so has limited our ability to study Graceland as an important and legitimate pilgrimage landscape. This paper has attempted to place the study of Graceland and its visitors within a more critical light.

Insummary, it is suggested that visitors to Graceland, rather than simply being passive consumers of Elvis' memory and religiosity, actually play an important role in representing the singer and his home as socially important. The memorialization of Elvis by ordinary people occurs at many different levels and at a variety of locations across the globe. However, an analysis of the Graceland wall allows us to witness the literal rewriting of Elvis' image beyond the immediate 'official' control of Elvis Presley Enterprises. The reading or interpretation of these writings has been facilitated, conceptually, by viewing pilgrimage landscapes as the product of multiple discourses about the meaning of religious heritage. While sometimes appearing odd or silly, the messages found on Graceland's wall represent the convergence of different and, at times, competing discourses about Elvis and the meaning of his image. Whether one is a fan, a curious observer, or an indifferent tourist, no one image of Elvis is embraced at Graceland. Rather, visitors represent and identify with many different Elvises. To some, Elvis is sex and romance. To others, he is the embodiment of success, wealth, and consumption (even over-consumption). For many ardent fans, the prevailing discourse is one of family.

Clearly, more discourses can be read from the inscriptions found on Graceland's fieldstone wall. The author of one wall message perhaps stated it best, "Elvis, I love you for 100,000 reasons." Graceland, like any pilgrimage landscape, can attract a complex range of visitors. Visitors can include those who come for spiritual reflection, those (like the author) who seek knowledge about the social and cultural aspects of Graceland, those who come for fun or spectacle, and even those who come for all these reasons. As suggested by Jackowski and Smith (1992) and Nolan and Nolan (1992), the categories of pilgrim and tourist conceal a rich diversity of motivations, expectations, and experiences. While visitors who love Elvis often write on the Graceland wall, those who are critical of the phenomenon also construct it. One inscription simply stated:"I preferred the Beatles." Future work should focus more closely on the politics of Elvis' memory and the role that Graceland plays in these struggles over the meaning and importance of his image.

Other pilgrimage landscapes exist where the role of fans as authors of Elvis' memory and religiosity can be examined. For example, in Tupelo, Mississippi, visitors can tour the small house where Presley was born as well as a chapel, park, museum, and street bearing his name. Realizing the profits to be made, the city of Tupelo is promoting its connection with Elvis more aggressively than in the past. Another, more minor pilgrimage site is Lauderdale Courts-a public housing project in Memphis where Elvis lived as a teenager from 1949 to 1953. While at "the Courts," Elvis attended high school and began performing publicly, using the laundry room in the basement and the front steps of the apartment building as practice areas. Incidentally, mobilized fans recently saved the apartment where Elvis grew up from demolition, thus further illustrating the need to view Elvis' faithful as important cultural and even political agents.

It is hoped that others will explore the authorship and agency of visitors within all pilgrimage landscapes, whether those landscapes support traditional religions or unconventional faiths such as that found at Graceland.The analysis of authorship should not be confined to the study of inscriptions and message writing. Rather, visitors author or construct pilgrimage landscapes through a variety of social practices. For example, something as seemingly innocent as walking can be interpreted as an important tourist performance (Edensor 2000).The Internet is another emerging forum for exploring pilgrim authorship. Users can visit, for example, a virtual Graceland and post a message on the King's on-line wall (http://books.dreambook.com/gracewall/king.sign.html). These inscriptions, like their terrestrial counterparts in Memphis, may reveal even more ways in which the faithful make and remake Elvis' image as part of their cultural religious heritage.

References

ARTHUR, C. 1989. Road trip: going to Graceland. *American Demographics* 11(5): 47–48.

BARNES, T. and J. DUNCAN (eds) 1992.*Writing Worlds: Discourse, Text & Metaphor in the Representation of Landscape.* New York, Routledge.

BERG, L.D.and R.A. KEARNS 1996.Naming as norming:'race,' gender, and the identity politics of naming places in Aotearoa/New Zealand. *Environment and Planning D: Society and Space* 14(1): 99–122.

BODNAR, R.E. 1992. *Remaking America: Public Memory, Commemoration, and Patriotism in the Twentieth Century.* Princeton, Princeton University Press.

CAMPO, J.E. 1998. American pilgrimage landscapes. *Annals of the American Academy of Political & Social Science* 558Quly): 40–56.

CHADWICK, V. (ed.) 1997. *In Search of Elvis.*Boulder, CO, Westview Press. COFFEY, F.1997. *The Complete Idiot's Guide to Elvis.* New York, Alpha Books.

DOSS, E. 1999. *Elvis Culture: Fans, Faith, & Image.*Lawrence, University Press of Kansas.

EDENSOR, T. 2000.Staging tourism: tourists as performers. *Annals of Tourism Research* 27(2): 322–344.

FOX NEWS 1997. Opinion dynamics poll. August 11. Available at Roper Center for Public Opinion Research, University of Connecticut (question ID: USODFOX.081197, R3).

GALBRAITH, M. 2000. On the road to Czestochowa: rhetoric and experience on a Polish pilgrimage. *Anthropological Quarterly* 73(2): 61–73. HAZEN, C. and M. FREEMAN 1997. *Memphis: Elvis-Style.*Winston-Salem, NC, John F. Blair Publisher.

JACKOWSKI, A. and V.L. SMITH 1992. Polish pilgrim-tourists. *Annals ofTourism Research* 19(1): 92–106. JACKSON, J.B.1984. *Discovering the Vernacular Landscape.* New Haven, Yale University Press.

KING, C. 1993. His truth goes marching on: Elvis Presley and the pilgrimage to Graceland. In J. Reader and T. Walter (eds) *Pilgrimage in Popular Culture.* London, Macmillan: 92–104.

LOS ANGELES TIMES 1997. 700,000 Graceland visitors a year can't be wrong. *Los Angeles Times* 03 August:8.

NOLAN, M.L. and S. NOLAN 1992. Religious sites as tourism attractions in Europe. *Annals of Tourism Research* 19(1): 68–78.

OLSON, M. and D. CRASE 1990. Presleymania: the Elvis factor. *Death Studies* 14(3): 277–282.

RODMAN, G.B.1994. A hero to most? Elvis, myth, and the politics of race. *Cultural Studies* 8(3): 457–483.

RODMAN, G.B.1996. *Elvis after Elvis: The Posthumous Career of a Living Legend.* New York, Routledge.

SEATON, A.V. 1999. War and thanatourism: Waterloo 1815–1914. *Annals of Tourism Research* 26(1): 130–158.

SCHEIN, R.H. 1997. The place of landscape: a conceptual framework for an American scene. *Annals of the Association of American Geographers* 87(4): 660–680.

WRIGHT, D.1996. *Dear Elvis: Graffiti from Graceland.* Memphis, Mustang Publishing.

Submitted: December 19, 2000
Accepted: May 20, 2001

Sacred World Heritage Sites: Balancing Meaning With Management

Myra Shackley

ABSTRACT

This article reviews the characteristics of the 166 World Heritage Sites (WHS) (29% of the List) which are of sacred or religious significance. They belong to many different religious traditions, but are predominantly (51%) of Christian affiliation and located in the northern hemisphere. Sacred sites are examined within the contextual framework of contemporary cultural tourism, highlighting the particular management challenges faced in balancing site conservation with enhanced visitor experience, while retaining a sense of the numinous.

Reverence for sacred places has existed ever since the world's earliest settled farming communities. Today, there is a great deal of interest in issues relating to sacred sites, both in their meaning and their management. In earlier times visiting a sacred site meant an encounter with the holy, and a visitor experience which might include a call to move beyond the self. The vast majority of today's cultural tourists do not have such motivations, but part of the challenge of managing visitors at sacred sites lies in difficulties associated with an interface between the sacred and the profane. The introduction of qualitative elements such as 1a sense of the numinous' or a 1spirit of place' makes the visitor experience difficult to classify.

Part of the diversification of the late 20th century quest for religious experience includes not only sites sacred to the great monotheistic religions but also contemporary pilgrimage and 'New Age' sites.Some indication of current levels of interest can be seen in the vastly increased interest in pilgrimage. Spain has now re-opened the medieval pilgrim hostels on way to Santiago de Compostela (a WHS) one of Europe's most important pilgrimage destinations, and it is estimated that 3.5m pilgrims/year make a visit (Bywater 1994). This is nothing to the 12 million/year visiting Notre Dame in Paris, the most popular tourism attraction in Europe. Managing such huge volumes of visitors to these sacred sites presents immense challenges, not the least in balancing the need to conserve the fabric of the site with the provision of a special experience for the visitor. The sacred site may function for the visitor as witness to a message or value system as well as a place of encounter with the numinous and an interesting artefact. The literature about sacred sites includes atlases of holy places (Wilson 1997), pilgrimage and travel guides (CCBA 1992) and academic works grounded in tourism management (Carmichael et al. 1994; Vukonic 1996), theology (Brown and Loades1995; Holm1994), and cultural geography (Park 1994).

Visitors to sacred sites may be divided into two basic groups; those whose primary purpose is to gain a religious experience (including pilgrims) and the potentially far larger group of those whose major motivation is visiting an element of the world's religious heritage. Patterns of systematic religious tourism and pilgrimage are constantly altering in response to factors such as political change, site accessibility, particular promotions and events in religious calendars. The most significant of these was the rise in such travel during the year 2000, when for example between 20-50 million tourists visited Rome. Purposive religious tourism (as opposed to the opportunistic visiting of a sacred site in the course of any leisure or business trip) is especially significant for the coach industry, and also in rail and air charter travel. Visiting sacred sites (particularly when there is a large concentration of such sites within an urban centre) is also a major motivation within the European short break market since a very high proportion of all visits to major religious centres last only for a single day or a two-night stopover. Domestic religious travel (often involving small, highly-specialized direct sale operators) is a growth market. Religious tourism also has merchandising implications - annual sales of religious souvenirs in Italy are estimated at US$ 255m (Bywater 1994) and 200 shops in Lourdes depend on sales of religious objects. The tourism industries of countries such as Israel and the Vatican City is entirely dependent on sacred sites;in Israel, for example, 95% of all visitors go to Jerusalem and more than 60% to Nazareth and Bethlehem. Such tourism is relatively recession resistant, but it is highly vulnerable to political uncertainties (Boniface 1995).

Classifying Sacred World Heritage Sites

There is an immense variety of sacred places: churches, temples, tombs, shrines, sites of apparitions or miracles, sites of particular significance in the natural world (including sacred mountains or rivers), and places associated with the life of a prophet or religious founder. Brockman (1997) has devised a useful classification:

Table 1. Classification of Sacred Sites (after Brockman 1997).

Site Category	Description
1	Places sanctified by events in the life of a prophet, saint or deity
2	Sites of miracles and healing
3	Places of apparition or visions
4	Locations dedicated to special religious rituals
5	Tombs of saints/ prophets/ founders
6	Shrines of a miraculous statue, icon or relic
7	The ancestral or mythical homes of the gods
8	Places that manifest the energy or mystical power of nature
9	Places commemorating great evil that have become a focus for remembrance

In practice many sacred sites combine several of these attributes and it is not difficult to find examples of each category in the World Heritage List. By far the majority of sacred WHS come under classes 4, 5, and 6 although there are several examples that combine the attributes of classes 7 and 8 in the cultural landscape category. The concentration camp of Auschwitz and the Peace Memorial at Hiroshima, both WHS, are excellent examples of category 9. Most of the locations of visions or apparitions (including famous Marian shrines such as Guadaloupe or Lourdes) do not conform to WH criteria, but some of the ancestral/ mystical homes of the gods do appear, either as elements in a cultural landscape or if they have acquired a significant built heritage component (as at Delos in the Aegean). Sacred sites may still be actively used today (as in the many temples cultures or religions that no longer exist, or do not exist in the same form, such as Machu Picchu in Peru, or Rapa Nui (Easter Island) (Shackley 1998a). The final category (9) also includes sites that are not sacred in the conventional sense but where events of great magnitude took place, which exert a powerful hold on our spiritual sensibilities (such as Isle du Goree, the slave fort in Senegal). The nature of the visitor experience at such sites is greatly influenced by their cultural background; a Jewish person visiting Auschwitz or an African-American visiting Goree is likely to have quite a different experience from an uninvolved tourist.

Sacred sites on the WH list are generally cultural, but do include some natural sites (such as Mt Taishan in China) as well as cultural landscapes such as the island of Rapa Nui, the Uluru-Kata Tjuta National Park (Ayers Rock, Australia) or Tongariro cultural landscape in New Zealand (Te Heuheu 1995; Layton and Titchen 1995). When calculating the percentage of the World Heritage list which is composed of sacred sites (29%), the writer has included rock-art sites and petroglyphs (such as Tassili b' Ajjer in Algeria) but not major urban centres with complexes of buildings (although these usually include sacred sites) unless such sites are included in the title (e.g. 'The Historic Mosque City of Bagerhat in Bangladesh'). Geographically, 92% of sacred sites are found in the northern hemisphere (which reflects a larger land mass, the presence of the major monotheistic religions and greater economic prosperity). Not all countries possess both cultural and natural WH sites. Some countries (e.g., Denmark, Egypt, Mali) have WH lists dominated exclusively by cultural sites.

Table 2. Religious Mfiliations of Sacred World Heritage Sites.

Religion	%of sites
Christian	51
Pre-Christian	14
Buddhist/ Confucian/Shinto	12
Islamic	7
Rock art	6
Hindu/Jain	4
Landscapes *	4
Other**	2

* includes natural and cultural landscapes
**Hiroshima Memorial, Auschwitz, Jerusalem, and slave fort of Goree in Senegal

The list of sacred sites is immensely varied, from central and south American pre-Columbian sites focused around major ceremonial centres such as Tikal or Chichen Itza to cultural landscapes of sacred value such as Sagarmatha National Park in Nepal (Everest). However, the World Heritage list does not conform dosely to a list of what many people might consider to be the world's most significant sacred sites.It omits, for example, the Black Hills of Dakota, the Mormon Temple in Salt Lake City, concentration camps like Dachau/Buchenwald, Oberammergau, Assisi, Lourdes, the Golden Temple at Amritsar, Varanasi and Nazareth.

Management and Visitors at Sacred World Heritage Sites

Figure 1 plots the relative volume of visitors to a selection of sacred sites against the perceived degree of sociopolitical instability. Visitor volumes are dearly relative,

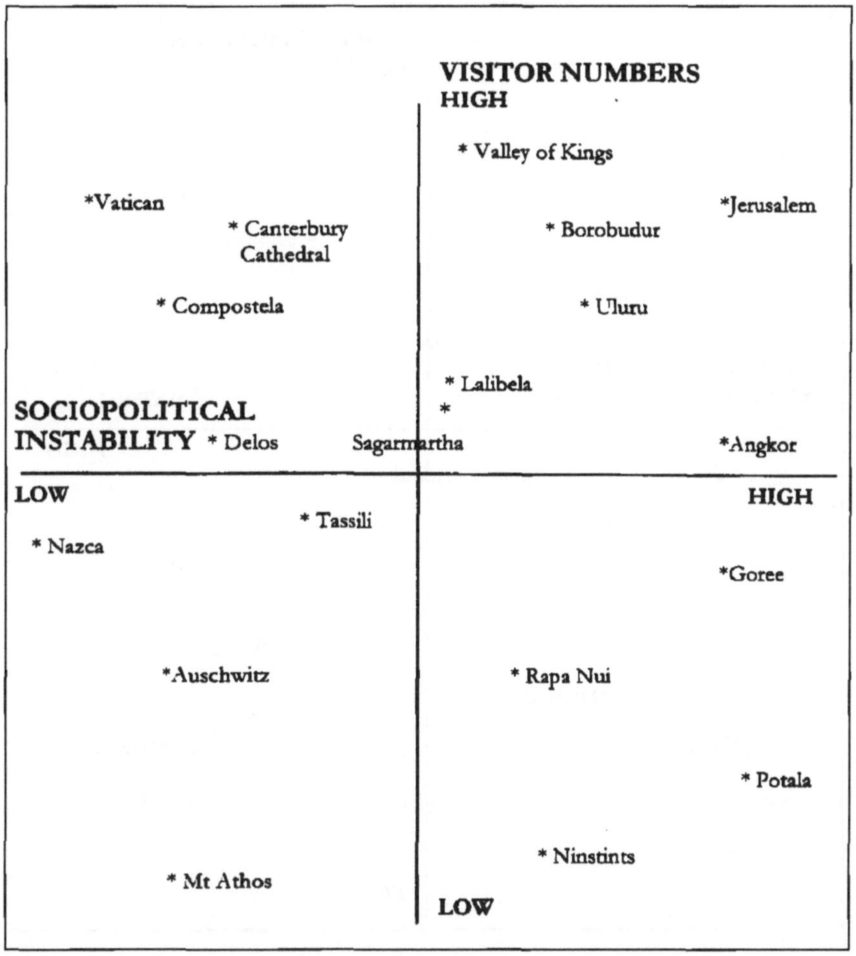

Figure 1. The relationship between visitor volume and socio-political stability at sacred World Heritage Sites.

and the term 'instability' is used here to mean anything from a distant threat from terrorism to disagreement between national government policies and a religious, social or political group with a vested interest in the site. Thus we

see that a site with very high visitor volume and high uncertainty inevitably has management problems (such as Jerusalem). The Valley of the Kings (Thebes) in Egypt also has high visitor volumes but a slightly less elevated uncertainty factor due to security improvements, whereas remote sites like the rock art of Tassili in Algeria not only have few visitors but are located in a political stable area. However, this combination does not necessarily eliminate adverse visitor impact; at Nazca in Peru, for example, there are many worries about vandalism of the famous Nazca Lines. High uncertainty may also be created if the visitors are differentially motivated, and may thus react to the site differently. A classic example occurs as the slave fort of Goree where unpleasant incidents have been reported where African-Americans emotionally affected by the site interacted aggressively with Europeans. On Rapa Nui (Easter Island), which is part of Chile, there is an emergent political movement to reclaim the island's Polynesian heritage from its present Chilean control (Shackley 1998a).

Native people throughout the world are becoming increasingly vocal about who has the responsibility for managing their sacred cultural heritage (Carmichael *et al.* 1994). Such locations often connect with what the western world calls natural features (mountain, rivers, caves). For many native people the notion of a sacred site carries with it a set of rules and regulations regarding visitor behaviour, and implies a set of beliefs to do with the non-empirical world.

A rather different view of site management issues is obtained if we plot visitor numbers against the emotive or spiritual quality of the site (Fig. 2). Sites which provide the highest quality of visitor experience are emotive (Goree, Ninstints), generating a powerful 'spirit of place' even if (perhaps especially if) visitors numbers are low. But the nature of the emotional or spiritual impact received by the visitor dearly varies with his/her

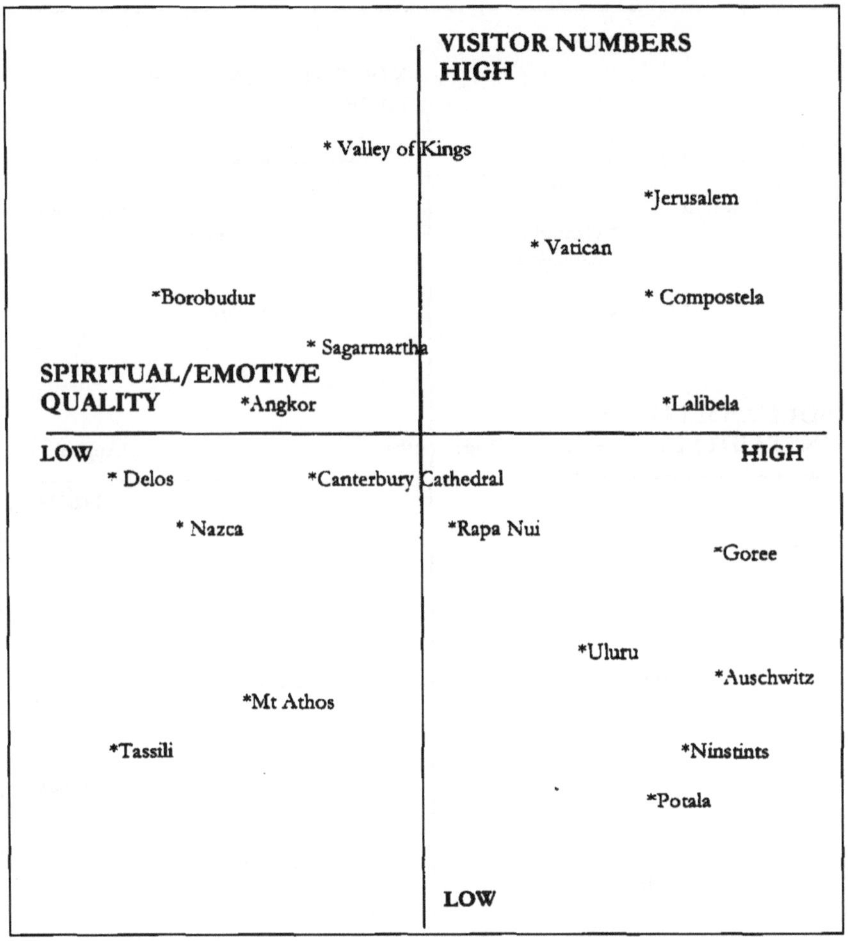

Figure 2. The relationship between visitor volume and experience quality at sacred World Heritage Sites.

motivation and cultural background. Tibetan Buddhists are more affected by the Potala than ethnic Chinese since it represents their own religious tradition. The rock-cut churches of Lalibela in Ethiopia (Carlisle 1998) are of outstanding religious significance to the thousands of Coptic pilgrims who throng them each year, but to the European visitor may appear dark and tawdry. The nature of experience obtained at Angkor Wat is affected by the persistent terrorist threat surrounding the site, and in any case can be described as more architectural than spiritual. The same is true for the Borobudur temple, where any spiritual experience is militated against by crowds of persistent hawkers, and by the site's location surrounded by artificially planted manicured green lawns. Archaeological sites like Delos have very little emotional appeal to the majority of visitors, since they are not 'working' spiritual sites. Places of high emotional impact also have problems with safety - a classic example is the famous 'Jerusalem effect' which regularly hospitalises several dozen people a year who are overcome by emotion and hysteria when visiting the city.

The easiest sites to manage are those where stability is high even if visitor numbers are high (such as the Vatican or Canterbury Cathedral) since this stability and control permits the development and implementation of effective visitor management systems. Borobudur has become Indonesia's major tourist attraction, but is difficult to approach as a place of pilgrimage. Many Asian Buddhist visitors who come in spirit of reverence resent the fact that the temple is not freely open for worship, and indeed

Borobudur is now an island of Buddhism in a country that is now 90% Muslim. This situation is analogous to the Christian monastery of St Katherine in Sinai (Shackley 1998c).

The management of sacred sites may need to be extremely sensitive, and difficult decisions must somehow be made. Auschwitz, for example, one of the largest Nazi extermination camps during the Holocaust, exemplifies the need to care for different ethnic and religious sensibilities. The site is not exclusively a memorial to Jewish dead (although these are in the majority) but is also recognised by Christians as a shrine to St Maximilian Kolbe, a Polish priest-martyr. The camp was restored by the Polish government after World War II but under Communism its guides were notorious for their anti-Semitic attitude, and displays in the museum still concentrate on Polish Christians. The building of a convent next door to the site by Carmelite nuns angered many Jews who felt it was another attempt to make Auschwitz a Christian Polish shrine, and the sisters were relocated in 1993 (Adelsberger 1996). Auschwitz now has the atmosphere of a museum complete with interpretation for visitors including a documentary film and signboard in five languages. But the site emotionally to grief and horror, pray, lightcandles and lay flowers. Historians, Jewish survivors and the Polish government are currently debating whether the site should be managed as a museum to educate the public or a religious memorial. If it is to be a museum then some people feel that this should include rebuilding to enhance the visitor experience, including offering visitors the opportunity to experience the horror of entering the notorious 'shower rooms' and then to see the crematorium. Others find this idea offensive, cheapen 'holy' ground.

Visitors to sacred sites often complain that the experiencing the numinous, and some site management strategies have been developed to address this issue. At Westminster Abbey in London, which receives 2.5 million visitors/year (up to17,000/ day at peak times), much anxiety had been felt about stress to both buildings and visitors, and a new policy, 'Recovering the Calm', was instituted in1998. Inan increasingly secular world visitors to a church do not necessarily know worshippers by loud chat and milling around. Westminster Abbey had become a convenient waiting room or meeting point in central London, and the new plan involved a raised admissions charge as a deterrent. Visitors were encouraged to be quiet and follow a well-marked route. Results seem to indicate that worshippers have returned in increased numbers, and criticism of the admissions charge has been addressed by creating a quiet area in the nave for prayer, accessed by a special door. The plan required capital investment and staff training for a new 'ministry of welcome', and not all sites have either the funds or a management system to operate such procedures. Nor do all sacred sites need silence to preserve a sense of the numinous. On the island of Mont-Saint-Michel in France, for example, the vast monastic structure is an intrinsic part of a noisy commercialized urban area full of souvenir sellers, probably much as it always was. Here the abbey was secularized after the French revolution but monastic life was resumed in 1966 for the 1000th anniversary of the community, providing an interesting and unusual temporal management reversal.

Discussion

Deciding what elements of the WH List consist of sacred sites is not an easy business. The definition of such sites needs to encompass how they can be recognised, who owns them, who has rights of access, what they mean to individuals and communities as well as what their relationship is to the living and the dead. The word'sacred' is derived from a Latin term whose technical meaning is 'restricted by belonging to the Gods', which implies that such sites come with restrictions and prohibitions on human behaviour at that location. This may not always be compatible with cultural tourism. It is frequently difficult to convey a concept of sacredness across cultural boundaries. In England, for example, the concept of sacredness is mostly derived from a Christian heritage with no idea of the sacredness of land, but this is diametrically opposed to the view of the sacred that might be expressed, for example, by a Native American. The nature of the management challenges faced by sacred WHS is not just related to difficulties with cross-cultural interpretation, but also depends on the physical structure of the site. Some are isolated, others form part of a complex of buildings, some constitute integral components of large urban centres. The historic WH listed centre of Cracow, for example, includes 300 historic town houses, 58 churches, 6000 historic buildings and 30 museums. Sacred sites are not only powerful cultural symbols but they need to present the visitor with an evocative experience by creating a visitor environment within which the original spirit of place is retained whilst still creating adequate facilities and providing sufficient information. This is notoriously difficult to do, unless the site is extremely remote. The inaccessibility of the Native American Haida site of Ninstints in Canada (Shackley 1998b) contributes both to its lack of serious management problems and high quality of visitor experience. At the other extreme the Valley of the Kings at Thebes (Egypt) sees 3000 visitors per day and has appalling management problems compounded by difficult conservation requirements and dire traffic congestion.

Throughout the world cultural tourism is experiencing a period of rapid growth, and it is inevitable that WHS should be viewed by national governments as major potential generators of funding from heritage tourism, and by private-sector tourism operators as having unlimited potential for merchandising, retail and service operations. However, entry to sacred WHS is quite likely to be free. Although fees are charged for admission to some sacred sites there are widespread anxieties about such policies, crossing religious traditions. Many of the problems encountered in coping with visitors to the world's sacred heritage are also worsened by difficulties with ownership and balancing the requirements of visitor and visited. At Lalibela in Ethiopia, for example, the site is owned by the Ethiopian Orthodox Church but without properly determined boundaries. Tourism provides vital revenues to support (amongst others) 560 priests and deacons, paying for salaries, festivals and the upkeep of churches (Carlisle 1998).

Many sites manage their cross-cultural religious heritage by consensus, but this is often contentious. Jerusalem is a particularly difficult issue since it contains Christian, Islamic and Jewish sites within a relatively small area presenting almost insurmountable difficulties. Management of the Church of the Holy Sepulchre is carefully controlled, with Roman Catholics having the largest share followed by Syrians, Greek Orthodox, Armenians, Syrian and Coptic Christians. Each group is jealous of the other's space and eager to claim right to it so that even the simplest repair can cause years of negotiation, meaning that the church has serious maintenance problems (Sizer 1999). For Muslims the city ranks second only in importance to Mecca and Medina as a pilgrimage site.Such political difficulties can only be minimized under repressive regimes. Visitors to the Potala Palace and associated Jokhang temple in Lhasa, including pilgrims and tourists, are kept under permanent surveillance by a small army of police (Thubten and Turnbull1983) as part of Chinese policy of politico-religious repression in Tibet. A different kind of political control occurs at Mt Athas in northern Greece, governed as a semi-autonomous monastic theocracy. Here female visitors are forbidden and male potential visitors are screened after producing reference letters and police registration documents. Only 10 non-Orthodox men are admitted each day.

It is interesting to speculate about the role that sacred WHS may play in the tourism of the future. One element that must be factored in is the severe abrasion that many such sites are receiving as a result of their popularity as visitor attractions. Some experts, for example, predict that the tombs of the Valley of the Kings in Thebes will be too fragile for visitation within the next century, casualties of the 661m visits/year expected by the year 2000 (Hawass 1998). It is a curious paradox that the contemporary resurgence of interest in a quest for the numinous, manifested in visiting sacred places, may eventually result in their closure. The concept of endangered cultural sites is already with us, although only 10% of the List of World Heritage in Danger consists of sacred sites. The very process of tourism in modem societies might be interpreted as possessing certain quasi-religious characteristics (Allcock 1988), by those subscribing to the theory that the apparent process of secularization evident today actually conceals sacralization of previously profane activities. This is a concept now being widely resurrected in management circles.Some academics have viewed the process of tourism itself as a modern pilgrimage, based on the ideas of MacCannell (1992), or as a secular substitute for organized religion with public places functioning as religious texts (after Durkheim 1915). The central challenge for today's sacred sites is therefore not the conflict between tourism and pilgrimage, or between religious and secular usage, but managing an escalation in the numbers of those seeking a sense of the numinous, by whatever means that act of pilgrimage is achieved.

References

ADELSBERGER, L. 1996. *Auschwitz: A Doctor's Story.* Boston, Northeastern University Press.

ALLCOCK, J.B. 1988. Tourism as a sacred journey. *Society and Leisure* 11: 33–38.

BONIFACE, P. 1995. *Managing Quality Cultural Tourism.* London, Routledge.

BROCKMAN, N. 1997. *Encyclopedia of Sacred Places.* Santa Barbara, AbC-Oio Inc.

BROWN, D. and A. LOA DES (eds) 1995. *The Sense of the Sacramental.* London, SPCK.

BYWATER, M. 1994. Religious travel in Europe. *Travel and Tourism Analyst* 2:39–52.

CARLISLE, S. 1998. Lalibela (Ethiopia): a religious town in rock. In M. Shackley (ed.) *Visitor Management: Case Studies from World Heritage Sites.* Oxford: Butterworth-Heinemann: 139–160.

CARMICHAEL, D., J. HUBERT and B. REEVES 1994. *Sacred Sites, Sacred Places.* London, Routledge.

COUNOL OF CHURCHES FOR BRITAIN AND IRELAND (CCBI) 1992. *Holy Land Pilgrimage: A Guide for Visitors.* London, CCBT.

DURKHEIM, E. 1915. *Elementary Forms of the Religious Life* (trs. J.W. Swain). London, Allen and Unwin.

HAW ASS, Z. 1998. Site Management: The Response to Tourism. *Museum International* 50(4): 31–37.

HOLM, J. 1994. *Sacred Place.* London, Pinter.

LAYTON, R. and S. TTTCHEN 1995. Uluru: An Outstanding Aboriginal Cultural Landscape. In B. von Drost, H. Plachter and M. Rossler (eds) *Cultural Landscapes of Universal Value.* Stuttgart, Gustav Fischer Verlang Jena/UNESCO: 174–182.

MACCANNELL, D. 1992. *Empty Meeting Grounds* London, Routledge.

PARK, C. 1994. *Sacred Worlds: An Introduction to Geography and Religion.* London, Longmans.

SHACKLEY, M. 1998a. The Cultural Landscape of Rapa Nui (Easter Island, Chile). In M. Shackley (ed.) *Visitor Management: Case Sh1dies from World Heritage Sites.* Oxford, Butterworth Heinemann: 66–82.

SHACKLEY, M.1998b. Ninstints (Canada). In M. Shackley (ed.) *Visitor Management: Case Studies from World Heritage Sites.* Oxford, Butterworth Heinemann: 182–194.

SHACKLEY, M. 199&. A Golden Calf in Sacred Space? The Future of St Katherine's Monastery, Mount Sinai (Egypt). *International Journal of Heritage Studies* 4(3/ 4): 123–134.

SIZER, S. 1999. The Ethical Challenges of Managing Pilgrimages to the Holy Land. *International Journal of Contemporary Hospitality Mar1agement* 11(2/3): 6–12.

TE HEUHEU, T. 1995. A Sacred Gift: Tongariro National Park, New Zealand. In B. von Drost, H. Plachter and M. Rossler (eds) *Cultural Landscapes of Universal Value.* Stuttgart, Gustav Fischer Verlang Jena/UNESCO: 170–174.

THUBTEN, J.N. and C. TURNBULL 1983. *Tibet: Tts History, Religion and People.* New York, Penguin.

VUKONIC, B. 1996. *Tourism and Religion.* Oxford, Pergamon Press.

WILSON, C. 1997. *The Atlas of Holy Places and Sacred Sites.* New York, Reed Publishing.

Submitted: November 10, 1999
Accepted: January 27, 2001

Complementarity: Bridging the Tourism Academic/Religion Divide

Stephen Schweinsberg

The opposite of a big truth is also a big truth (Niels Bohr)

Friedrich Nietzsche argued that

> the whole of nature is the conception of religious men a sum of actions by conscious and volitional beings ... we men of today feel precisely the opposite; the richer a man feels within himself, the more polyphonic his subjectivity is, the more powerfully is he impressed by the uniformity of nature (Nietzsche, 1977, p. 167).

The idea that theistic and mono-theistic beliefs are misguided is long standing and based on the 'reasoned' arguments of scientific luminaries such as Charles Darwin, Stephen Hawking, and Bertrand Russell, etc. (Dixon & Shapiro, 2022). Scientific knowledge, they argued is premised on the basis of observations and incremental reasoning of the natural world (Gower, 1997). The sense of what it means to be human – freedom, ethics, personal and societal values and an effort to understand our place in the universe – not only do not require the presence of a God or Gods, but in arguing that we can devolve responsibility for our rationale thinking to religious belief, we are trusting in a 'religious method [that] is refuted by its failure' (Stenger in INCH, 2014).

Science and the scientific method have long been recognised as one of the foundational disciplines of tourism knowledge (Tribe & Liburd, 2016; Urano et al., 2022). Schweinsberg (2023) has built on this disciplinary framework in this special issue, arguing that religious faith can serve as a values-based meta narrative governing an academic's sense of self as they seek to engage with different disciplinary perspectives. What he did not consider, however, was how academics who have a religious faith might engage constructively with tourism practices, which conflict with their own faith-based perspective. Could I as a person of faith argue that abortion tourism should be developed in a particular destination, so long as it adheres to a particular political regulatory framework? Similarly, could I recommend the establishment/ expansion of a red-light district in a township if I could prove its economic value for the region? Hollinshead (2010) once asked the Academy to look beyond its traditional colonialist and economic framings and embrace a truly post-disciplinary interpretation of knowledge. This paper asks a follow-on question; how can new knowledge producers reconcile their idea of 'truth' to the rationalist socio-political and economic imperatives upon which the tourism industry still ultimately relies?

Tourism scholars are frequently asked to critically investigate forms of tourism, which can be seen by many in society to be ethically suspect including sex tourism, child sex tourism, suicide tourism, fertility/ reproductive tourism, abortion tourism, and slum tourism. The economic importance of such industries is well documented in particular destination localities (e.g. Brooks & Heaslip, 2019; Guiney & Mostafanezhad, 2015). However, underpinning all of the afore mentioned tourism sectors are also wider societal issues regarding: the sanctity of life; cultural stigmatisation; the marginalisation of the 'other', respect or disrespect of privacy, and unequal power relations; all of which religious institutions have strong, often negative opinions on (e.g. Huschke & Schubotz, 2016; Masci, 2016). Carr (2016) suggests with respect to sex tourism that academics should find ways to engage with contentious tourism in all its forms as a means of illuminating the darkest parts of the world. But, to what end? If our goal is to advocate for principles of love, respect, generosity, and forgiveness for all people then these are laudable goals. After all, 'We love because he [God] first loved us' (1 John 4:19). However, if we are being asked to put aside our own moral principles and advocate for a developmental outcome for a form of tourism that we know goes against our own religious beliefs; what does this mean for the development of truly integrative post-disciplinary perspectives in tourism? Are we not at-risk of falling back into the trap of short-sightedness that the call for this special edition has characterised the Academy to-date?

Even before tourism research took on a greater focus on research engagement and impact as a foundation of its work (Font et al., 2019; Schweinsberg et al., 2022) many readers would have been uncomfortable with the idea that scholarship on issues like sex tourism was focused largely on academic naval gazing. As Carr (2016, p. 188) has argued, shouldn't the focus instead be on 'working in a real world context to help people' (Carr, 2016, p. 188)? However, how might one achieve this while being faithful to their own religious beliefs? The paragraphs that follow will introduce the notion of complementarity as an ontological mindset for how people of faith might engage with views they disagree with. Niels Bohr developed the theory of complementarity in the early twentieth century to describe the probabilistic positioning of electrons in the quantum world (Gribbin, 2012). Bohr acknowledged that

> the epistemological lesson we have received from the new development in physical science, where the problems enable a comparatively concise formulation of principles, may also suggest lines of approach in other domains of knowledge where the situation is of essentially less accessible character (Bohr, 1950, p. 54).

In such contexts, complementarity theory asks human beings to recognise that 'there can be several different ways of describing a single system, each useful and internally consistent, but mutually incompatible' (Wilcezek, 2016 in Goetz, 2018, p. 199).

It has long been recognised that an absence of complementary thinking has the potential to drive a wedge between scientific and religious worldviews (Reich, 1989). As a Christian, my responsibility is not to approach science and tourism's other paradigms from a dogmatic referential perspective, but rather as a relational construct. The following analogy will illustrate the challenges and opportunities of such a goal. Imagine you are listening to a piece of dissonant polyphonic music; a piece of music where a particular combination of notes is causing you to feel uncomfortable; at the extreme you might describe it as like hearing fingernails scrape on a chalkboard. For some listeners such dissonance is not problematic, a reflection of their gradual exposure to new tonal structures (Johnson-Laird et al., 2012). For others, as was the case in an early performance of Igor Stravinsky's *The Rite of Spring* (subtitled *Pictures from Pagan Russia*) in 1913 at the Théâtre des Champs-Elysées in Paris, the musical dissonance and its description of a 'pagan ritual in which a sacrificial virgin dances herself to death' led to what has popularly been described as a riot amongst members of the Parisian audience (Classic FM, 2018). Complementarity perspectives suppose that when assessing the quality of the music, the application of its melodies, tonal structures and instrumentation exist on the one hand. On the other you have the point of view and expectations of the audience. As Wilczek (2021, p. 217) argues – 'Each is a meaningful way to interact with the music. You can switch between them. But you can't really do both at once'. For the record, Stravinsky's reaction to seeing the audience's disquiet was decidedly anti-complimentary, storming out of the theatre in disgust saying '"Go to hell" … [noting that] they were very naïve and stupid people' (Classic FM, 2018).

Tourism academics of religious faith often find themselves operating in professional environments that can seem dissonant, contrary to one's sense of purpose. Should we seek to influence others in the Academy or those in the wider 'world of tourism' to adopt our faith-based views, clinging to a narrow view of the world dictated by dogmatic religious observance? Or alternatively, should we recognise that our ability to understand alternative ways of thinking through complementarity thinking does not require us to agree with everyone (Wilczek, 2021)? Christian academics must recognise that when they seek to bring faith into a discussion of sex tourism, they will be inherently aligned with most in the community around issues to advocate for the ending of sexual slavery (Daniel-Hughes, 2018; Miles et al., 2014). At the same time, however, such an alignment of views must be seen against the backdrop of more divisive debates in various church denominations over various gender and power issues such as rights for LGBTQI peoples (Human Rights Campaign, 2023), abortion rights (Masci, 2016) and the rights of women in the church for ordination and other leadership positions (Riches & Jennings, 2016). Gender-based power and discrimination is often seen as an underlying theme behind in many darker manifestations of commercial forms of sex tourism (Berdychevsky, 2022).

When I as a Christian academic come to an understanding of a divisive form such as sex tourism, I am reminded of the fact that my beliefs do not match perfectly with the everyday experiences of the world I am observing. At the same time, however, I am cognisant of the fact that as a scholar I have access to a diverse range of knowledge domains to describe tourism (see Tribe & Liburd, 2016). No form of tourism can be 'embraced in its entirety in one narrative, statistic or discourse analysis, so tourism research is not accounted for or told in one consistent way, from one position, with one voice' (Ren et al., 2010, p. 899). It is therefore incumbent on me to keep talking collaboratively with different knowledge domains, as per the principles of complementarity articulated by Bohr (McGrath, 2009). This is not always easy but if we do there is the opportunity to truly understand the nature of the whole.

Disclosure statement

No potential conflict of interest was reported by the author(s).

ORCID

Stephen Schweinsberg http://orcid.org/0000-0002-8820-8302

References

Berdychevsky, L. (2022). Sex in tourism. In D. Buhalis (Ed.), *Encyclopedia of tourism management and marketing* (pp. 68–71). Edward Elgar Publishing.

Bohr, N. (1950). On the notions of causality and complementarity. *Science*, *111*(2873), 51–54. https://doi.org/10.1126/science.111.2873.51

Brooks, A., & Heaslip, V. (2019). Sex trafficking and sex tourism in a globalised world. *Tourism Review*, *74*(5), 1104–1115. https://doi.org/10.1108/TR-02-2017-0017

Carr, N. (2016). Sex in tourism: Reflections and potential future research directions. *Tourism Recreation Research*, *41*(2), 188–198. https://doi.org/10.1080/02508281.2016.1168566

Classic FM. (2018). This is what REALLY happened at The Rite of Spring riot in 1913. https://www.classicfm.com/composers/stravinsky/news/rite-and-the-riot/.

Daniel-Hughes, C. (2018). Evangelical women are shaping public attitudes about sex work. https://theconversation.com/evangelical-women-are-shaping-public-attitudes-about-sex-work-89129.

Dixon, T., & Shapiro, A. (2022). *Science and religion: A very short introduction*. Oxford University Press.

Font, X., Higham, J., Miller, G., & Pourfakhimi, S. (2019). Research engagement, impact and sustainable tourism. *Journal of Sustainable Tourism*, *27*(1), 1–11. https://doi.org/10.1080/09669582.2019.1560673

Goetz, S. (2018). *CS lewis*. John Wiley & Sons.

Gower, B. (1997). *Scientific method: An historical and philosophical introduction*. Psychology Press.

Gribbin, J. (2012). *In search of Schrodinger's cat: Updated edition*. Random House.

Guiney, T., & Mostafanezhad, M. (2015). The political economy of orphanage tourism in Cambodia. *Tourist Studies*, *15*(2), 132–155. https://doi.org/10.1177/1468797614563387

Hollinshead, K. (2010). Tourism studies and confined understanding: The call for a "new sense" postdisciplinary imaginary. *Tourism Analysis*, *15*(4), 499–512. https://doi.org/10.3727/108354210X12864727693669

Human Rights Campaign. (2023). Stances of faiths on LGBTQ Issues: Roman Catholic Church. https://www.hrc.org/resources/stances-of-faiths-on-lgbt-issues-roman-catholic-church.

Huschke, S., & Schubotz, D. (2016). Commercial sex, clients, and Christian morals: Paying for sex in Ireland. *Sexualities*, *19*(7), 869–887. https://doi.org/10.1177/1363460716638094

INCH. (2014). Debate: Can religion and science co-exist. https://www.ineos.com/inch-magazine/articles/issue-7/debate/.

Johnson-Laird, P. N., Kang, O. E., & Leong, Y. C. (2012). On musical dissonance. *Music Perception: An Interdisciplinary Journal*, *30*(1), 19–35. https://doi.org/10.1525/mp.2012.30.1.19

Masci, D. (2016). Where major religious groups stand on abortion. https://www.pewresearch.org/fact-tank/2016/06/21/where-major-religious-groups-stand-on-abortion/.

McGrath, A. (2009). *Science and religion: A New introduction*. John Wiley and Sons.

Miles, G., Crawford, C. F., DoCarmo, T., & Velazco, G. A. (2014). *Stopping the traffick: A Christian response to sexual exploitation and trafficking*. 1517 Media/Fortress Press.

Nietzsche, F. (1977). *A Nietzsche reader*. Penguin Classics.

Reich, H. (1989). Between religion and science: Complementarity in the religious thinking of young people. *British Journal of Religious Education*, *11*(2), 62–69. https://doi.org/10.1080/0141620890110202

Ren, C., Pritchard, A., & Morgan, N. (2010). Constructing tourism research: A critical inquiry. *Annals of Tourism Research*, *37*(4), 885–904. https://doi.org/10.1016/j.annals.2009.11.006

Riches, T., & Jennings, M. (2016). Explainer: why some churches teach that women are 'separate but equal'. https://theconversation.com/explainer-why-some-churches-teach-that-women-are-separate-but-equal-64305.

Schweinsberg, S. (2023). Religion, spirituality, and the formation of tourism knowledge. *Tourism Recreation Research*, 1–12. https://doi.org/10.1080/02508281.2023.2188353

Schweinsberg, S., Sharpley, R., & Darcy, S. (2022). Competitive positioning of tourism academic knowledge. *Tourism Management*, *91*, 104502. https://doi.org/10.1016/j.tourman.2022.104502

Tribe, J., & Liburd, J. (2016). The tourism knowledge system. *Annals of Tourism Research*, *57*, 44–61. https://doi.org/10.1016/j.annals.2015.11.011

Urano, D. G., Mathe, A. J., Coutinho, A. C. A., & Marques Junior, S. (2022). Scientific production in tourism: An analysis of the methodological approaches. *Current Issues in Tourism*, *26*(3), 380–391.

Wilczek, F. (2021). *Fundamentals: Ten keys to reality*. Penguin.

Sacred Journeys: Religious Heritage and Tourism

Dallen J. Timothy

For millennia people have travelled in search of Deity or deification in one form or another. Before the advent of formalized religion, ancient peoples felt the need to worship and seek personal connections to divinity and spiritual connections to nature. Often this aspiration (inherent or cultural) led people to travel as individuals or groups in search of a spiritual association with a power greater than themselves. Practices among some indigenous peoples even today are evidence of this long-established pattern.

With the development of formal religious associations, such journeys and wanderings became formalized as demonstrations of faithfulness or righteousness. As a result, in many religions the trip and the destination have become of paramount importance, even to the extent that religious adherents will not gain salvation in the afterlife if they fail to make a journey to sacred places in this life. Less stringent are faiths that encourage visitations at places deemed sacred as a way of acquiring blessings, supplicating God, or receiving enlightenment. Likewise, in purely a secular sense, leisure travellers, in an attitude of curiosity, commonly visit places that people of other faiths hold sacred. Thus, travel to sacred sites may be viewed on a motivation continuum where at one end it is required by formal religion as a means of gaining salvation. On the other end of the spectrum is simple curiosity by secular tourists about places of interest (Cohen 1992; Morinis 1992; Turner 1973).

Russell (1999: 46) defines religious travel as"a journey to one or more sacred places, undertaken for religious motives." Some of these motives include visiting a site where a miracle took place or where one is expected to occur in the future; attending a family religious ceremony; fulfilling a commandment or religious requirement; obtaining forgiveness for sins; praying and seeking a cure for illness; attending a prayer meeting with a religious leader; witnessing a religious ceremony or performance; and participating in worship services (Timothy and Boyd 2002).

The majority of papers in this collection focus on the more secular end of the spectrum, where pilgrimage is not a religious requirement, but where people travel for spiritual enlightenment. Jutla, for example, examines the role of pilgrimage in Sikhism and considers the scriptural foundations for why it is not actively encouraged as it is in other South Asian religions. Despite the fact that scriptural references and practices do not officially encourage visits to sacred sites, many Sikhs do believe it is important to visit holy places, but more for the purpose of learning about their religious heritage than for any kind of saving ordinance or practice. This is also the case among American Jews, according to Ioannides and Ioannides. While much American Jewish travel does involve visits to sites of spiritual importance, a great deal is also of a more secular nature to visit sites of historical importance. American Jews appear to be more inclined to visit destinations that have a Jewish presence than places that do not.

Similarly, the papers by Davis and Austin, and Olsen and Timothy both examine the dimensions of tourism and religious heritage among a fairly recent Christian faith founded in North America-the Church of Jesus Christ of Latter-day Saints, which does not require formal pilgrimage, but whose members travel extensively to sites of historical and sacred importance in an effort to learn about their heritage and to receive inspiration from both history and a present spiritual consciousness. Attix, in her contribution, examines a spiritually motivated form of tourism that also differs significantly from the traditional views of pilgrimage and religious travel. New Age tourism, where people travel in search of their connections to Mother Earth, for purposes of healing and enlightenment through metaphysical means, is increasing rapidly as society in the Western world becomes older but more interested in outdoors, nature, and inner self enlightenment.

Religiously motivated travel has grown so significant!yin recent years, involving millions of international travelers, that some travel agencies and tour operators have begun to specialize specifically in this form of travel (Olsen and Timothy 1999; Russell1999;San Filippo 2001). Attix demonstrates this increase in interest by drawing attention to the rampant growth of New Age spiritual literature, including guidebooks to places of supernatural significance, as well as the rapid expansion in recent years of New Age centred travel agencies and tour operators.

Given the importance and magnitude of religiously motivated travel, it is not surprising that so many researchers of cultural and migration studies have become interested in this phenomenon as a field of study (e.g., Bhardwaj 1998; Collins-Kreiner 1999; Fleischer 2000; Morinis 1992; Noble 1998), although scholars traditionally have tended to avoid bringing religious travel, or pilgrimage, into the realm of tourism. A distinction has long been maintained by some scholars and religious organizations between tourists and pilgrims and tourism and pilgrimage, which according to the argument papers by Vukonic and Santos, has been propagated in part by religious organizations which do not wish to associate sacred practices and spaces with more hedonistic forms of travel, profane space, and the negative impacts that commonly result from tourism. Thus, while some people argue that differences exist between pilgrims and tourists or that they are two distinctively different groups, the distinction remains unclear, and researchers have been unsuccessful in drawing clear boundaries between them.

Despite this contention, more and more tourism researchers have begun considering pilgrimage and all forms of religious travel as tourism. This is so because pilgrims travel from their home environments, they usually stay a certain length of time in the destination, and they utilize tourism services (e.g., lodging, eating establishments, etc.) in transit and at the destination. Simply stated, from a tourism perspective, pilgrims can be considered tourists who are motivated by religious purposes, and therefore, pilgrimage can be seen as a form of tourism.The distinction (if there needs to be one at all) probably does not lie within the pilgrim's motivation, for motivation has rarely been considered a part of defining who is a tourist or what is tourism. Instead, perhaps the distinction should be between leisure-based tourists and religious-based tourists (which can also overlap) rather than between pilgrims and tourists. In support of this argument, Gupta (1999: 91) argues

> *the fact that [pilgrimage] is often carried out on foot, is an older form and has many religious connotations, has made people overlook it as a form of tourism. Apart from the devotional aspect, looked at from the broader point of view, pilgrimage involves sightseeing, traveling, visiting different places and, in some cases, voyaging by air or sea etc., and buying the local memorabilia, almost everything a tourist does.*

The pilgrim - tourist debate aside, it is clear that tourism to sacred sites, whether by adherents to a particular religion or another, is a major global phenomenon, and many destinations are beginning to broaden their attraction and service base to reflect this growth (Olsen and Timothy 1999).

Through official recognition, or even less formal endorsements, secular places can become sacred. Alderman's analysis of visitor inscriptions on the wall outside Elvis Presley's Graceland provides an interesting and useful look into symbolic pilgrimages among fans of popular culture legends. In the case of Elvis and Graceland, the visitors who seek to connect to the singer demonstrate many of the same characteristics that formal pilgrims exhibit while on a serious trek of spiritual growth. That the home of Presley is a pilgrimage landscape results in the transformation of a profane space into one that is sacred and highly revered. In some situations, the route to the destination is as much a sacred space (and perhaps more so) than the actual destination itself, as in the case of the route leading to Santiago de Compostela in northwestern Spain.

Places of national and personal heritage, such as war cemeteries and sites related to one's own familial past, also exert a significant appeal to travellers, who often view their experience at these places as sacred (Timothy 1997). Seaton provides an interesting historical explanation of the appeal that cemeteries and sites of darkness can exert on visitors as visitations to such places become a form of religious experience.

This special issue, *Sacred Journeys*, is an attempt to highlight many of the forms of religious travel that exist throughout the world and to bring the scope of religious travel more into the mainstream tourism literature. The papers included in this issue are timely and shed light on a variety of religions, religious movements, and definitional arguments in the realm of tourism studies.

References

BHARDWAJ, S.M. 1998. Non-Hajj pilgrimage in Islam: a neglected dimension of religious circulation. *Journal of Cultural Geography* 17(2): 69–87.

COHEN, E. 1992. Pilgrimage centers: concentric and excentric. *Annals of Tourism Research* 19:33–50.

COLLINS-KREINER, N. 1999. Pilgrimage holy sites: a classification of Jewish holy sites in IsraeL *Journal of Cultural Geography* 18(2): 57–78.

FLEISCHER, A. 2000. The tourist behind the pilgrim in the Holy Land. *International Journal of Hospitality Management* 19: 311–326.

GUPTA, V. 1999. Sustainable tourism: learning from Indian religious traditions. *International Journal of Contemporary Hospitality Management* 11(2/3): 91–95.

MORINIS, A. 1992. Introduction: The territory of the anthropology of pilgrimage. In A. Morinis (ed.), *Sacred Journeys:* The *Anthropology of Pilgrimage.* Greenwood Press, Westport, CT: 1–30.

NOBLE, A.G. 1998. South Asian sacred places. *Journal of Cultural Geography* 17(2): 1–3.

OLSEN, D.H. and TIMOTHY, D.J. 1999. Tourism 2000: selling the millennium. *Tourism Management* 20(4): 389–392

RUSSELL, P. 1999. Religious travel in the new millennium. *Travel & Tourism Analyst* 5: 39–68.

SAN FIUPPO, M. 2001. The religious niche. *Travel Weekly* 60(18): 12.

TIMOTHY, D.J. 1997. Tourism and the personal heritage experience. *Annals of Tourism Research* 34(3): 751–754.

TIMOTHY, D.J. and BOYD, S.W. 2002. *Heritage Tourism.* London: Prentice Hall.

TURNER, V. 1973. The center out there: Pilgrim's goal. *History of Religion* 12(3): 191–230.

Part II
Spiritual Dimensions of Religious Tourism

Exploring the Restorative Benefits of Spiritual Retreats: The Case of Clergy Retreats in Australia

Chelsea Gill, Jan Packer and Roy Ballantyne

ABSTRACT

Despite being one of the oldest forms of tourism, religious tourism is a growing and increasingly diverse sector. Spiritual retreat tourism is one type of religious tourism that has been found to have a range of benefits for participants, particularly in relation to restorative benefits such as rest and recovery of cognitive capacity. This mixed-methods study applies Attention Restoration Theory to explore the ways in which participants of spiritual retreats attain restorative outcomes, and to investigate whether different types of retreats result in different benefits. The analysis of data from 152 clergy who participated in spiritual retreats in Australia demonstrated that restoration was a key outcome, amongst other benefits. The practical implications of the findings for both organisers and attendees of spiritual retreats are discussed, and a research agenda for future exploration of the value of retreats is proposed.

Introduction

Spiritual retreats offered within Christian, Buddhist and secular contexts have been found to provide opportunities for reflection (Rodrigues & McIntosh, 2014), renewal (Chun & Chong, 2011), restoration (Ouellette, Kaplan, & Kaplan, 2005), personal development (Schutte & Dreyer, 2006), transformation (Voigt, Brown, & Howat, 2011) and healing (Nelson-Johnson, 2015). Research attention is increasing into Buddhist and secular retreats, yet there is minimal investigation and evaluation of Christian spiritual retreats, despite their frequent occurrence. As restorative experiences have been found significant in religious tourism research in Christian contexts (Bond, Packer, & Ballantyne, 2015; Ouellette et al., 2005; Rodrigues & McIntosh, 2014), this article applies Attention Restoration Theory (ART) to explore the restorative outcomes of spiritual retreats for Christian clergy in Australia.

Literature review

Religious tourism is one of the oldest forms of tourism (Rinschede, 1992) and represents a significant, evolving, growing and increasingly diverse sector of the global tourism market (Sharpley, 2009). Common examples of religious travel and tourism noted by Stausberg (2011) include retreats, seminars, conferences, festivals and pilgrimages. According to McKercher's (2016) taxonomy of tourism products, religious tourism is situated within the Personal Quest category, which refers to travel for personal reasons associated with self-development and/or learning, and includes pilgrimage, sacred travel, missionary safaris and spiritual retreats. An estimated 300 million travellers, engaging in both traditional and modern experiences of religion and spirituality, contribute to the $18-billion religious tourism industry (Wright, 2008). According to Raj, Griffin, and Blackwell (2015, p. 105) religious tourism includes 'a range of spiritual sites and associated services, which are visited for both secular and religious reasons'. A review of religious tourism literature provided by Timothy and Olsen (2006) demonstrates the diversity of the industry with case studies from the perspectives of Christianity, Judaism, Buddhism, Hinduism and Islam, as well as more recent new-age spirituality movements.

Recently, the term 'spiritual tourism' has gained increased use in both academic and non-academic settings. Stausberg (2014, p. 356) explains:

> Spiritual tourism can be used as a synonym for religious tourism but alternatively it can be used as a specific form of religious tourism, one that is more individual, more prone to connectedness, personal meaning and quest, or it can refer to a travel process of identity-work, voyages in search of meaningful experiences, not articulated in religious language.

Norman (2012) identified five types of spiritual tourism experiences, namely spiritual tourism as healing; quest; retreat; collective and experiment. These

Table 1. McKercher's (2016) personal quest need family and Norman's (2012) classification of five spiritual tourist experiences.

McKercher's (2016) tourism product taxonomy – Personal quest need family	Norman's (2012) classification of five spiritual tourist experiences
Personal history: diaspora; genealogy/ roots; personal memory	Quest
Religious: pilgrimage; sacred travel; religious; missionary safaris; spiritual retreats	Retreat/Collective
Medical/wellness: procedures; health; wellness	Healing/Experiment
Learning: cultural exchange; study tour; service learning; immersion programmes; personal development	Quest/Experiment

types align with McKercher's (2016) Personal Quest category, as illustrated in Table 1. Within Norman's (2012) classification, spiritual tourism as healing refers to both physical and psychological healing, with wellness activities often being practiced in an attempt to repair or maintain self. Spiritual tourism as quest refers to tourists seeking new knowledge or fresh discovery of self. Spiritual tourism as a retreat (Norman, 2012) is characterised by renewal or an escape from the everyday routine. Spiritual tourism as collective (Norman, 2012) recognises that some experiences are more significant because they are encountered with others and not merely as an individual. Spiritual tourism as an experiment (Norman, 2012) refers to tourists seeking alternatives as a substitute for something that has become unsustainable or unhelpful in their normal lifestyle and existence, for instance travelling to India to backpack or attend a yoga course. According to Raj and Griffin (2015), the notion of spiritual tourists refers to contemporary 'wayfarers' who are not driven by religious motives but a more general spiritual search for meaning, outside of the confines of dogma and traditional religion.

Clearly, there is an overlap between religious, spiritual and wellness tourism (McGettigan, 2016), and research into spiritual retreats might be positioned within religious tourism (McKercher, 2016), spiritual tourism (Norman, 2012), wellness tourism (Voigt et al., 2011) or transformational tourism (Fu, Tanyatanaboon, & Lehto, 2015; Heintzman, 2013). As this paper focusses on Christian clergy retreats in Australia, it is positioned within the religious tourism field.

Shackley (2004) suggests that most retreat centres in the developed world are either Buddhist or Christian. There has been some research into Buddhist and meditation retreats (e.g. Carter et al., 2005; Emavardhana & Tori, 1997; Walach, Buchheld, Buttenmuller, Kleinknecht, & Schmidt, 2006), with a particular focus on Temple Stay, which is a form of religious tourism whereby Buddhist monasteries provide accommodation as well as cultural, spiritual, religious and leisure activities to tourists (Chun, Roh, & Spralls, 2017). The motivations for participating in Temple Stays include experiencing tranquillity, mental relaxation and reduced stress (Choe, Blazey, & Mitas, 2015) as well as being with nature, learning and self-growth (Chun et al., 2017). Participants reported inner healing and renewal resulting from the silence, meditation and self-reflection while at Temple Stays (Chun & Chong, 2011). From ancient times, Christian retreats have been associated with the notion of 'desert spirituality' – a way to draw closer to God in silence and solitude by leaving behind concerns from everyday life (Tullio & Rossiter, 2009). Research into spiritual retreats has also been undertaken within an orthodox Christian context at a Benedictine monastery in Canada (Ouellette et al., 2005), a Catholic monastery in New Zealand (Rodrigues & McIntosh, 2014), a sacred shrine on Mt Athos, Greece (Andriotis, 2009) as well as Roman Catholic weekend retreats (Nelson-Johnson, 2015). These weekend retreats were found to be transformative, healing, revitalising and calming due to the sense of community, being outside one's normal environment and the hospitality provided. Trono (2015, p. 19) also note that spiritual retreats can be offered in a secular retreat (non-religious) context, for 'those who seek a holiday of the soul that meets the need for growth of the person, spirituality and creativity'.

In practice, spiritual retreats take numerous forms and are marketed in a variety of ways. A retreat involves leaving the everyday routine and structure behind so as to enter into a new temporary experience that includes separation from everyday life and a fresh experience of some form of community in the new environment (Schutte & Dreyer, 2006). The word 'retreat' incorporates notions of escape, relaxation, renewal, refreshment, rejuvenation, stillness, solitude, rest, recharging physical/ mental batteries, reflection, finding self, and exploring new possibilities upon return to ordinary life (Huszar, 2016; Tullio & Rossiter, 2009). Retreats can provide opportunities for healing, companionship, personal transformation, receiving increased strength to live out one's vocation, and becoming more experientially aware of the divine (Schutte & Dreyer, 2006). While outcomes of attending a spiritual retreat include restoration (Ouellette et al., 2005), spiritual transformation and well-being (Andriotis, 2009; Heintzman, 2013; Klimova, 2011), and eudemonic experiences (Voigt et al., 2011), this research will focus on restorative outcomes, specifically in the context of Christian clergy retreats.

Restoration and religious tourism experiences

Previous research has shown that places of worship and monasteries have the potential to restore participants' ability to focus attention and reflect (Herzog, Gray, Dunville, Hicks, & Gilson, 2011; Ouellette et al., 2005). In studying the restorative experiences of 521 visitors to a Benedictine monastery, Ouellette et al. (2005) found visitor outcomes included satisfaction, feeling relaxed, positive, rested and clear-headed and that the destination itself often afforded personal transformation. Similarly, visitors to a Catholic monastery in New Zealand noted reflection and personal growth as valuable experiences (Rodrigues & McIntosh, 2014) while spiritual retreat visitors in the study by Voigt et al. (2011) appreciated feelings of renewal, peace and calm as well as the opportunity for contemplation and meditation. In a study of visitor experiences at three Christian pilgrimage sites in England, Bond et al. (2015) found that restorative experiences were more important to participants than the spiritual or cognitive benefits often associated with religious tourism experiences.

Sharpley (2009) suggested that further research into the experiential aspects of religious tourism is needed to 'inform the management, development and promotion of religious/sacred tourism sites and experiences' (p. 241). Similarly, more targeted research is needed in order to develop, manage and promote effective retreat practices. This paper addresses this need by applying ART to explore the restorative benefits of different types of spiritual retreats and better understand the processes that lead to restorative outcomes. This theoretical framework will be used to shape the research questions and interpret the research findings.

Attention Restoration Theory

ART was developed by Stephen Kaplan and Rachel Kaplan (1989) within the field of environmental psychology and attempts to explain the process by which physical, psychological and social capabilities can be replenished or renewed. The theory maintains that the capacity to focus attention and inhibit distractions becomes fatigued with overuse. In order to recover from mental fatigue, the individual's attention needs to be effortlessly, rather than intentionally, engaged. When effort is not required to inhibit distractions, the capacity for directed attention is able to recover (Kaplan, 1995). Environments that effortlessly engage attention are referred to as restorative environments. The theory has been widely used to study restorative benefits and experiences in contexts such as natural environments (Berto, 2005; Herzog, Colleen, Maguire, & Nebel, 2003; Laumann, Garling, & Stormark, 2001), museums (Packer & Bond, 2010), zoos (Pals, Steg, Siero, & van der Zee, 2009), monasteries (Ouellette et al., 2005), houses of worship (Herzog, Ouellette, Rolens, & Koenigs, 2010), favourite places (Korpela & Hartig, 1996), vacations (Lehto, 2013) and workplaces (Kaplan, 1993).

According to ART, there are four characteristics of a restorative environment (Kaplan, 1995): (1) being away (a physical or psychological break from routine); (2) extent (an environment with sufficient size and scope to occupy the mind without calling upon directed attention); (3) fascination (an environment that holds attention without any effort) and (4) compatibility (alignment between a person's goals and what the setting provides). Natural environments are purported to be the most restorative as they contain all four characteristics in abundance (Berto, 2005; Herzog, Black, Fountaine, & Knotts, 1997; Kaplan, 2001).

Four benefits gained from exposure to restorative environments (Kaplan & Kaplan, 1989) have been identified and were later grouped into two categories of 'attention recovery' and 'reflection'. Attention recovery includes clearing the mind and recovering directed attention; reflection includes considering immediate and unresolved personal problems, as well as reflecting on priorities and one's position in the world (Herzog et al., 1997). Clearly, having the capacity to focus attention is necessary in order to function effectively (Kaplan, 1995). Thus spending time in restorative environments to replenish attention faculties is a valuable endeavour for anyone struggling with mental fatigue.

ART has previously been applied to examine the causes and consequences of mental fatigue in clergy and offer practical interventions to restore cognitive well-being, with spiritual retreats being one suggestion (Gill, Packer, & Ballantyne, 2017). However, the theory has not been previously used to study the extent to which different types of spiritual retreats are actually able to achieve restorative outcomes. This is the focus of the present research.

Clergy context

The clergy face a unique working environment (Blaikie, 1979) and are particularly susceptible to mental fatigue. They are expected to fulfil multiple professional roles such as preacher, ritualist, pastor, teacher and administrator (Carroll, 2006), as well as handle conflict, allocate resources and disseminate information (Kuhne & Donaldson, 1995). Clergy provide organisational leadership, manage volunteers and staff and are required to respond appropriately to the constant demands of congregational members, church hierarchies and the wider community (Davey, 1996; Proeschold-Bell et al., 2015; Rowatt, 2001). The many and varied stressors associated with the clergy role have been categorised as work-related stressors and boundary-related stressors (Wells, Probst, McKeown, Mitchem, & Whiejong, 2012). Work-related stressors include multiplicity of roles, role conflict, unrealistic time expectations, low pay and interpersonal criticism. Boundary-related stressors impact on the clergy's family life and include lack of time for family due to the constant work demands; isolation due to a lack of support and close relationships that often lead to a clergy's own needs going unaddressed; and the challenges of fitting in with a congregation and managing difficult personalities (Hill, Darling, & Raimondi, 2003). Such stressors can not only increase clergy's mental fatigue but also limit their opportunities for mental recovery and restoration (Gill et al., 2017). The consequences of directed attention fatigue include inappropriate responses to situations, a greater likelihood of making errors, feeling irritable, being easily distracted and unable to focus attention, as well as being unable to solve problems (Staats, Kieviet, & Hartig, 2003). In a major study of Australian Christian clergy, Blaikie (1979) concluded that research was needed in order to propose potential strategies to resolve the issues and stressors associated with the clergy profession. In this article, we explore the role of spiritual retreats in providing restorative experiences that support clergy well-being and assist them to effectively live out their vocation.

It is common for denominational hierarchies across various Christian religious institutions to provide an opportunity for clergy to experience refreshment and renewal in the form of an annual retreat away from the normal work environment within work time. Clergy Renewal Programs in America 'provide an opportunity for pastors to step away briefly from the persistent obligations of daily parish life and to engage in a period of renewal and reflection. Renewal periods are not vacations but times for intentional exploration and reflection' (Clergy Renewal Program, 2017). Spiritual enrichment retreats have been identified as a potential opportunity to provide support to clergy and their families (Darling, Hill, & McWey, 2004), and are valued by clergy and their spouses as a component of holistic health programming to enhance their well-being (Proeschold-Bell et al., 2012). Some retreats emphasise reflective experiences and are designed to provide 'interior spiritual development through a prolonged period of reflection' (Catholic Retreats, n.d.). Others emphasise educational experiences and are designed to impart information and facilitate discussion amongst attendees, away from the normal place of work and in an environment conducive to learning new things (Rice, 2014). Although clergy health programmes in America are often retreat-based, there has been minimal evaluation of the outcomes of such experiences (Wallace et al., 2012). Empirical research is required to explore the effectiveness and restorative benefits of the various retreat experiences offered by religious organisations.

Aims and focus of this study

This study presents a first attempt to investigate the restorative potential of spiritual retreat experiences, in the context of clergy retreats in Australia. The research aims to explore the extent to which different types of spiritual retreats (reflective and educational) contribute to restorative outcomes for participants, and compare the types of benefits gained from a retreat experience. Specifically, this article will address the following research questions:

(1) Does a spiritual retreat have a measurable and durable impact on participants' recovery from mental fatigue?
(2) Do different types of retreat elicit different restorative outcomes and benefits?
(3) How do participants perceive the impact of the retreat on their mental state?
(4) How do participants perceive the unfolding of the restorative process during the retreat?

The research findings will enable retreat facilitators to more effectively manage, develop and promote appropriate experiences. Research in this field could also benefit other professions that are considering the provision of annual retreats for their employees.

Method

This research is situated within a post-positivist paradigm, which recognises that all research methods are flawed and that it is impossible to provide an objective account of reality as data are already being interpreted at the same time that it is being observed (Mottier, 2005). Thus in order to produce stronger inferences and increase diversity of findings, inquiry was undertaken using quantitative methods coupled with qualitative approaches (Guba, 1990). This mixed methods approach was considered important to most effectively address the research problem (Creswell & Plano-Clark, 2011) as it enabled the ART quantitative findings to be more fully understood alongside qualitative insights describing how restorative benefits and experiences unfold.

As clergy retreats often consist of a small number of participants, several retreats were included in this research. The Queensland head offices of seven religious organisations were approached to identify whether they provided retreats for their clergy. Four organisations (Anglican, Roman Catholic, Lutheran and Uniting Church) offered retreats during the period of data collection and were willing to be involved in this research. This resulted in a total sample of 17 clergy retreats. Retreats ranged in size from 7 to 52 attendees and were held in various locations throughout Queensland, Australia. Although all were considered to be spiritual retreats, two distinct types were offered: reflective retreats (RR) ($n = 10$) and educational retreats (ER) ($n = 7$). For the purpose of this research, RR were defined as providing at least 3 hours per day of quiet time when participants were encouraged to privately reflect. Some RR were completely or predominantly silent (silent reflective retreats, SRR) while others imposed no or limited expectations regarding silence (non-silent reflective retreats, NSRR). ER were more focussed on imparting information to participants, to help them develop professionally, personally, emotionally, mentally, or spiritually with topics such as personal self-care and well-being, managing conflict, maintaining personal, professional and pastoral boundaries as well as information about new initiatives or updates from the local/state/national hierarchies. The RR ranged from 4–5 nights in duration, while the ER were 2–3 nights. A summary of the 17 retreats across the four denominations is provided in Table 2.

Participants and procedures

Organisers disseminated an email one week prior to the retreat to encourage attending clergy to participate in the research. The researcher then invited the 388 clergy via email (which included a 2-minute video outlining the research purpose), to participate in the study.

Participants were asked to complete three questionnaires. They were asked to provide an email address so that all of their survey responses could be matched. All responses were combined into a single data file and were subsequently de-identified. The first questionnaire was conducted online, one week prior to the retreat (Time 1). A total of 251 questionnaires were completed (response rate = 65%). Before leaving the retreat (Time 2), participants were invited to complete a hard-copy questionnaire, which produced a total of 200 completed questionnaires (response rate = 80% of those who completed the first questionnaire). Two weeks after the retreat (Time 3), participants received another email with the final online questionnaire, resulting in a total of 152 participants who completed all three questionnaires (response rate = 60% of those who completed first questionnaire or 39% of all eligible participants). In the final questionnaire (Time 3), participants were asked if they would be willing to take part in a follow-up telephone interview. Subsequently, semi-structured interviews were conducted with 30 participants (21 from RR and 9 from ER; 22 males and 8 females) with an average duration of 20–30 minutes. The interviews were audio recorded with participants' permission and transcribed for analysis. Interviewees were asked to elaborate on the process and benefits of the retreat experience.

Descriptive data regarding the 152 participants who completed all three questionnaires are presented in Table 3. Chi-square tests comparing those who completed all three questionnaires ($n = 152$) and those who completed only the first and/or second questionnaire ($n = 99$) showed no significant differences in age, χ^2 (4) = 7.0, $p = .136$, or repeat attendance of the same retreat, χ^2 (1) = .011, $p = .916$, or a similar retreat χ^2 (1) = 3.209, $p = .073$. There were statistically significant differences in gender, χ^2 (1) = 11.870, $p = .001$, with a higher proportion of women completing all three questionnaires; and denomination, χ^2 (3) = 12.920, $p = .005$, with a lower proportion of Catholics completing all three questionnaires.

Table 2. Type of retreat by denomination.

	Anglican	Catholic	Lutheran	Uniting
Retreats offered	7	6	1	3
Reflective (RR)	4	6	0	0
Silent (SRR)	2	2		
Non-silent (NSRR)	2	4		
Educational (ER)	3	0	1	3

Instruments

The first questionnaire (Time 1) contained basic demographic questions and a pre-visit measure of participants'

Table 3. Participant demographic information.

	Anglican	Catholic	Lutheran	Uniting
Gender				
Male	52 (58%)	22 (100%)	9 (64%)	15 (56%)
Female	37 (42%)	0	5 (36%)	12 (44%)
Total	89	22	14	27
Age				
20–39	8 (9%)	2 (9%)	1 (7%)	0 (0%)
40–49	8 (9%)	4 (18%)	3 (21%)	6 (22%)
50–59	33 (37%)	3 (14%)	9 (65%)	6 (22%)
60+	40 (45%)	13 (59%)	1 (7%)	15 (56%)
Total	89	22	14	27
Attended this retreat previously[a]				
Yes	61 (69%)	15 (68%)	12 (86%)	13 (72%)
No	27 (31%)	7 (32%)	2 (14%)	5(28%)
Total	88	22	14	18
Attended similar retreat previously[a]				
Yes	79(90%)	16 (76%)	9 (64%)	20 (74%)
No	9(10%)	5 (24%)	5 (36%)	7 (26%)
Total	88	21	14	27

[a]Indicates some missing data.

Relaxed mental state (RMS) (see below). The second questionnaire (Time 2) contained an onsite measure of RMS, participants' level of attention recovery and reflection, and open-ended questions about the retreat experience. The third questionnaire (Time 3) contained a follow-up measure of RMS, asked questions about the benefits gained, and provided space for participants to further describe what they found valuable from the retreat experience.

Relaxed mental state

An adapted and extended Restored Mental State scale (Packer & Bond, 2010) was used across the three time periods to measure an individual's level of mental fatigue and thus gauge whether any changes had occurred as a result of the retreat experience. The original scale (six items) covered physical and mental aspects of relaxation, such as feeling refreshed, calm, thoughtful, tranquil, alert and positive. For the current research, three additional items were included from the Swedish Occupational Fatigue Inventory (Åhsberg, Gamberale, & Kjellberg, 1997) to measure whether participants felt exhausted, uninterested and tired. A 7-point Likert scale (1 = not at all and 7 = completely) was used for all nine items. A factor analysis was used to test whether the nine items represented a single factor. The analysis confirmed one factor, thus a single composite variable was created (fatigue items were reverse scored; Cronbach's alpha = .891) and is referred to as the RMS scale.

Attention recovery and reflection

The Attention Recovery and Reflection Scale (Staats et al., 2003) containing 13 items relating to attention recovery and reflection was included in the onsite questionnaire. (An additional six items relating to social stimulation were not used as they were not relevant to the aims of this research). A 7-point Likert scale (1 = not at all, 7 = completely) was used, as in the original research. A principal axis factor analysis was performed to verify that Attention Recovery (seven items) and Reflection (six items) constructs were distinguishable from the onsite (Time 2) measure of RMS (nine items) described above. The analysis confirmed three orthogonal factors: relaxation, recovery and reflection (Cronbach alphas were .905 for the Attention Recovery scale and .878 for the Reflection scale).

Benefits of retreat

Bond et al.'s (2015) scale that measures the benefits associated with visits to religious tourism sites was adapted for inclusion in the follow-up questionnaire. The benefits included social benefits, personal growth, spiritual growth, cognitive/learning benefits and restorative benefits and were measured using 20 items (four in each subscale) each rated on a 7-point scale (1 = most important, 7 = least important). Principal axis factor analysis with oblique rotation identified three related factors: the first included personal growth, spiritual growth and cognitive benefits, while social and restorative benefits formed the second and third factors, respectively. The same subscales were used as in the previous research to aid interpretation of the findings. Composite variables were created for each of the five subscales (Cronbach alpha's were .919, .923, .894, .860, .922, respectively).

Data analysis

Repeated measures ANOVA was used to determine the extent to which the retreats had a measurable and durable impact on participants' recovery from mental fatigue (Aim 1). Two-way ANOVA with repeated measures, and multivariate analysis of variance (MANOVA) were used to compare the impact of different types of retreat (reflective vs. educational; and within the reflective category, silent vs. non-silent) on restorative outcomes and perceived benefits (Aim 2). All statistical analyses were conducted with a sample of 152 participants who had completed all three questionnaires. The open-ended questionnaire and interview responses were analysed qualitatively to identify themes in relation to the perceived impact of the retreat on participants' mental state (Aim 3) and their perceptions of the unfolding of the restorative process during the retreat (Aim 4).

Results

Results are presented for each of the four research questions in turn.

RQ1: Does a clergy retreat experience have a measurable and durable impact on participants' mental state?

A repeated measures ANOVA was used to compare the RMS means based on Time 1, Time 2 and Time 3 composite scores. Mauchly's test indicated that the assumption of sphericity had been violated (χ^2 (2) = 16.74, $p < .001$), therefore degrees of freedom were corrected using Greenhouse–Geisser estimates of sphericity ($\varepsilon = .90$). The results demonstrate that RMS mean scores increased significantly from Time 1 to Time 2, and decreased significantly from Time 2 to Time 3. However, Time 3 scores remained significantly higher than Time 1 scores, F (1.809, 273.152) = 37.275, $p < .001$ indicating

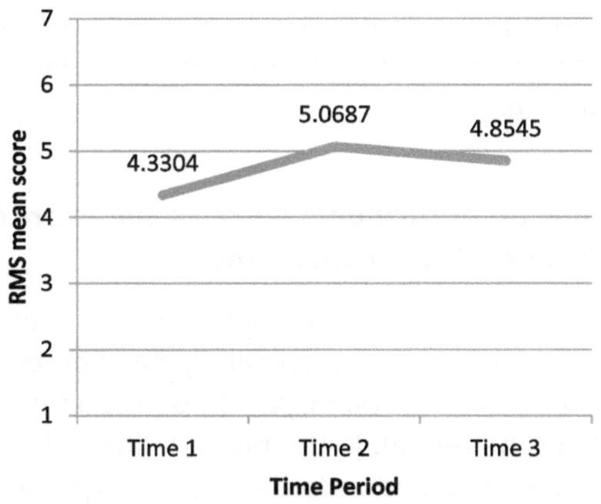

Figure 1. Changes in relaxed mental state over time.

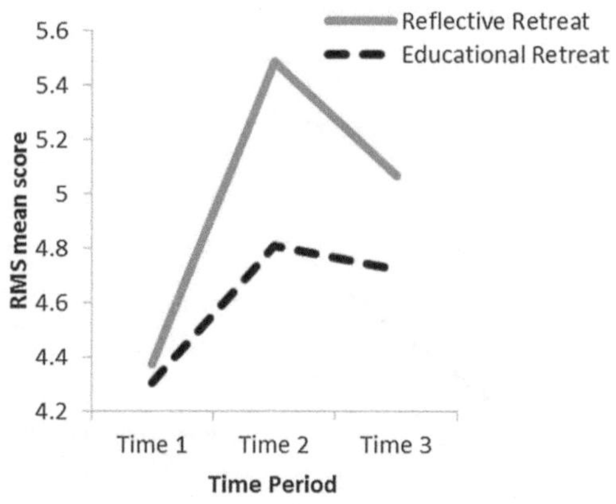

Figure 2. Restorative outcomes (relaxation) for reflective vs educational retreats.

that the retreats had an overall positive impact on participants' mental state (Figure 1) and a partial effect persisted for at least two weeks after the retreat.

RQ2: Do different types of retreats elicit different restorative outcomes and benefits?

RR were compared with ER in terms of their impact on three dependent variables: changes in RMS over time (from Time 1 to Time 2 to Time 3); reported levels of Recovery and Reflection immediately after the retreat (Time 2); and ratings of the perceived benefits of the retreat (Time 3). Within the reflective retreat category, silent RR were compared with NSRR.

Impact of type of retreat on RMS

A two-way ANOVA with repeated measures was conducted to investigate whether retreat type (reflective vs. educational) had an impact on the extent to which RMS scores changed over time. Within the reflective retreat category, a further analysis was conducted to test the difference between retreats with and without periods of silence (SRR vs. NSRR).

Reflective vs. educational retreats. An initial *t*-test comparing RR and ER participants' RMS scores showed no significant difference between groups at Time 1. A two-way ANOVA with repeated measures was performed with type of retreat (RR or ER) as the between-subjects variable and RMS as the repeated measure (Time 1, Time 2, Time 3). Mauchly's test indicated that the assumption of sphericity had been violated (χ^2 (2) = 16.08, $p < .001$); therefore degrees of freedom were corrected using Greenhouse–Geisser estimates of sphericity ($\varepsilon = .90$). The two main effects and the interaction effect were all statistically significant (see Figure 2). Of particular interest for this research question is the interaction effect, $F(1.814, 272.14) = 5.791$, $p = .005$ indicating that type of retreat had an influence on the way RMS scores changed over time. Specifically, RR resulted in higher RMS at both Time 2 and Time 3.

Silent reflective retreats vs. non-silent reflective retreats. A further two-way ANOVA with repeated measures was performed with silent vs. NSRR as the between-subjects variable (see Figure 3). An initial *t*-test comparing SRR and NSRR participants' RMS scores showed no significant difference between groups at Time 1. Mauchly's test indicated that the assumption of sphericity had been violated ($\chi^2(2) = 14.65$, $p < .05$); therefore degrees of freedom were corrected using Greenhouse–Geisser estimates of sphericity ($\varepsilon = .91$). The interaction effect was significant, $F(1.83, 274.31) = 8.14$, $p = .001$, indicating that silence did have an impact on the way in which RMS changed over time. Participants in silent retreats reported a higher RMS when measured onsite (Time 2), but this advantage was not maintained in the weeks after the retreat (Time 3).

Impact of types of retreat on recovery and reflection

To explore whether the type of retreat impacted on the level of recovery and reflection (measured in the onsite questionnaire at Time 2), a MANOVA was conducted with recovery and reflection as the dependent variables, and the type of retreat (RR vs. ER) as the independent

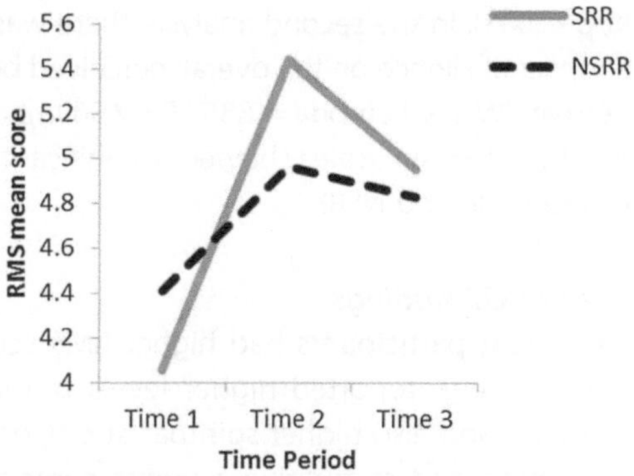

Figure 3. Restorative outcomes (relaxation) for silent vs. NSRR.

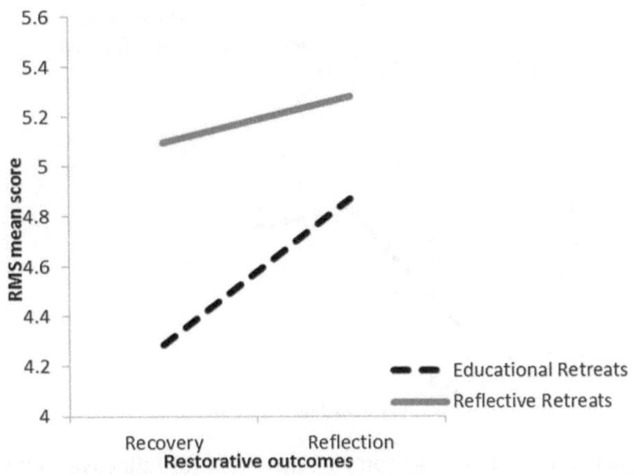

Figure 4. Restorative outcomes (recovery and reflection) for reflective vs educational retreats.

variable (see Figure 4). The type of retreat had a significant overall impact (Wilk's Lambda = .837, $F = 24.759$, $p < .001$), with significant effects on both recovery ($F(1, 256) = 49.347$, $p < .001$) and reflection ($F(1, 256) = 16.248$, $p < .001$). The effect size was higher for recovery than for reflection ($eta^2 = .162$, .060, respectively, see Figure 4). A second MANOVA was conducted using silence (SRR vs. NSRR) as the independent variable, and found no significant effect of silence on recovery and reflection (Wilk's Lambda = 1.00, $F = .008$, $p = .992$).

Impact of type of retreat on perceived benefits

Two MANOVAs were conducted with perceived benefits of the retreat (measured in the follow-up questionnaire at Time 3) as the dependent variables, and type of retreat as the independent variables (reflective vs. educational in the first analysis; silent reflective vs. non-silent reflective in the second analysis). The type of retreat (RR or ER) had a significant impact on the perceived benefits of the retreat (Wilk's Lambda = .801, $F = 7.480$, $p < .001$), with RR producing significantly higher perceived benefits on all five subscales: spiritual benefits ($F(1, 155) = 23.101$, $p < .001$); social benefits ($F(1, 155) = 5.312$, $p = .023$); personal growth benefits ($F(1, 155) = 22.879$, $p < .001$); learning benefits ($F(1, 155) = 9.285$, $p = .003$); restorative benefits ($F(1, 155) = 19.706$, $p < .001$). In the second analysis, there was a significant impact of silence on the overall perceived benefits of the retreat (Wilk's Lambda = .835, $F = 2.569$, $p = .035$), but none of the five subscales showed a significant difference between SRR and NSRR.

Summary of RQ2 findings

Reflective retreat participants had higher RMS scores at Time 2 and Time 3, reported higher levels of recovery and restoration and also higher spiritual, social, personal growth, learning and restorative benefits compared to educational retreat participants. Silent reflective retreats produced higher RMS scores at Time 2, but no advantage was noted at Time 3 compared to the non-silent reflective retreat participants.

RQ3: How do participants perceive the impact of the retreat on their mental state?

Qualitative data, collated from interviews and responses to open-ended questions, were analysed in order to gain further insight into participants' perceptions of the benefits of the retreat. In the presentation below, interviewees are identified according to the type of retreat they had attended: ER, SRR or NSRR. Pseudonyms have been used to protect anonymity. (Open-ended comments extracted from the questionnaire data have not been classified or identified).

When asked about the value of the retreat experience, restorative and social benefits were the most commonly cited themes, followed by spiritual benefits and cognitive benefits.

Restorative benefits

Getting away 'from the daily obligations of work to slow down and recuperate', 'having some free time to re-create, time to rest', having 'space to be me' and 'time to spend just being, time to spend just thinking' out of the normal work environment and routine were very significant for retreat participants. James {ER} noted how relaxing and unwinding at a retreat is 'a critical thing'.

Jim {SRR} and Simon {NSRR} appreciated being away from the world:

> It enabled him 'just to be. To be quiet. Stop and pause … some of the day-to-day pressure is off, so you end up restored, energized, you're not worried about tomorrow';

> It was refreshing … it was so good just to stop. It was like having a holiday. I do the same thing on my holidays. I have that time to sit and do nothing and not be responsible;

Nick {ER} also experienced restorative benefits from the retreat 'because it is a break from normal routine, and I find that refreshing', while Matt {ER} 'came away with a real sense of renewal'.

Another restorative benefit identified by Amon {NSRR} was that of reflection:

> [the retreat was] time away from the ordinary demands of my life, so that I have time to reflect.

Social benefits

In terms of social benefits, participants appreciated: 'the like-minded people at the dinner table', 'time for deep and honest sharing/discussion with my colleagues';

'good companionship among participants' and 'catching up with people, meeting new colleagues'. The notion of sharing information and ideas was a valuable outcome of the social connection at retreat.

> I think being with other priests is very valuable, listening to their stories, because you know we are all not just keepers of the truth, everyone else can share information with us which is really important. {Mike, NSRR}

> You're just busy doing your own work and you don't have a lot of opportunity to feed off what other people are thinking and doing so this is an opportunity for that to happen. {Nick ER}

Jack {NSRR} reported a deeply significant social interaction that was very important to him:

> I was able to get some clarity around this particular friendship that had frayed at the edges ... call it reconciliation, healing, transforming a misunderstanding, talking things through with this mate of mine.

Spiritual benefits

In the written responses, participants appreciated 'the opportunity to be led closer to God and encouraged in ministry', 'being informed on different aspects of the life of the [region]', 'seeing the wider scope of ministry, experiencing the great range of theological positions of attendees', 'the chance to worship regularly with others' and 'the experience of the Scriptures – presented and then reflected upon'.

Surprisingly, in the interviews, few participants overtly mentioned specific spiritual benefits. For some, reflection (i.e. a restorative benefit) facilitated a growth in relationship with God, however, this was not often verbalised as the main benefit of the retreat experience. Nathan {ER}, for instance, notes 'I felt just refreshed spiritually ... I came away I think thankful and walking more closely with God'.

When spiritual benefits were articulated or implied, it was usually in association with RR and some degree of silence. Amon {NSRR}, for example, appreciated that 'for those 5 days, [reflection] it's on my spiritual life and my relationship with God ... [and this led to] a deepening of my own faith'.

Cognitive benefits

Some participants noted that cognitive benefits had emerged from the input session content. Fresh insights were gained and some resources provided that they could then share with others. For instance, participants appreciated 'the retreat giver – I've gained much from his talks/reflections', the 'opportunity for learning' and 'the chance to think and be stimulated by the retreat conductor'. Mike {NSRR} elaborated on these ideas in greater detail by explaining:

> Some of the input has been really valuable. Because one of the things in ministry, you just kind of do your own thing and sometimes just that little bit of an input can make you realize 'oh I didn't consider this angle or I didn't consider those options', so it means you go away with new knowledge.

However, Kate {ER} suggested that the mental stimulation resulting from discussion and debate may have limited the opportunity to experience mental restoration:

> The topics, the teaching we had was very mentally stimulating and there were some people who disagreed and so I found myself coming away thinking even more. My thoughts were even more scattered or heightened. Hearing people talk – again my thoughts were stimulated. I wasn't given – my mind didn't feel at rest. My thoughts increased if anything.

Summary of RQ3 findings

Regardless of whether participants attended a reflective retreat or educational retreat, there is clear qualitative evidence of restorative, social, spiritual and cognitive benefits resulting from the experience. Although format, content, location and attendees varied for each retreat, these four beneficial outcomes were consistently noted by participants.

RQ4 How do participants perceive the unfolding of the restorative process during the retreat?

Qualitative data were also analysed to explore the *process* of restoration as it unfolded during the retreat. This analysis provided some support for Kaplan and Kaplan's (1989) conceptualisation of a four-step process of clearing the mind of cognitive clutter; recovering directed attention; reflecting on immediate and unresolved personal problems and reflecting on life's larger problems and goals. The language used by participants provides rich insight and nuances as to the reality and outworking of restorative benefits as experienced by this professional group.

Clearing the mind of cognitive clutter

Interviewees' responses suggested that the process of clearing the mind depended on two related elements: being away, and stopping. Even though these terms were often used alongside each other, there was a clear acknowledgment that mentally stopping was required, in addition to simply being away. This confirmed Kaplan's (1995) notion that a conceptual shift is more important than being physically away, although the two may or may not coexist.

The retreat provided an opportunity to get away from busyness and ministry demands. This was considered essential as it enabled participants to slow down, recharge, recover and reflect.

> Its value for me is always that I stop from what I am doing normally and, you know, change of location, change of everything really. Stopping helps to recharge my batteries. A lot of the time I'm responding to people. I often find retreats a chance for myself, to slow down, and the process of the retreat – it nourishes me. {Jack, NSRR};

> Retreats are a protected quiet space where people can recover from day-to-day work {Jim, SRR}

While Jim {SRR} claimed that retreats 'are a protected quiet space', it appears this is a result of intentional decision-making and is not guaranteed. For instance, Karen {NSRR} described how setting boundaries to protect the retreat time and space is an intentional decision and necessary for her self-care:

> This sort of dedicated time, where I could – I didn't have my phone, everything was switched off, and I chose that. That was already arranged with my workplace and everyone ... I kind of put in place that self-care framework ... I don't know if I would use the word disconnect because that has certain connotations, but it's a protection and I believe taking it seriously.

Other participants reported that it often takes intentional choice and effort to stop and be still in order to experience a conceptual shift while away. Simon {NSRR} noted how

> it takes a while just to calm down. You're rushing for all this time and then all of a sudden you stop. You've got to get rid of all that excess adrenaline and just breathe, and that takes at least 24 hours.

Indeed the act of stopping is not always easy, as Bob {NSRR} commented: 'usually it's quite heavy-going to actually try to be still.' When probed if it is tiring to try to be still, he responded 'yes, it's probably more mentally tiring in some sense because I'm needing to discipline myself. But then once you get into it, it's reenergizing'.

In contrast to those who deliberately put boundaries in place to protect the retreat time and space from distractions and interruptions, some participants mentioned they did check their emails so that there would not be a backlog on return to work. Bill {SRR} admitted to feeling a tinge of guilt for having checked emails, which implies an acknowledgement that getting away and stopping normal work is important:

> I maybe have a little sense of guilt looking at my emails because I was concerned that I would be stressed when I got to the end of the week and found there were all these things that needed to be attended to.

Likewise, Jack {NSRR} noted that constant phone interruptions affected his retreat experience: 'I knew that I wasn't as relaxed as I have been on previous retreats because my main secretary kept phoning me. I understand why, but that was particularly distracting'. Thus even though the retreat participants were physically away from the normal work environment, the interruptions and distractions limited their ability to experience a total conceptual shift.

Jill {SRR} noted why the physical, mental and emotional 'stopping' while at retreat is an important aspect of 'being away' for this professional group:

> The stopping physically is incredibly important. Most of us get sick on retreat because it's been so long since we've stopped. The physical stopping is important. But more important is the emotional stopping, the mental stopping, the stopping so that spirituality can flow back into our lives in ways that has been pushed out because to some extent, we are sort of like a CEO, or a manager of a company.

Recovering directed attention

Jim {SRR} articulated how being away led to the recovery of his directed attention and enabled him to 'forget about stuff, to not be under the pressure to do this, think about that – you're able to focus I suppose'.

Bob {NSRR} also explained that as his mind cleared from the usual noise, demands and cognitive clutter, he was able to focus attention and synthesise the past and present:

> There's a strong sense of reality without static – not the noise going on. And the pace that I'm going at slows down and a certain clarity of where I've come from, not so much where I'm going, but there's a sense that I'm able to move forward easier because I've left the past behind.

Recovering the ability to direct attention was also described in terms of 'receiving' spirituality, clarity and refreshment instead of operating out of a reactive mode that constantly 'gives' to others and results in mental fatigue. It thus facilitated a sense of nourishment, energisation and restoration.

> We're very much kind of in a mode of reaction as priests – so we react to situations. So the clarity comes because you don't have to be creatively thinking spontaneously, it just gives your mind an opportunity to relax. So in that relaxing, that's where your clarity comes back. {Mike, NSRR}

> [Refreshment] ... usually it's physical – feeling a bit more energy and not as tired, but also mental – ready to get back into things, and spiritual – feeling, you know sometimes you get spiritually drained by things that are happening, so for me it's all of those. It enables me to get back in and keep going. {Nick, ER}

Reflecting on immediate and unresolved problems

Consistent with Kaplan's (1995) model of the restorative process, after recovering the ability to focus attention, participants were able to reflect on current personal problems or work issues. For example, the retreat enabled Frank {SRR} 'the opportunity to repair things and get things back into perspective'. Similarly, Roger *{ER}* valued having the time and space to reflect while at retreat because at other times, responding to work demands takes priority:

> Because it was a different situation [to normal environment], I didn't need to reflect on the things of business, I could reflect on the things of God – it took primary place.

Several participants noted the importance of relaxation in relation to reflection. Mike {NSRR} found that relaxation enabled him to 'go a little bit deeper in self-reflection', while Jim {SRR} suggested '[relaxation] is integral to being able to get all the benefit from the retreat that you can. I think that's the first thing that has to happen and then all the other things flow from it'.

An outcome of reflection, according to Amon {NSRR}, is 'greater personal integration, being more at home with myself, with my own faults, failings, foibles, eccentricities, but also with my personal faith and my commitments'. Thus the third component appears to unfold in two stages: relaxation and then reflection.

Reflecting on life's larger problems and goals and priorities

The final component of restoration is being able to reflect not just on past or present issues, but also reflect on larger realities, goals and priorities for the future. Karen {NSRR} described this process:

> [retreat] was gentle and clarifying. I don't know if it's a direct linear thing because life isn't linear. I think it's a bit more back and forth and all around really. Because as you're reflecting, you're going back on your life and then you're going forward. So it's not a linear, straight line. It's not messy and confusing in any way, but maybe meandering is probably the word.

In synthesising the past, present and future, several participants noted a change in their outlook. For example, Bob {NSRR} likened the retreat experience to New Year's Eve, due to a shift in perspective from newly formed resolutions:

> Having a clearer picture of where I am at, emotionally, spiritually, physically. There's been space to be able to look at relationships, so it is very much across the board. Now I might come out with a whole lot of, its like New Year's Eve, and come out with a whole lot of resolutions, so that I can be more productive and more at peace. [Some] aspects of it do stay. So I think there tends to be a shift that does happen and that shift tends to be fairly permanent.

Reflection engendered a shift in perspective, which according to Bob, 'tends to be fairly permanent'. Frank {SRR} mentioned a similar notion by identifying that the retreat provides an opportunity to create strategies in order to remain 're-centred' and 'balanced' once back in the normal environment:

> It's really about coming back to base all the time. I see a retreat as being a chance to get re-centred, and to look at strategies for when you're back to keep that more in place where its fallen away over the past 12 months for this or that reason.

Similarly, Amon {NSRR} used the term 're-orienting' along with the illustration of restoring a painting to describe how integrating the past, present and future adds value and focus to future goals:

> Restore, revive, revitalize would be words that I would use. Reorienting would be what I would say, taking the material on board that the retreat leader was giving and doing an examination of one's own life and seeing areas where you need to repent and convert again. Your own personal focus can go off course so you need to reorient – how I am tracking? I then add value to restoring because restoring at times can have a sense of loss or depletion. If you're restoring something, I'm looking at a painting here on the wall and if it was in need of restoration, the dirt would have to be taken off and all of that, so there would be something of that process going on to us personally when we involved ourselves in the annual retreat. But also there's this reorienting so that you're coming back to your goals, your commitments.

Likewise, the higher level process of reflecting on the future is implied by Karen {NSRR} as a key outcome of the spiritual retreat experience, when she said 'I don't know why it's called retreat, why it's not called advance, because it's a moving forward, so even the word retreat is an interesting word'.

Evidently, participants see the retreat as an opportunity to reflect, refocus, re-centre and move forward. Frank {SRR} astutely notes that this benefits not only self, but also others due to improved interactions and relationships:

> I think when you don't have the discipline of prayer and silence you don't deal with [some] people very nicely, you fob them off or ignore them or whatever. But it gives you the ability somehow, because you've gone deeply into yourself and seen your own vulnerabilities, your own weaknesses, you can be better in how you deal with people.

Summary of RQ4 findings

The restorative process that unfolded during the retreat experience, as described by participants, supports

Kaplan and Kaplan's (1989) four-step process. The qualitative data demonstrates that being away does not automatically lead to recovery and reflection. Rather, intentional choices to stop – in a physical, emotional and mental sense – are critical. As participants' minds were cleared of cognitive clutter and the usual demands to respond and give to others, they recovered the ability to focus attention. Consequently, they had the mental space to reflect on immediate issues as well as take a longer-term outlook and assess their goals and priorities. This resulted in fresh perspectives, re-orienting and re-focusing, and moving forward in a more positive state of mind.

Discussion

The aims of this study were to investigate whether a spiritual retreat experience has a measurable restorative impact on participants' mental state, whether different types of retreat elicit varying restorative outcomes, how participants perceive the impact of the retreat on their mental state and how the restorative process unfolds while at retreat. Data from 152 participants were analysed in this mixed-methods study and provide evidence to illustrate the positive impact of retreats.

Spiritual retreats were shown to have a significant and positive impact on participants' mental state. Although the level of restoration decreased in the two weeks following the retreat, there was still a measurable improvement compared to baseline levels. Thus, not only do clergy desire retreats to enhance their well-being (Proeschold-Bell et al., 2012), the findings of this study suggest that retreats have a positive impact on participants' mental state. Hence, annual spiritual retreats could be utilised both as a preventative technique to guard against mental fatigue and burnout, as well as an opportunity to support those whose mental well-being requires intervention.

Different types of retreat elicit varying levels of restoration. Although participants of both educational and RR reported improvements in their RMS while on retreat, reflective retreat participants experienced a greater increase than educational retreat participants. This is not surprising given that ER include several hours of delivered content each day which requires significant amounts of directed attention. As one ER participant noted, the mental stimulation limited the opportunity to experience restoration.

Silent reflective retreats were more effective than NSRR in achieving an RMS during the retreat. Silence enables a quietening of the mind (Huszar, 2016), facilitates greater understanding, deep thoughtfulness and absence of mental activity (Jensen, 1973). External silence can encourage inner stillness and thereby awaken and encourage the spiritual life (Mantzarides, 2011). However, there was no evidence that silence resulted in higher levels of recovery or reflection, and the impact on relaxation was not sustained over time.

Follow-up measures indicated that the RMS experienced during the retreat decreased within two weeks, across all retreat types (ER, SRR, NSRR). Such fade-out effects are consistent with observations by de Bloom et al. (2009) who notes vacation effects are usually temporary and fade-out within two to four weeks of vacation. Nonetheless, as all Time 3 scores were higher than Time 1 scores, the retreats did have a positive impact on participants' perceived mental state.

Interviewees' responses suggest that restorative outcomes from the retreat experience are at least as important as the social, spiritual and cognitive benefits. This is consistent with Bond et al.'s (2015) findings regarding religious tourism experiences. The restorative outcomes described in the present research are related to some of the changes noted by Thailand retreat participants who reported a more positive perspective after reflecting on their lives, enabling them to feel better able to handle stressful situations in daily life (Fu et al., 2015). The social benefits acknowledged by some participants of this study align with Bone's (2013) findings that community and connection with others and God were often sought by retreat attendees.

The unfolding process of restoration described by retreat participants gives further insight into the ways in which restorative outcomes develop. Relaxation was a crucial component of the retreat experience for many participants and usually preceded deep reflection. This is consistent with the notion of attention recovery and reflection being separate constructs (Herzog et al., 1997; Kaplan & Kaplan, 1989; Staats et al., 2003). Key aspects of the retreat experience such as being away from the normal work environment, avoiding distractions and resting from the mental effort (Kaplan & Kaplan, 1989) enabled a clearing of cognitive noise (Kaplan, 1995). Being away on retreat enabled participants to separate from their normal routine and enter into a liminal space where communitas evolved and restoration unfolded, which is akin to Turner's (1969) findings regarding rituals and pilgrimage. Nelson-Johnson (2015) observed that the experience of liminal space and communitas on a weekend retreat within Roman Catholicism led to healing and transformation for many attendees. In terms of the current study, in quietness and stillness participants found greater clarity, insight, fresh perspective as well as a reorienting and realigning of their values, purpose and future direction. These findings are similar to the restorative experiences gained from retreating to a Benedictine Monastery, which include feeling rested, relaxed, clear-headed and positive, with visitors appreciating the opportunity to reflect and pray (Ouellette et al., 2005).

The research findings have practical implications for retreat organisers and attendees, as well as offering theoretical contributions. Relaxation was shown to provide an important basis for the emergence of deeper levels of restorative benefits. Placing boundaries around the retreat time and space allows these processes to unfold, and thus enhances restorative benefits. As noted by Hartig (2007, p. 164), 'a restorative environment promotes, not merely permits, restoration'. If a person is distracted (for instance, due to interruptions stemming from their normal working environment), their potential to experience restoration is hindered. In addition, retreats that incorporate some degree of reflection into the programme will be more restorative for participants. It would be helpful to educate retreat participants regarding the process of restoration and the factors that support and facilitate deeper benefits.

This study has contributed to the niche area of religious tourism and restoration by demonstrating the restorative benefits of spiritual retreats. While previous research into retreats has hinted at the transformative and restorative potential of retreats, the current research has utilised a more comprehensive research design to show comparisons across different types of spiritual retreats. Although the qualitative and quantitative data have provided rich insights into the process and benefits of restoration, the study was subject to a number of limitations. These need to be addressed in future research. For instance, the length and type of retreat could not be disentangled in this study due to the limited range of retreat sites that were able to be included. Similarly, due to the sample size, denomination was unable to be used as a dependent variable in the analyses. Future research could explore denominational differences as well as a wider variety of retreats, of different lengths, in various settings. As silent retreats were shown to positively impact on-site mental state measures, it would be useful to further explore the process, impact and effectiveness of silent retreats in achieving restorative and other benefits. The transition from retreat into normal routine would also be a worthwhile topic for further research, with the aim of identifying strategies for increasing the longevity of restorative benefits after participating in a spiritual retreat.

Disclosure statement

No potential conflict of interest was reported by the authors.

ORCID

Chelsea Gill http://orcid.org/0000-0002-9001-5593
Jan Packer http://orcid.org/0000-0003-2250-1175

References

Åhsberg, E., Gamberale, F., & Kjellberg, A. (1997). Perceived quality of fatigue during different occupational tasks: Development of a questionnaire. *International Journal of Industrial Ergonomics, 20*, 121–135. doi:10.1016/S0169-8141(96)00044-3

Andriotis, K. (2009). Sacred site experience: A phenomenological study. *Annals of Tourism Research, 36*(1), 64–84.

Berto, R. (2005). Exposure to restorative environments helps restore attentional capacity. *Journal of Environmental Psychology, 25*, 249–259. doi:10.1016/j.jenvp.2005.07.001

Blaikie, N. H. (1979). *The plight of the Australian clergy: To convert, care or challenge?* Brisbane: University of Queensland Press.

Bond, N., Packer, J., & Ballantyne, R. (2015). Exploring visitor experiences, activities and benefits at three religious tourism sites. *International Journal of Tourism Research, 17* (5), 471–481. doi:10.1002/jtr.2014

Bone, K. (2013). Spiritual retreat tourism in New Zealand. *Tourism Recreation Research, 38*(3), 295–309. doi:10.1080/02508281.2013.11081755

Carroll, J. W. (2006). *God's potters: Pastoral leadership and the shaping of congregations.* Grand Rapids, MI: William B. Eerdmans.

Carter, O. L., Presti, D. E., Callistemon, C., Ungerer, Y., Liu, G. B., & Pettigrew, J. D. (2005). Meditation alters perceptual rivalry in Tibetan Buddhist monks. *Current Biology, 15*(11), R412–R413.

Catholic Retreats. (n.d.). Retrieved from http://www.itmonline.org/bodytheology/

Choe, J., Blazey, M., & Mitas, O. (2015). Motivations of non-Buddhists visiting Buddhist temples. *Current Issues in Tourism, 18*(1), 70–82. doi:10.1080/13683500.2013.771627

Chun, B., Roh, E. Y., & Spralls, S. A. (2017). Living like a monk: Motivations and experiences of international participants in Templestay. *International Journal of Religious Tourism and Pilgrimage, 5*(1), 39–55.

Chun, B. G., & Chong, Y. J. (2011). An investigation of effect and outcome of post-temple stay experience. *Journal of Tourism Science, 35*(10), 73–96.

Clergy Renewal Program. (2017). Retrieved from http://www.cpx.cts.edu/renewal

Creswell, J., & Plano-Clark, V. (2011). *Designing and conducting mixed methods research* (2nd ed.). Thousand Oaks, CA: SAGE.

Darling, C., Hill, E., & McWey, L. (2004). Understanding stress and quality of life for clergy and clergy spouses. *Stress and Health, 20*(5), 261–277. doi:10.1002/smi.1031

Davey, J. (1996). *Burnout: Stress in ministry.* Leominster: Gracewing.

de Bloom, J., Kompier, M., Geurts, S., de Weerth, C., Taris, T., & Sonnentag, S. (2009). Do we recover from vacation? Meta-analysis of vacation effects on health and well-being. *Journal of Occupational Health, 51*, 13–25. doi:10.1539/joh.K8004

Emavardhana, T., & Tori, C. D. (1997). Changes in self-concept, ego defense mechanisms, and religiosity following seven-day Vipassana meditation retreats. *Journal for the Scientific Study of Religion, 36*(2), 194–206.

Fu, X., Tanyatanaboon, M., & Lehto, X. Y. (2015). Conceptualizing transformative guest experience at retreat centers. *International Journal of Hospitality Management, 49*, 83–92. doi:10.1016/j.ijhm.2015.06.004

Gill, C., Packer, J., & Ballantyne, R. (2017). *Applying Attention Restoration Theory to understand and address clergy's need to restore cognitive capacity* (Unpublished manuscript). University of Queensland.

Guba, E. (1990). *The paradigm dialog.* California: SAGE.

Hartig, T. (2007). Three steps to understanding restorative environments as health resources. In C. Ward-Thompson & P. Travlou (Eds.), *Open space: People space* (pp. 163–179). London: Taylor & Francis.

Heintzman, P. (2013). Retreat tourism as a form of transformational tourism. In Y. Reisinger (Ed.), *Transformational tourism: Tourist perspectives* (pp. 67–81). Wallingford: CAB International.

Herzog, T. R., Black, A. M., Fountaine, K. A., & Knotts, D. J. (1997). Reflection and attentional recovery as distinctive benefits of restorative environments. *Journal of Environmental Psychology, 17*, 165–170. doi:10.1006/jevp.1997.0051

Herzog, T. R., Colleen, T. R., Maguire, C. P., & Nebel, M. B. (2003). Assessing the restorative components of environments. *Journal of Environmental Psychology, 23*, 159–170. doi:10.1016/S0272-4944(02)00113-5

Herzog, T. R., Gray, L. E., Dunville, A. M., Hicks, A. M., & Gilson, E. A. (2011). Preference and tranquillity for houses of worship. *Environment and Behavior, 45*(4), 504–525. doi:10.1177/0013916511410422

Herzog, T. R., Ouellette, P., Rolens, J. R., & Koenigs, A. M. (2010). Houses of worship as restorative environments. *Environment and Behavior, 42*(4), 395–419. doi:10.1177/0013916508328610

Hill, E. W., Darling, C. A., & Raimondi, N. M. (2003). Understanding boundary-related stress in clergy families. *Marriage and Family Review, 35*(1/2), 147–166. doi:10.1300/J002v35n01_09

Huszar, O. (2016). The role of silence at the retreats of a Buddhist community. *KOME – An International Journal of Communication Inquiry, 4*(2), 59–73. doi:10.17646/KOME.2016.25

Jensen, J. V. (1973). Communicative functions of silence. *ETC: A Review of General Semantics, 30*(3), 249–257.

Kaplan, R. (1993). The role of nature in the context of the workplace. *Landscape and Urban Planning, 26*, 193–201. doi:10.1016/0169-2046(93)90016-7

Kaplan, S. (1995). The restorative benefits of nature: Toward an integrative framework. *Journal of Environmental Psychology, 15*, 169–182. doi:10.1016/0272-4944(95)90001-2

Kaplan, S. (2001). Meditation, restoration and the management of mental fatigue. *Environment and Behavior, 33*(4), 480–506. doi:10.1177/00139160121973106

Kaplan, S., & Kaplan, R. (1989). *The experience of nature: A psychological perspective*. New York, NY: Cambridge University Press.

Klimova, J. (2011). Pilgrimages of Russian Orthodox Christians to the Greek Orthodox monastery in Arizona. *Tourism, 59*(3), 305–318.

Korpela, K., & Hartig, T. (1996). Restorative qualities of favourite places. *Journal of Environmental Psychology, 16*, 221–233. doi:10.1006/jevp.1996.0018

Kuhne, G. W., & Donaldson, J. F. (1995). Balancing ministry and management: An exploratory study of pastoral work activities. *Review of Religious Research, 37*(2), 147–163. doi:10.2307/3512398

Laumann, K., Garling, T., & Stormark, K. M. (2001). Rating scale measures of restorative components of environments. *Journal of Environmental Psychology, 21*, 31–44. doi:10.1006/jevp.2000.0179

Lehto, X. (2013). Assessing the perceived restorative qualities of vacation destinations. *Journal of Travel Research, 52*(3), 325–339. doi:10.1177/0047287512461567

Mantzarides, G. (2011). *Monasticism*. Retrieved from https://www.stanthonysmonastery.org/monasticism.php

McGettigan, F. (2016). Walking labyrinths: Spirituality, religion and wellness tourism. *International Journal of Religious Tourism and Pilgrimage, 4*(5), 37–50. doi:10.21427/D7V7C7D

McKercher, B. (2016). Towards a taxonomy of tourism products. *Tourism Management, 54*, 196–208. doi:10.1016/j.tourman.2015.11.008

Mottier, V. (2005). The interpretive turn: History, memory and storage in qualitative research. *Forum: Qualitative Social Research, 6*(2), doi:10.17169/fqs-6.2.456

Nelson-Johnson, I. J. (2015). *The beloved retreat: An examination of a transformative experience* (Doctoral dissertation). Retrieved from ProQuest Digital Dissertations (#10008989).

Norman, A. (2012). The varieties of the spiritual tourist experience. *Literature and Aesthetics, 22*, 20–32.

Ouellette, P., Kaplan, R., & Kaplan, S. (2005). The monastery as a restorative environment. *Journal of Environmental Psychology, 25*, 175–188. doi:10.1016/j.jenvp.2005.06.001

Packer, J., & Bond, N. (2010). Museums as restorative environments. *Curator: The Museum Journal, 53*(4), 421–436. doi:10.1111/j.2151-6952.2010.00044.x

Pals, R., Steg, L., Siero, F. W., & van der Zee, K. I. (2009). Development of the PRCQ: A measure of perceived restorative characteristics of zoo attractions. *Journal of Environmental Psychology, 29*, 441–449. doi:10.1016/j.jenvp.2009.08.005

Proeschold-Bell, R. J., LeGrand, S., Wallace, A., James, J., Moore, H., Swift, R., & Toole, D. (2012). Tailoring health programming to clergy: Findings from a study of United Methodist clergy in North Carolina. *Journal of Prevention and Intervention in the Community, 40*(3), 246–261. doi:10.1080/10852352.2012.680423

Proeschold-Bell, R. J., Smith, B., Eisenberg, A., LeGrand, S., Adams, C., & Wilk, A. (2015). The glory of god is a human being fully alive: Predictors of positive versus negative mental health among clergy. *Journal for the Scientific Study of Religion, 54*(4), 702–721. doi:10.1111/jssr.12234

Raj, R., & Griffin, K. (2015). Introduction to sacred or secular journeys. In R. Raj & K. Griffin (Eds.), *Religious tourism and pilgrimage management: An international perspective* (2nd ed., pp. 1–15). Wallingford: CAB International.

Raj, R., Griffin, K., & Blackwell, R. (2015). Motivations for religious tourism, pilgrimage, festivals and events. In R. Raj & K. Griffin (Eds.), *Religious tourism and pilgrimage management: An international perspective* (2nd ed., pp. 103–117). Wallingford: CAB International.

Rice, K. (2014). *Discovery channel founder launches educational retreats*. Retrieved from http://www.travelweekly.com/North-America-Travel/Discovery-Channel-founder-launches-educational-retreats

Rinschede, G. (1992). Forms of religious tourism. *Annals of Tourism Research, 19*(1), 51–67. doi:10.1016/0160-7383(92)90106-Y

Rodrigues, S., & McIntosh, A. (2014). Motivations, experiences and perceived impacts of visitation at a Catholic monastery in New Zealand. *Journal of Heritage Tourism, 9*(4), 271–284. doi:10.1080/1743873X.2014.899602

Rowatt, W. (2001). Stress and satisfaction in ministry families. *Review and Expositor, 98*(4), 523–543. doi:10.1177/003463730109800404

Schutte, C. H., & Dreyer, Y. (2006). Monastic retreat and pastoral care in the Dutch reformed tradition. *Theological Studies, 62*(4), 1453–1468. doi:10.4102/hts.v62i4.402

Shackley, M. (2004). Accommodating the spiritual tourist: The case of religious retreat houses. In R. Thomas (Ed.), *Small firms in tourism: International perspectives* (pp. 225–237). Amsterdam: Elsevier Science BV.

Sharpley, R. (2009). Tourism, religion and spirituality. In T. Jamal & M. Robinson (Eds.), *The SAGE handbook of tourism studies* (pp. 237–254). London: SAGE.

Staats, H., Kieviet, A., & Hartig, T. (2003). Where to recover from attentional fatigue: An expectancy-value analysis of environmental preference. *Journal of Environmental Psychology, 23*, 147–157. doi:10.1016/S0272-4944(02)00112-3

Stausberg, M. (2011). *Religion and tourism: Crossroads, destinations and encounters*. Oxon: Routledge.

Stausberg, M. (2014). Religion and spirituality in tourism. In A. A. Lew, C. M. Hall, & A. M. Williams (Eds.), *The Wiley Blackwell companion to tourism* (pp. 349–360). Oxford: John Wiley & Sons.

Timothy, D., & Olsen, D. (2006). *Tourism, religion and spiritual journeys*. Abingdon: Routledge.

Trono, A. (2015). Politics, policy and the practice of religious tourism. In R. Raj & K. Griffin (Eds.), *Religious tourism and pilgrimage management: An international perspective* (2nd ed., pp. 16–36). Wallingford: CAB International.

Tullio, R., & Rossiter, G. (2009). Critical issues for the future of senior class retreats in Australian Catholic schools: Part 1 – major theoretical and educational issues. *Journal of Religious Education, 57*(4), 57–70.

Turner, V. (1969). *The ritual process*. Chicago: Aldine.

Voigt, C., Brown, G., & Howat, G. (2011). Wellness tourists: In search of transformation. *Tourism Review, 66*(1/2), 16–30. doi:10.1108/16605371111127206

Walach, H., Buchheld, N., Buttenmuller, V., Kleinknecht, N., & Schmidt, S. (2006). Measuring mindfulness – the Freiburg Mindfulness Inventory (FMI). *Personality and Individual Differences, 40*(8), 1543–1555. doi:10.1016/j.paid.2005.11.025

Wallace, A., Proeschold-Bell, R. J., LeGrand, S., James, J., Swift, R., Toole, D., & Toth, M. (2012). Health programming for clergy: An overview of protestant programs in the United States. *Pastoral Psychology, 61*, 113–143. doi:10.1007/s11089-011-0382-3

Wells, C. R., Probst, J., McKeown, R., Mitchem, S., & Whiejong, H. (2012). The relationship between work-related stress and boundary-related stress within the clerical profession. *Journal of Religion and Health, 51*, 215–230. doi:10.1007/s10943-011-9501-9

Wright, K. (2008). *Religious tourism: A new era, a dynamic industry*. Retrieved from http://tourism-review.com/travel-torusim-magazine-religious-tourism-a-new-era-a-dynamic-industry-article695

Religion, Spirituality, and the Formation of Tourism Knowledge

Stephen Schweinsberg

ABSTRACT

Over the last decade, tourism academia has been increasingly characterised as working in a knowledge domain that is sociologically framed. To date, however, there has been no explicit consideration in the literature as to whether and how an academic's spiritual and/ or religious beliefs should influence their engagement with knowledge formation. Drawing on Barkathunnisha et al.'s article "Towards a spirituality-based platform in tourism higher education" published in Current Issues in Tourism that proposed the notion of the 7th platform of tourism knowledge based on spirituality and the underlying four pillars – a spirit of knowing; a spirit of doing; a spirit of being; and a spirit of becoming – this exploratory paper will seek to propose some preliminary thoughts on some of the challenges and opportunities that a belief in spirituality and religion might have for tourism academia. Drawing on autobiographical insights from the author's own Christian faith, the paper concludes by discussing issues that the tourism academy will have to face in order to find ways to bring such a spiritual/ religious perspective into the formulation of tourism knowledge .

Introduction

> Do whatever it takes to avoid fooling yourself into believing that something is true when it is false, or that something is false when it is true (Ibn al-Haytham in Tyson, 2022)
>
> Science is vitally important, but it is only one figure on the hand of humanity, and its greatest potential can be actualised only so long as we careful to remember this (His Holiness the Dalai Lama Lama, 2005, p. 11)

This exploratory paper will seek to propose some preliminary thoughts on some of the challenges and opportunities that a belief in spirituality and religion might have for tourism academia, drawing on autobiographical insights from the author's own Christian faith. Over recent decades tourism knowledge has been increasingly framed as a sociological construct, one where a scholar's own 'knowledge force field' (position, person, ends, rules and ideology) mediates between the 'theorised world of tourism (the canon of its knowledge) and its phenomenal world' (Tribe, 2006, p. 360). At the same time, the increasingly commercialised nature of tourism academia has led tourism scholars to critically reflect on how best to manage the interplay between their own moral position and the notions of abstract and objective scientific understanding that characterise the 'dominant institutional and historical structures we are embedded in' (Hales et al., 2018). As a tourism scholar with an interest in social and environmental sustainability issues who is also a Christian; my own moral position stems from my belief in God and the idea that through a relationship with God through the death of his son Jesus Christ. Because of this and the salvation that it affords, I have a responsibility to use my research and teaching to equip future generations to safeguard the beauty of his creation now and into the future. As Hay (2009) has argued, we cannot assume a right to 'discount the well-being of future generations of human beings, each of which will be loved by God as we are' (p. 64).

Recent scholarship has considered how tourism academics can potentially implement a spiritual and religious dimension into their knowledge building and educational activities (see Bakar, 2020; Barkathunnisha et al., 2019; Barkathunnisha et al., 2017). One mechanism through which a spiritual dimension can be legitimised in the study of tourism is through the work of Barkathunnisha et al. (2019) who proposed an extension of earlier tourism knowledge models including those suggested by Jafari (2001) and Macbeth (2005). The so-called 7th knowledge platform argued that there should be a 'focus on the spiritual development and consciousness that emphasises the inter-relatedness and mutual dependency among human beings and the world around them' (Barkathunnisha et al., 2019, p. 2148).

When Barkathunnisha et al. (2019) proposed the notion of the 7th platform of tourism knowledge based on spirituality they argued that it was underpinned by four pillars – a spirit of knowing; a spirit of doing; a spirit of being; and a spirit of becoming. Each pillar has the potential to influence the recipients of tourism knowledge (see Barkathunnisha et al., 2019). However, while there is an increasing level of understanding of research into the religious dimensions of tourism knowledge formation (Collins-Kreiner, 2020; Kim et al., 2020), what has not been considered to-date is the impact of the pillars on the belief of the scholar who advocates them and the link to a particular organised religious faith. Therefore the purpose of this paper will be to offer comment on the four pillars, drawing on insights from the author's own spiritual journey. The challenge for the reader to consider is on what basis is your own explanation of tourism knowledge based?

Pillar 1: a spirit of knowing

Tourism academics have access to a range of disciplinary perspectives (business studies, social science, arts, humanities and science) and inter-disciplinary perspectives (problem centred knowledge, values based knowledge, indigenous knowledge and web 2.0 knowledge) (Tribe & Liburd, 2016) with which to conceptualise and influence the world of tourism. For Barkathunnisha et al. (2019) the spirit of knowing does not propose that there is one definitive knowledge framework; nor does it encourage the imposition of any knowledge perspective on practice. Rather, it asks how the knower can incorporate different knowledge frameworks in order to 'gain meaningful interpretation(s) of the tourism phenomenon' in question (Barkathunnisha et al., 2019, p. 2147 & 2149). Such arguments are in line with growth in critical tourism scholarship with its concern for understanding 'intuitive aspirations for inclusivity, diversity, partnership, sacredness and quality of life, deep play, sustainability, universal human rights, the rights of nature and peace on Earth' (Ateljevic, 2011, p. 509). Representing a concerted pushback against the principles of modernity, which characterised the late nineteenth and early twentieth century (Rojek, 1995), critical tourism has challenged scholars to move away from traditional rationalist perspectives of reality and embrace a more nuanced and self-critical perspective (Jamal & Everett, 2004). Often such perspectives have also resisted a focus on any 'authoritarian imposition of religious certainty… rediscover[ing] the sacred as a dimension of life and of our societies' (Ateljevic, 2011, p. 504).

Thomas and Ormerod (2017, p. 380 in Font et al., 2022) 'suggest that, as a complex social phenomenon, tourism is best understood by combining the knowledge(s) produced by academics with that produced by other actors, so that it might have greater traction with practitioners'. Spirituality refers to the 'human search for meaning, purpose, and connection with self, others, the universe, and ultimate reality' and religion as the 'systematic body of beliefs and practices related to a spiritual search' (Sheridan et al., 1994, p. 365 in Abu Bakar et al., 2021, p. 7). As academics we should all have our own informed perspectives as to whether, and if so how, the knowledge provided by spiritual or religious perspectives informs our academic work. To-date, however, there is limited evidence regarding what such concepts mean for the study of tourism (exception Bakar, 2020). Drawing on a small sample of twenty-two academics representing a mixture of Christian, Buddhist, Jewish, Hindu, Muslim, Tao and atheist beliefs, Bakar (2020) observed that tourism scholars with an interest in spirituality and religion often had challenges integrating such concepts into materialistic driven Business School environments. A common refrain was whether they be accepted by their academic colleagues, or would students view anything on the study of religion as being 'too weird, too hippy, too new age' (Bakar, 2020)? The motivation of the participants to focus on religion came from a desire to focus on what is the meaning in wicked problems like sustainability (Murphy, 2012); if one was forced to 'teach sustainability without a sense of spirituality … it would be almost inhumane, robotic and shallow, and it wouldn't have the full-bodiedness of sustainability. So, spirituality helps give a form of understanding to what sustainability means' (Agnes in Bakar, 2020). Even the atheist participant in the Bakar (2020) study, whilst not believing in religion, acknowledged the value in encouraging students to interrogate the meaning behind different values-based perspectives.

The inclusion of values-based perspectives has implications for the application of the scientific method. The scientific method is, at its most fundamental level, the practice of asking 'a question and suggesting an explanation (hypothesis) based on an observation, followed by the careful design and execution of controlled experiments and finally validation, refinement and rejection of the hypothesis' (Carrol & Goodstein, 2009, p. 237). Often implemented haphazardly in tourism academia (Urano et al., 2022), the strength of such scientific processes is its ability to allow knowledge to transcend 'all kinds of frontiers: race, ideology, religion, political conviction and the myriad of other things that divide people from one another' (Lennox, 2009, p. 31). Tribe (2018) has argued with respect to tourism that early in one's academic career we learn the need to assess theories in the world 'and those that are supported by empirical evidence and survive attempts at falsification are accepted into the canon of a subject' (p. 15). Tribe (2018) then builds on this, suggesting that what will ultimately underpin the legitimacy of research is the process whereby it is created and how it is accepted or curated by others. He frames such processes according to organised frameworks and benchmarks including: significance; rigour; theoretical basis; and originality (Tribe, 2018). It is against such frames that interpretive perceptions of 'truth' will be determined. However, what of the notion in 'each of the great Monotheistic traditions [that] sees God as truth, love and knowledge' (Miller, 2007, p. 267). In this paper I wish to extend Miller's (2007) argument around evolutionary biology, which argued that 'if faith and reason are both gifts from God, then ... [should they not] play complementary, not conflicting, roles in our struggle to understand the world around us [in the world of tourism]' (p. 267)? I would argue in line with the views of Gooding and Lennox (2019) that the voices of science, religion, history, and philosophy can variously inform our understanding of knowledge and truth. No perspective should be forced on another by dogmatic self-interest. However, neither should one who propagates religious or scientific beliefs be naive as to the limits of their understanding. As Einstein once observed with respect to the relative limits of religion and science:

> Accordingly, a religious person is devout in the sense that he has no doubt of the significance and loftiness of those super-personal objects which neither require nor are capable of rationale foundation. They exist with the same matter-of-factness as he himself. In this sense religion is the age-old endeavour of mankind to become clearly and completely conscious of these values and goals, and constantly to strengthen and extend their effects. If one conceives of religion and science according to these definitions, then conflict between them appears impossible. For science can only ascertain what is, but not what should be, and outside its domain value judgements of all kind remain necessary. Religion, on the other hand, deals only with evaluations of human thought and action; it cannot justifiably speak of facts and relationships between facts (Einstein, 1940, p. 605)

Einstein (1940) was arguing that rather than being in direct conflict, science and religion have the potential to be self-supporting when each perceives its own limitations. Religion is reliant on the knowledge gained from science to conceptualise what it means to be part of God's creation. At the same time, however, science can only be fashioned by someone who has the sufficient understanding of what is truth that a wider world view might provide (Einstein, 1940). As far back as the 1540s, John Calvin, the French theologian who was an influential figure in the Protestant Reformation, argued that the apparent orderliness observed by science was evident for a role for religion in its understanding. Also, he argued that the Bible should not be misread as a scientific textbook; it is not a statement of ecology, anthropology, physics, biology or chemistry; it is solely the word of God testifying to the birth and resurrection of Jesus Christ (McGrath, 2013a).

Because tourism is an interdisciplinary field of academic inquiry it has access to a range of scientific framings and methodologies. Scholars from disciplinary backgrounds will apply different perspectives to the study of a tourism phenomenon in order to shed light on specific issues at hand. A recent work on the interface of ideas of spirituality with tourism education noted that trans-modernism 'posits a new approach to tourism education, one that is transformative and multi-dimensional, emphasising cognitive, affective, moral, social, cultural and spiritual developments in students' (Brown 1998 in Barkathunnisha et al., 2017, p. 179). For the reader of this paper, what is it that serves as your foundation for developing your own affective, cognitive, social, moral, cultural and spiritual thinking? Tourism knowledge has grown exponentially in recent decades (Schweinsberg et al., 2022). We know so much about what tourism is and how it works, but at the same time struggle to reconcile sometimes mutually exclusive considerations from our ever-evolving academic brethren, the industry that serves as the canvas for our research and the community at large. Disagreements often manifest themselves most acutely in circumstances where 'wicked' or socially constructed problems impact on the industry (Schweinsberg et al., 2021). When this occurs what foundation should a scholar fall back on? Critics of the religious view would respond to debates such as this with arguments that whilst there is current debate over what constitutes the limits of scientific understanding, we cannot reasonably place our absolute faith in a theistic/ deistic entity that is by definition empirically unknowable (Dawkins, 2006). It is ok, some would suggest, that we don't yet understand how to scientifically conceptualise wicked problems like tourism's sustainable management. Science is by its nature open to the idea that is comfortable with the idea that something can remain unknowable, which the celebrated physicist Richard Feynman articulated in these terms:

> The scientist has a lot of experience with ignorance and doubt and uncertainty, and this experience is of very great importance ... When a scientist doesn't know the

> answer to a problem, he is ignorant. When he has a hunch as to what the result is, he is uncertain. And when he is pretty darn sure of what the result is going to be, he is in some doubt. We have found it of paramount importance that in order to progress we must recognise the ignorance and leave room for doubt (Feynman, 1955, p. 14)

Being religious does not for the present author devalue the merits of the scientific approach around 'wicked' problems like sustainability. Quite the opposite, it is only through the reasoned analysis of information that we can hope to further understand God's creation. At the same time, however, scientific understanding as currently defined by the disciplines of tourism should always be open to other knowledge lenses. For myself, I am reminded that how we know what we know about aspects of the world 'is influenced by the characteristics and positionality of the knower, who is enmeshed in a complex social web of roles, heritages, experiences, and self-understandings' (Belhassen & Caton, 2009, p. 338). Such cultural perspectives are evident in the historical evolution and interplay between religion and science (see Craig, 2022b; Fara, 2009; McGrath, 1999; Meyer, 2021 for historical summaries). From my own perspective, rather than being in conflict, religion and science exist in a symbiotic relationship, asking complimentary questions around what, how, and why that jointly inform our collective spirit of knowing. When I reflect on what this means for the study of tourism, I am reminded of two things. First, I am reminded of the fact that tourism is not an industry as much it is a sociological framing of society of what it means to responsibly experience the world around us. As an academic, I must strive to make a positive difference in the world through my research. However, as Hales et al. (2018) argue, it is impossible to completely separate the impacts we aim to achieve through our research from the values we have as people. As a Christian, my own values-based perspectives are framed by my faith, which in turn grounds a sense of reality in my perception of what I observe in the world through my research; it guides how I perceive ontological and epistemological disputes, how I perceive a value for empathy with those who I deal with; and forces me to critically consider my role as a researcher. In doing so, I am conscious of the fact that scientists, including social scientists, are often associated more with secular impulses, using the knowledge they create to decipher 'the mysteries of the natural order without resource to supernatural aid or guidance' (Ecklund & Long, 2011, p. 2).

A second issue that flows from the complimentary interface between religion and science is the issue of 'validity' of the knowledge they respectively impart. Dixon and Sharpio (2022) note that we 'generally derive knowledge of the world from four sources: our senses, our powers of rationale thought, the testimony of others, and our memory' (p. 19). The role of science has been described as an attempt to 'weave these individually feeble threads into a more resilient web of knowledge' (Dixon & Shapiro, 2022, p. 19). Tourism has adopted many of the ideas of traditional science disciplines, privileging 'technical, rational and scientific [perspectives]' (Font et al., 2022) as a way of ensuring its own pedagogical legitimacy. This has led some to speculate whether the inclusion of a religious perspective limits rather than enhances our ability to engage with the complexities of tourism's place in the world. Such observations are bound in long standing debates over the merits of a secularisation of society (Berger, 1967) and more specifically the interplay of science and religion in the formation of knowledge (Dixon & Shapiro, 2022; Ecklund et al., 2019; Lennox, 2009). However, for the present author it misses the point. Religion will never win a battle with science if the aim is to narrowly ask, which viewpoint 'gives you more predictive power and the ability to change an outcome' (John Templeton Foundation, n.d.). Religion is not competing with science on what can be proved to be falsifiable or not. Rather, it is a statement of faith and a launching pad to understanding the ecstasy that one might experience in study of nature that underpins much of my own research around sustainability in tourism. As C. S. Lewis wrote:

> Nature never taught me that there exists a God of glory and of infinite majesty. I had to learn that in other ways. But nature gave the word glory a meaning for me. I still do not where else I could have found one (in Yancey, 2002, p. 30)

Pillar 2: a spirit of doing

The Spirit of Doing calls upon academics to consider carefully the relationship between one's ethics and values, and one's praxis in engaging with tourism in the real world (Barkathunnisha et al., 2019). Over recent decades, one of the dominant themes in the study of tourism has been whether a focus on neoliberalism is commensurate with the cultivation of moral and ethical values (Mura & Wijesinghe, 2022). Such deliberations have manifested themselves in debates over the merits regarding tourism growth versus degrowth (Butcher, 2021a, 2021b; Everingham & Chassagne, 2020; Fletcher et al., 2019); as well as discussions over the ethicality of specific traveller practices including elephant treks (Cui & Xu, 2019), sex tourism (Jeong & Lee,

2022) and medical tourism (De la Hoz-Correa et al., 2018). There are numerous religious and non-religious perspectives that an academic may bring to an ethical study of the natural world (see Jenkins et al., 2017; Zagonari, 2021). Religious perspectives should, I would argue, have an important place in how we view ethics as a society; different religious traditions espousing core principles including the importance of stewardship, love of one's neighbours, and the importance of the equal dignity of humans and non-humans (Zagonari, 2020). For some, such perspectives are not absolute ethical standards in their own right, but rather expressions of one's own standards presented as though derived from an absolute authority (Blackburn, 2001 in Fennell, 2018). Others in contrast argue that a belief in God is an essential metanarrative for all moral values, without which 'morality is wholly subjective and non-binding' (Craig, 2022a). As Griffiths (2015) has argued, metanarratives 'frame all other narratives and can be framed and explained by none' (p. 15). As a Christian, my own metanarrative is entwined with my ethical perspectives. At the heart of my metanarrative is a belief in God's power to create the universe, produce humankind in his image, and his decision to manifest himself in his son Jesus Christ to save us from our sins. Such a belief represents an important lesson for my understanding of tourism neoliberalism and in particular its relationship to its social and environmental setting.

In Barkathunnisha et al. (2019), a spirit of doing is about the interplay between one's values and praxis. When seen in relation to tourism education, this means dovetailing a reflexive appreciation of ethics and moral values with the development of vocational skills. In relation to tourism research, it is about the ability to critically reflect on one's role in tourism academia and one's relationship with other stakeholders in the world of tourism. Schweinsberg and Darcy (2022, p. 28) have suggested that the role of a tourism academic is through research, teaching, and engagement, to develop the next generation of tourism professionals whilst being prepared 'to ask tough questions of industry practice and to advocate without fear or favour for the development of a more "sustainable" and "responsible" (i.e. "Responsustable") tourism sector'. Our epistemic authority as academics exists based on the confidence placed in us by other stakeholders within the world of tourism; industry, government, media, and the community at large (Schweinsberg, 2022). Because of this, tourism scholars have a moral obligation 'to lead in a way that is inclusive, empowering, and democratic' (Schweinsberg, 2022), engaging 'constantly and reflexively with the ethical questions that surround … [our] own identity, power and responsibility as academics' (Khoo-Lattimore, 2018, p. 239). It is therefore only with a personal critical reflection on what constitutes a moral 'good' that an academic will be able to grapple with important questions facing the tourism industry.

Tourism is a form of human activity that begets a number of fundamental ethical debates. Whether it is questions over the merits or costs in pursuing tourism as a form of capitalist expansion (Fletcher et al., 2021); or questions over the commodification of humans and the natural world to serve as the playthings of hedonistic travellers (Cole, 2007; Ni'am et al., 2021; Tickle & von Essen, 2020); tourism is a microcosm of wider societal challenges. Whether something is bad or good, it depends on the perspective of the beholder (Hall, 2015). Such viewpoints can manifest themselves through personal spiritual or organised religious beliefs that have their basis in Western and Eastern philosophical traditions (see Armstrong, 2022). As human beings we have the tendency to 'assign meaning and value' in a way that transcends science (Miller, 2007, p. 267). From a Christian perspective this means that I must wrestle with challenging ethical questions relating to how my own position of authority as a tourism scholar may lead to a subjugation of the 'insignificant other' through my interpretation of the results of my research. How might this be at odds with my own Christian ideal – You should love your neighbour as yourself (Mathew's Gospel 22:39)? In doing so, however, I must always remember that I am part of a system that prioritises competition, academic performance benchmarks, grant attainment etc., which can lead to a corruption of the work-life balance (Rosa, 2022) and a loss of personal integrity and academic freedom (Schweinsberg et al., 2021). I must also, however, be open to a critical examination of the ethical frameworks of my own metanarrative. Harari (2018) made the important point that any religion's claim of moral absolutisms should be treated with caution, lest we may mistakenly assume that every contemporary morality debate can have its solution found in only one source of knowledge. Complementarity thus forms an important part of instilling a religious perspective into my spirit of doing; the ability to consider other points of view simultaneously and what they reveal in relation to a specific issue or situation (Wilczek, 2022).

Pillar 3: a spirit of being

The ability to reconcile contradictions in one's academic praxis – to engage in complimentary thinking – is dependent on the ability to critically reflect on one's own sense of self, which is the focus of a spirit of being (Barkathunnisha et al., 2019). Over the course of our time as academics we are influenced by a range of institutional and personal forces, including academic mentors, colleagues, and industry collaborators who in different ways help us to identify our place in the academy (Doorne et al., 2007; Schweinsberg et al., 2018). For some in academia such influencers extend to incorporate religion, which is sometimes seen as a pathway towards interconnectedness, transcendence, and a personal quest for meaning (Bakar, 2020). While religion is only a marginal influence in many parts of the academy owing to concerns over stigmatisation and allegations of unsophisticated thinking by colleagues given the pre-dominance of relativist thinking throughout much of the social sciences (Ecklund et al., 2019; Ecklund & Scheitle, 2007; Spencer & Waite, 2022), in many respects a religious scholar's faith is not dissimilar to an atheist's considered position that there is no God. There will always things that can be empirically proven and other things that must be taken on faith, be it faith in oneself, a deistic or theistic entity or something else. Barkathunnisha et al. (2019) talk of one's spirit of being in relation to tourism including the idea 'that values and ethics cannot be taught for practical wisdom but must be experienced through an engagement with the inner consciousness' (p. 2151). In an educational context this involves educators guiding students 'toward self-discovery and authenticity, develop their sense of the transcendence and shape their experience in ways that empower the individual spirit to take good actions' (p. 2151). When, however, seen from the perspective of an individual academic one should explore how their religious faith or lack thereof has evolved and juxtaposed with their growing experience in the study of tourism.

Like many before him, the present author is a lapsed human geographer who got into the study of tourism initially as a pathway to transition from a Bachelor of Arts to the study of business. Over the last decade or so my research interests have built on these human geography roots (Schweinsberg & McManus, 2020), evolving into a focus on tourism sustainability (Schweinsberg et al., 2012; Schweinsberg et al., 2020; Schweinsberg et al., 2013). Areas of specific research interest has been in relation to tourism's place-based setting, with a specific focus on tourism development in tourism's national park-based settings (Schweinsberg et al., 2015; Wearing et al., 2016). Reflecting back over the period of my academic life the one constant of my research into sustainability was the understanding that the more I know of the world, the less I know. When I was a child, my understanding of nature came from documentaries, family holidays and reading. Children's literature including seminal works such as *The Lorax*, *The Tale of Peter Rabbit*, and *Wind in the Willows* have played a role throughout the western world in bonding generations of children to the natural world and instilling in them the power of non-human agency in environmental management (Lawrence, 2014). The mechanisms by-which such binding occurs include a mixture of works e.g. Anna Sewall's *Black Beauty*, which explores non-human (often animal) perspectives on human–environmental interaction with the aim of exploring the obvious limits of human kind's concern for the natural world in which they are co-inhabit (Buell, 2014). Perhaps not surprisingly my early understanding of sustainability, such as it was in primary school in the 1980s, was often limited to a focus on environmental preservation, with little appreciation of the wicked and contested social problems that underpin sustainability debates today.

Such perspectives were also evidenced in my perspectives on Christianity and the Bible as a child. I knew that God created the heavens and the earth, and that humankind has a responsibility to be stewards the natural world (e.g. Genesis 1:28-30 & Psalm 24:1). In addition to attending Sunday school and church youth groups I was influenced heavily as a child by writings including C. S Lewis's *The Narnia Chronicle's* (see Hooper, 1996 for a discussion of plot points and key themes). Written by C. S. Lewis from 1950 to 1956, the Chronicles have become canons of twentieth century English literature – often being compared to other masterpieces by Lewis's great friend, fellow Christian and colleague J. R. R. Tolkien, *The Hobbit* and *The Lord of the Rings* (see Zaleski & Zaleski, 2015). Being also overtly Christian in their subject matter (Bruner & Ware, 2005) the Chronicles also represent a useful point of comparison to understand the relationship between one's religious perspectives and contemporary sustainability considerations. The *Narnia Chronicles* have since their initial publication been questioned by readers for the underlying message(s) they convey – the presence of its potential medieval symbolism (Ward, 2008); the presence of a deliberate Christian allegory in Lewis's writing (Babu & Vishnuvardhan, 2022) etc. When I read the Chronicles as a child, I was enthralled with the imagery of children entering secret parallel worlds, the grandeur of its environmental descriptions, the presence of talking beasts and battles between the lion Aslan and the White Witch. Over time, however, I began to appreciate the deeper religious symbolism and apologetic nature of the work and more recently as an adult the challenges that the books present over Lewis's treatment of issues that are central to social sustainability issues including gender and race. In recent years, there have also been attempts made to argue for and against the presence of environmental and sustainability messages in the *Narnia Chronicles* (e.g. Dickerson & O'Hara, 2009; Echterling, 2016a, 2016b; Morris, 2009; Noëlle O'Connor & Bolan, 2008). From a pro-sustainability perspective, the books have been argued to have made important arguments in favour of the positive ethical engagement with animals. McGrath (2013b) has observed that when Lewis wrote the *Narnia Chronicles* in the mid-twentieth century, he was writing in an age where vivisection was still seen as accepted practice in society. Morris (2009) has observed that Lewis's writings in children's literature thus sought to emphasise the importance of animal autonomy and offering criticisms of human/animal experimentation, themes that have been important foundations of arguments around sustainable consumption (Suphasomboon & Vassanadumrongdee, 2022) and the rights of animals in tourism contexts across current and future generations (Fennell, 2018, 2022).

The use of a children's novel to examine the interplay between faith and sustainability realities has parallels to the role of science fiction writing and tourism; specifically the manner in which fictitious scenarios can be used to test the boundaries of imagination and reality for the reader (Yeoman & McMahon-Beattie, 2021). From a literary perspective Pocock argued that 'literary landscapes would be described as landscapes of imaginative reflection' (1981 in Noelle O'Connor & Kim, 2014, p. 3). One of the best examples of an expert crafter of imaginative reflection was the environmental scientist Rachel Carson, author of the work *Silent Spring*. Carson was a Christian, although not an evangelical Christian with respect to her scientific work. For Carson, nature was to be seen 'through a scientific lens of cause and effect' (Wadsworth, 2016, p. 111). At the same time, however, the imagery e.g.: 'life is a miracle beyond our comprehension, and we should reverence it even where we have to struggle against it' (Wadsworth, 2016, p. 115) contain overt Christian overtones.

As I have grown in my faith, however, my childish ignorance has been tested. Through my reading on the relationship between tourism and sustainability and experiences in the field, I have been asked to reflect on many complex issues including tourism's marginalisation of the 'other'; the injustices inflicted by tourism on host communities, wildlife, local workers and at-risk members of the community; and the inherent ambiguity in what constitutes responsible tourism practice etc. Issues such as these can cause a number of operational opportunities and challenges for global tourism, which is forced to respond in a way that is reflective both of societal values and international/local laws. As a Christian I also have a responsibility to reflect on the words of the Bible and to pray. In doing so I must remember 'that what is perfectly lawful, and perfectly permissible, for everyone else – even our very close non-Christian friends – is not necessarily lawful and permissible for us' (Scalia, 2017, p. 31). What this demonstrates is that no tourism scholar can deny their own values, nor the links between one's faith and evolving trends in the philosophy of tourism, including the development of post humanist perspectives (Cohen, 2019). Post humanism has been recently observed as playing an important role in the development of space tourism (Cohen & Spector, 2020).

Pillar 4: a spirit of becoming

Tourism academia is a product of the late nineteenth and early twentieth century, developing initially to support hospitality and catering degrees in North America and Europe, before being formalised and the first tourism journals appearing in the 1930s and 40s (Schweinsberg et al., 2022). Because of this, the history of tourism academia was developing at the same time as the rise of secularism (Zinnbauer et al., 2015) and the first instances of open warfare between religion and science. This trend is evidenced in the work of Andre Dickson's *A History of the Warfare of Science with Theology under Christendom* in 1896 (Craig, 2022b). John William Draper argued that

> the history of science is not a mere record of isolated discoveries; it is a narrative of the conflict of two contending powers, the expansive force of the human intellect on one side, and the compression arising from traditionary faith and human interests on the other. (1874 in McGrath, 1999, p. 45)

Subsequent to this there has been a progressive shift towards scientific materialism, methodological naturalism, and a reduction in the effect of the religious perspective (Meyer, 2021). In the early twenty-first century, higher levels of societal interest in scientism is often linked to the new atheist movement, which was popularised by the writings of Richard Dawkins, the late Christopher Hitchens and others (Hitchens, 2008; Hitchens et al., 2007; Zenk, 2013).

In 1965, the Catholic Church announced 'a "General directory for the pastoral ministry to tourism", which recognised the importance of tourism and of its pastoral challenges in the contemporary world' (Cohen, 2018, p. 371). To-date, however, no formal theological tradition has seen fit to develop a 'theology of tourism' and until the last decade academics were often 'hesitant in developing their understanding of the philosophical

... process that underpins knowledge production and practices' (p. 361). Working with partner disciplines in organisational and management theory, which have historically viewed religion as a taboo subject (Van Buren et al., 2020), tourism academics have a responsibility to consider how religion can be incorporated into the corpus of the tourism knowledge domain. Barkathunnisha et al. (2019) note that the incorporation of spiritual dimensions in the teaching of tourism involves looking beyond traditional positivist Western paradigms and adopting an ontological perspective which 'emphasises the connection between nature, culture and society. The learner transcends, moving beyond one's own self into the consciousness of a larger universe' (p. 2142). This is the spirit of becoming.

Thoughts for the future

In the space remaining this paper will consider two issues that the tourism academy will have to face in order to find ways to bring such a spiritual/religious perspective into the formulation of tourism knowledge. First, we must remember that the tourism academy is a broad community, one that is subject to a myriad of intellectual interests with members drawn from all over the world and thus encompassing a wide variety of secular and religious perspectives. Future work should start by seeking to understand empirically what religious affiliations exist within the tourism academy. At the same time, however, we must learn from the earlier experiences from the disciplines including physics and biology, in relating with religion – nobody wins through conflicts between religion and science. Whether it was the ill-conceived attempt of the Catholic Church to bring Galileo before the inquisition in 1633 (see Hummel, 1986) or more recently Richard Dawkins criticising not only the ethics of the Hebrew old testament God but comparing the act of educating children to believe in Jesus as tantamount to abuse (Dawkins, 2006), pursuing divisions between religion and science means that each side misses out on the opportunity for a larger perspective than otherwise would have been available to them through their own narrow parochial viewpoints. One of the central aspects of Tribe and Liburd's (2016) tourism knowledge system was the idea that the relationship between different paradigms and non-paradigmatic ways of knowing are fluid (networked), subject to the possibility of forming, dissolving, and reforming over time. A question that I would therefore ask the reader to consider is this. Do you ascribe to the view of Keat (1981 in Ayikoru, 2009, pp. 68–69) that everything that is 'presumed to be non-scientific, such as the discourses of religion, metaphysics, ideology, politics and ethics or tourism for that matter... [should be] bracketed as inferior, cognitively meaningless or even nonsensical'? Or do you in fact ascribe to Keat's other point that you can 'uphold a positivist idea of what counts as scientific knowledge, without having to believe that science entails the only form of genuine knowledge' (1981 in Ayikoru, 2009, p. 69)? For the present author who is a Christian, Miller summed it up well when he said that for a scientist to be threatened by God assumes that the almighty is 'nothing more than a place holder for human ignorance' (John Templeton Foundation, ND, p. 39).

Second, I would ask the reader to critically consider the nature of the tourism academy of which we are all a part. Scientists have for over a century often been said to be less religious on average than the general population (Ecklund et al., 2019; Ecklund & Long, 2011; Ecklund & Scheitle, 2007). Secularisation theory, drawing on perspectives from Weber, Marx and others in the mid twentieth century espoused the notion that there was an association between the modernisation of the economy and society, which tourism has been an essential manifestation of, and the decline of religious relevance (Gross & Simmons, 2009). This paper will not be naive and suggest that religious people are currently in the majority with respect to the composition of the tourism academy. At the same time, however, the recent growth in post-secular tourism has drawn attention to the need to consider the link between the subjectivities of tourism scholars and their work to understand the world of tourism (Nilsson & Tesfahuney, 2018).

Conclusion

Globally it has been estimated that there are over 4300 religions dominated by Christianity, Islam, Hinduism, Buddhism and Judaism (Travel Almanac, 2022). The diversity of religions and by implication the potential diversity of religious perspectives in the tourism academy makes it inappropriate to generalise on how and to what degree a form of spirituality or religion might influence our engagement with tourism's scientific and other paradigms. It is for this reason that this preliminary exploratory investigation has drawn on the autobiographical perspectives of one academic, using the experiences of the author as the impetus to ask questions of the wider academy.

In 2022, the Theos think-tank explored UK societal and academic perspectives on the relationship between 'science' and 'religion' where it was acknowledged that both are 'highly complex, contestable [and] "polyvalent" terms' that are often debated to narrowly (Spencer & Waite, 2022, p. 3). The debates between 'science' and 'religion' are 'like the proverbial swimming pool where all the noise is up at the shallow end' (p. 26), which has often tended to discount the value that both perspectives bring to our understanding of the world. All epistemologies, 'or ways of knowing, as implemented in a pedagogy, or way of teaching and learning, tends to become an ethic, or way of living' (Palmer and Zajonc, 2010, p. 98 in Barkathunnisha et al., 2019, p. 2151). For the present author this means that it is not possible to distinguish the study of tourism from my Christian faith. However, as an insightful anonymous reviewer for this paper noted, the degree to-which I should allow my faith to impinge directly on my academic work does depend on the topic and approach I am taking. For myself, believing in God does not devalue the merits of the scientific approach to the study of tourism. Rather, God 'is the ground for all explanation: it is his existence which gives rise to the very possibility of explanation, scientific or otherwise' (Lennox, 2009, p. 48). The purpose of the paper has been to challenge the reader to consider, on what basis is your own explanation of tourism knowledge based?

Disclosure statement

No potential conflict of interest was reported by the author(s).

ORCID

Stephen Schweinsberg http://orcid.org/0000-0002-8820-8302

References

Abu Bakar, B., Price, A., Lee, D., & Wilson, E. (2023). Spirituality and pedagogy in tourism higher education: Voices of tourism educators. *Journal of Hospitality & Tourism Education*, *35*(1), 1–15. https://doi.org/10.1080/10963758.2021.1963745

Armstrong, K. (2022). *Sacred nature: How we can recover our bond with the natural world*. The Bodley Head.

Ateljevic, I. (2011). Transmodern critical tourism studies: A call for hope and transformation. *Turismo em Análise, 22*(3), 497–515.

Ayikoru, M. (2009). Epistemology, ontology and tourism. In J. Tribe (Ed.), *Philosophical issues in tourism* (pp. 62–79). Channel View Publications.

Babu, R. K., & Vishnuvardhan, V. (2022). The impact of fantasy and symbolism in CS Lewis: The Chronicles of Narnia. *Journal of Language and Linguistic Studies, 17*(2), 1210–1214.

Bakar, B. (2020). Integrating spirituality in tourism higher education: A study of tourism educators' perspectives. *Tourism Management Perspectives, 34*, 100653. https://doi.org/10.1016/j.tmp.2020.100653

Barkathunnisha, A., Diane, L., Price, A., & Wilson, E. (2019). Towards a spirituality-based platform in tourism higher education. *Current Issues in Tourism*, *22*(17), 2140–2156. https://doi.org/10.1080/13683500.2018.1424810

Barkathunnisha, A., Lee, D., & Price, A. (2017). Transcending towards a spirituality-based platform in tourism higher education: A contemplation of the pedagogical implications. *. Journal of Hospitality, Leisure, Sport & Tourism Education*, *21*, 174–184. https://doi.org/10.1016/j.jhlste.2016.11.003

Belhassen, Y., & Caton, K. (2009). Advancing understandings: A linguistic approach to tourism epistemology. *Annals of Tourism Research*, *36*(2), 335–352. https://doi.org/10.1016/j.annals.2009.01.006

Berger, P. L. (1967). *The sacred canopy: Elements of a sociological theory of religion*. Anchor Books.

Bruner, K. D., & Ware, J. (2005). *Finding God in the land of Narnia*. Tyndale House Publishers, Inc.

Buell, L. (2014). Environmental writing for children: A selected reconnaissance of heritages, emphases, horizons. In G. Garrad (Ed.), *Oxford companion to ecocriticism* (pp. 408–422). Oxford University Press.

Butcher, J. (2021a). COVID-19, tourism and the advocacy of degrowth. *Tourism Recreation Research*, 1–10. https://doi.org/10.1080/02508281.2021.1953306

Butcher, J. (2021b). Debating tourism degrowth post COVID-19. *Annals of Tourism Research, 89*, 103250. https://doi.org/10.1016/j.annals.2021.103250

Carrol, S., & Goodstein, D. (2009). Defining the scientific method. *Nature Methods, 6*(4), 237. https://doi.org/10.1038/nmeth0409-237

Cohen, E. (2018). The philosophical, ethical and theological groundings of tourism–an exploratory inquiry. *Journal of Ecotourism*, *17*(4), 359–382. https://doi.org/10.1080/14724049.2018.1522477

Cohen, E. (2019). Posthumanism and tourism. *Tourism Review*, *74*(3), 416–427. https://doi.org/10.1108/TR-06-2018-0089.

Cohen, E., & Spector, S. (2020). Transhumanism and cosmic travel. *Tourism Recreation Research*, *45*(2), 176–184. https://doi.org/10.1080/02508281.2019.1679984

Cole, S. (2007). Beyond authenticity and commodification. *Annals of Tourism Research*, *34*(4), 943–960. https://doi.org/10.1016/j.annals.2007.05.004

Collins-Kreiner, N. (2020). Religion and tourism: A diverse and fragmented field in need of a holistic agenda. *Annals of Tourism Research*, *82*, 102892. https://doi.org/10.1016/j.annals.2020.102892

Craig, W. (2022a). The Indispensability of Theological Meta-Ethical Foundations for Morality. https://www.reasonablefaith.org/writings/scholarly-writings/the-existence-of-god/the-indispensability-of-theological-meta-ethical-foundations-for-morality.

Craig, W. (2022b). What is the Relation between Science and Religion. https://www.reasonablefaith.org/writings/popular-writings/science-theology/what-is-the-relation-between-science-and-religion.

Cui, Q., & Xu, H. (2019). Situating animal ethics in Thai elephant tourism. *Asia Pacific Viewpoint, 60*(3), 267-279. https://doi.org/10.1111/apv.12221

Dawkins, R. (2006). *The God delusion*. Black Swan.

De la Hoz-Correa, A., Muñoz-Leiva, F., & Bakucz, M. (2018). Past themes and future trends in medical tourism research: A co-word analysis. *Tourism Management, 65*, 200–211. https://doi.org/10.1016/j.tourman.2017.10.001

Dickerson, M., & O'Hara, D. (2009). *Narnia and the fields of arbol: The environmental vision of C. S. Lewis.*: University Press of Kentucky.

Dixon, T., & Shapiro, A. (2022). *Science and religion: A very short introduction*. Oxford University Press.

Doorne, S., Cary, S. H., Graham Brown, J.-A., Lester, K. B., Pernecky, T., Curtin, S., … Morgan, N. (2007). Processes of becoming: Academic journeys, moments and reflections. In I. Ateljevic, A. Pritchard, & N. Morgan (Eds.), *The critical turn in tourism studies* (pp. 387). Routledge.

Echterling, C. (2016a). How to save the world and other lessons from children's environmental literature. *Children's Literature in Education, 47*(4), 283–299. https://doi.org/10.1007/s10583-016-9290-6

Echterling, C. (2016b). Postcolonial ecocriticism, classic children's literature, and the imperial-environmental imagination in "The chronicles of Narnia". *Journal of the Midwest Modern Language Association, 49*(1), 93–117. https://doi.org/10.1353/mml.2016.0016

Ecklund, E. H., Johnson, D. R., Vaidyanathan, B., Matthews, K. R., Lewis, S. W., Thomson, J. R. A., & Di, D. (2019). *Secularity and science: What scientists around the world really think about religion*. Oxford University Press.

Ecklund, E. H., & Long, E. (2011). Scientists and spirituality. *Sociology of Religion, 72*(3), 253–274. https://doi.org/10.1093/socrel/srr003

Ecklund, E. H., & Scheitle, C. P. (2007). Religion among academic scientists: Distinctions, disciplines, and demographics. *Social Problems, 54*(2), 289–307. https://doi.org/10.1525/sp.2007.54.2.289

Einstein, A. (1940). Science and religion. *Nature*, 146(3706), 605–607. https://doi.org/10.1038/146605a0

Everingham, P., & Chassagne, N. (2020). Post COVID-19 ecological and social reset: Moving away from capitalist growth models towards tourism as Buen Vivir. *Tourism Geographies, 22*(3), 555–566. https://doi.org/10.1080/14616688.2020.1762119

Fara, P. (2009). *Science: A four thousand year history*. Oxford.

Fennell, D. (2018). *Tourism ethics* (2nd ed.). Channel View Publications.

Fennell, D. (2022). An animal welfare literacy framework for tourism. *Annals of Tourism Research, 96*, 103461. https://doi.org/10.1016/j.annals.2022.103461

Feynman, R. P. (1955). The value of science. *Engineering and Science, 19*(3), 13–15.

Fletcher, R., Blanco-Romero, A., Blázquez-Salom, M., Cañada, E., Murray Mas, I., & Sekulova, F. (2021). Pathways to post-capitalist tourism. *Tourism Geographies*, 1–22. https://doi.org/10.1080/14616688.2021.1965202

Fletcher, R., Murray Mas, I., Blanco-Romero, A., & Blázquez-Salom, M. (2019). Tourism and degrowth: An emerging agenda for research and praxis. *Journal of Sustainable Tourism, 27*(12), 1745–1763. https://doi.org/10.1080/09669582.2019.1679822

Font, X., Montano, L., Wu, J., Coghlan, A., Woosnam, K., & Li, S. (2022). The purpose of a sustainable tourism journal. *Journal of Sustainable Tourism*. https://doi.org/10.1080/09669582.2022.2119244

Gooding, D., & Lennox, J. (2019). *Questioning our knowledge: Can we know what we need to know*. Myrtlefield House.

Griffiths, P. (2015). Imposible pluralism. In R. Reno & B. McClay (Eds.), *Religion and the social sciences* (pp. 9–17). Cascade Books.

Gross, N., & Simmons, S. (2009). The religiosity of American college and university professors. *Sociology of Religion, 70*(2), 101–129. https://doi.org/10.1093/socrel/srp026

Hales, R., Dredge, D., Higgins-Desbiolles, F., & Jamal, T. (2018). Academic activism in tourism studies: Critical narratives from four researchers. *Tourism Analysis, 23*(2), 189–199. https://doi.org/10.3727/108354218X15210313504544

Hall, C. (2015). What's tourism ever done for us? Depends where you are looking and who's looking. In T. V. Singh (Ed.), *Challenges in tourism research* (pp. 346–353). Channel View Publications.

Harari, Y. N. (2018). *21 lessons for the 21st century*. Signal.

Hay, D. (2009). Responding to climate change: How much should we discount the future? In R. White (Ed.), *Creation in crisis: Christian perspectives on sustainability* (pp. 53–66). Society for Promoting Christian Knowledge.

Hitchens, C. (2008). *God is not great: How religion poisons everything*. McClelland & Stewart.

Hitchens, C., Dawkins, R., Harris, S., & Dennett, D. (2007). The Four Horsemen: Hour 2 of 2 - Discussions with Richard Dawkins, Ep 1. https://www.youtube.com/watch?v = TaeJf-Yia3A&t = 2098s.

Hooper, W. (1996). *C. S. Lewis: The companion and guide*. Harper Collins.

Hummel, C. (1986). *The Galileo connection: Resolving conflicts between science and the Bible*. InterVarsity Press.

Jafari, J. (2001). The scientification of tourism. In V. Smith & M. Brent (Eds.), *Hosts and guests revisited: Tourism issues of the 21st century* (pp. 28–41). CABI.

Jamal, T. B., & Everett, J. (2004). Resisting rationalisation in the natural and academic life-world: Critical tourism research or hermeneutic charity? *Current Issues in Tourism, 7*(1), 1–19. https://doi.org/10.1080/13683500408667970

Jenkins, W., Tucker, M., & Grimm, J. (Eds). (2017). *Routledge handbook of religion and ecology*. Routledge.

Jeong, J. Y., & Lee, K.-Y. (2022). Is sex tourism intention uncontrollable? The moderating effects of ethics and law. *Journal of Travel Research, 62*(3), 00472875221093771. https://doi.org/10.1177/00472875221093771

John Templeton Foundation. (n.d.). *Does science make belied in God obsolete?* https://andrewjmonaco.files.wordpress.com/2012/10/templeton-science-religion.pdf.

Khoo-Lattimore, C. (2018). The ethics of excellence in tourism research: A reflexive analysis and implications for early career researchers. *Tourism Analysis, 23*(2), 239–248. https://doi.org/10.3727/108354218X15210313504580

Kim, B., Kim, S., & King, B. (2020). Religious tourism studies: Evolution, progress, and future prospects. *Tourism Recreation Research, 45*(2), 185–203. https://doi.org/10.1080/02508281.2019.1664084

Lama, D. (2005). *The universe in a single atom: The convergence of science and spirituality*. Harmony.

Lawrence, B. (2014). Environmental writing for children: A selected reconnaissance of heritages, emphases, horizons. In G. Garrad (Ed.), *Oxford companion to ecocriticism* (pp. 408–422). Oxford University Press.

Lennox, J. (2009). *God's undertaker: Has science buried God?* Lion Books.

Macbeth, J. (2005). Towards an ethics platform for tourism. *Annals of Tourism Research, 32*(4), 962–984. https://doi.org/10.1016/j.annals.2004.11.005

McGrath, A. (1999). *Science and religion: An introduction*. Blackwell.

McGrath, A. (2013a). *Christian history: An introduction*: Wiley-Blackwell.

McGrath, A. (2013b). *CS Lewis–a life: Eccentric genius, reluctant prophet*. Tyndale House Publishers, Inc.

Meyer, S. (2021). *Return of the God hypothesis: Three scientific discoveries that reveal the mind behind the universe*. HarperOne.

Miller, K. (2007). *Finding Darwin's God: A scientist's search for common ground between God and evolution*. Harper.

Morris, M. (2009). Middle earth, Narnia, Hogwarts, and animals: A review of the treatment of nonhuman animals and other sentient beings in Christian-based fantasy fiction. *Society & Animals, 17*(4), 343–356. https://doi.org/10.1163/106311109X12474622855183

Mura, P., & Wijesinghe, S. N. (2022). Ph. D. Students' experiences and emotions in neoliberal tourism academia. *Tourism Management Perspectives, 44*, 101022. https://doi.org/10.1016/j.tmp.2022.101022

Murphy, R. (2012). Sustainability: A wicked problem. *Sociologica, 6*(2), 1–23. https://doi.org/10.2383/38274

Ni'am, L., Koot, S., & Jongerden, J. (2021). Selling captive nature: Lively commodification, elephant encounters, and the production of value in Sumatran ecotourism, Indonesia. *Geoforum, 127*, 162–170. https://doi.org/10.1016/j.geoforum.2021.10.018

Nilsson, M., & Tesfahuney, M. (2018). The post-secular tourist: Re-thinking pilgrimage tourism. *Tourist Studies, 18*(2), 159–176. https://doi.org/10.1177/1468797617723467

O'Connor, N., & Bolan, P. (2008). Creating a sustainable brand for northern Ireland through film-induced tourism. *Tourism Culture & Communication, 8*(3), 147–158. https://doi.org/10.3727/109830408786177515

O'Connor, N., & Kim, S. (2014). Pictures and prose: Exploring the impact of literary and film tourism. *Journal of Tourism and Cultural Change, 12*(1), 1–17. https://doi.org/10.1080/14766825.2013.862253

Rojek, C. (1995). *Decentring leisure: Rethinking leisure theory* (Vol. 35). Sage.

Rosa, R. (2022). The trouble with 'work–life balance'in neoliberal academia: A systematic and critical review. *Journal of Gender Studies, 31*(1), 55–73. https://doi.org/10.1080/09589236.2021.1933926

Scalia, A. (2017). *Scalia speaks: Reflections on Law, faith, and life well lived*. Crown Forum.

Schweinsberg, S. (2022). The epistemic authority of tourism academics. *Annals of Tourism Research, 93*. https://doi.org/10.1016/j.annals.2022.103351

Schweinsberg, S., & Darcy, S. (2022). Tourism academic. In D. Buhalis (Ed.), *Encyclopaedia of tourism management and marketing* (pp. 27–30). Edward Elgar.

Schweinsberg, S., Fennell, D., & Hassanli, N. (2021). Academic dissent in a post COVID-19 World. *Annals of Tourism Research, 91*, 1–10.

Schweinsberg, S., Heizmann, H., Darcy, S., Wearing, S., & Djolic, M. (2018). Establishing academic leadership praxis in sustainable tourism: Lessons from the past and bridges to the future. *Journal of Sustainable Tourism, 26*(9), 1577–1586. https://doi.org/10.1080/09669582.2018.1489397

Schweinsberg, S., & McManus, P. (2020). Coal seam gas: A space-based perspective. *Geographical Research, 58*(1), 24–33. https://doi.org/10.1111/1745-5871.12375

Schweinsberg, S., Sharpley, R., & Darcy, S. (2022). Competitive positioning of tourism academic knowledge. *Tourism Management, 91*, 104502. https://doi.org/10.1016/j.tourman.2022.104502

Schweinsberg, S., Wearing, S., & Darcy, S. (2012). Understanding communities' views of nature in rural industry renewal: The transition from forestry to nature-based tourism in Eden, Australia. *Journal of Sustainable Tourism, 20*(2), 195–213. https://doi.org/10.1080/09669582.2011.596278

Schweinsberg, S., Wearing, S., & Lai, P.-H. (2020). Host communities and last chance tourism. *Tourism Geographies, 23*(5–6), 945–962. https://doi.org/10.1080/14616688.2019.1708446.

Schweinsberg, S., Wearing, S., & McManus, P. (2013). Exploring sustainable tourism education in business schools: The honours program. *Journal of Hospitality and Tourism Management, 20*, 53–60. https://doi.org/10.1016/j.jhtm.2013.05.003

Schweinsberg, S., Wearing, S., & Wearing, M. (2015). Transforming nature's value – cultural change comes from below: Rural communities, the 'othered' and host capacity building. In Y. Reisinger (Ed.), *Transforming tourism: Host perspectives* (pp. 102–113). CABI.

Spencer, N., & Waite, H. (2022). *'Science and Religion': Moving away from the shallow end*. https://www.theosthinktank.co.uk/cmsfiles/Science-and-Religion-Moving-away-from-the-shallow-end_REPORT.pdf.

Suphasomboon, T., & Vassanadumrongdee, S. (2022). Toward sustainable consumption of green cosmetics and personal care products: The role of perceived value and ethical concern. *Sustainable Production and Consumption, 33*, 230–243. https://doi.org/10.1016/j.spc.2022.07.004

Tickle, L., & von Essen, E. (2020). The seven sins of hunting tourism. *Annals of Tourism Research, 84*, 102996. https://doi.org/10.1016/j.annals.2020.102996

Travel Almanac. (2022). *Religions of the world*. https://www.thetravelalmanac.com/religions/.

Tribe, J. (2006). The truth about tourism. *Annals of Tourism Research, 33*(2), 360–381. https://doi.org/10.1016/j.annals.2005.11.001

Tribe, J. (2018). Creating and curating tourism knowledge. *Annals of Tourism Research, 73*, 14–25. https://doi.org/10.1016/j.annals.2018.08.004

Tribe, J., & Liburd, J. (2016). The tourism knowledge system. *Annals of Tourism Research, 57*, 44–61. https://doi.org/10.1016/j.annals.2015.11.011

Tyson, N. (2022). *Starry messenger: Cosmic perspectives on civilisation*. Harper Collins.

Urano, D. G., Mathe, A. J., Coutinho, A. C. A., & Marques Junior, S. (2022). Scientific production in tourism: An analysis of the methodological approaches. *Current Issues in Tourism, 26*(3), 380–391. https://doi.org/10.1080/13683500.2021.2014790.

Van Buren III H. J., Syed, J., & Mir, R. (2020). Religion as a macro social force affecting business: Concepts, questions, and future research. *Business & Society, 59*(5), 799–822. https://doi.org/10.1177/0007650319845097

Wadsworth, A. (2016). Carson's Christianity and environmental crises. *Criterion: A Journal of Literary Criticism, 9*(1), 109–119.

Ward, M. (2008). *Planet Narnia: The seven heavens in the imagination of CS Lewis*. Oxford University Press.

Wearing, S., Schweinsberg, S., & Tower, J. (2016). *The marketing of national parks for sustainable tourism*. Channel View.

Wilczek, F. (2022). *Fundamentals: Ten keys to reality*. Penguin.

Yancey, P. (2002). *Finding God in unexpected places*. Hodder and Stoughton.

Yeoman, I., & McMahon-Beattie, U. (Eds). (2021). *Science fiction and the future of tourism*. Channel View.

Zagonari, F. (2020). Comparing religious environmental ethics to support efforts to achieve local and global sustainability: Empirical insights based on a theoretical framework. *Sustainability, 12*(7), 2590. https://doi.org/10.3390/su12072590

Zagonari, F. (2021). Religious and secular ethics offer complementary strategies to achieve environmental sustainability. *Humanities and Social Sciences Communications, 8*(1), 1–13. https://doi.org/10.1057/s41599-021-00802-0

Zaleski, P., & Zaleski, C. (2015). *The fellowship: The literary lives of the inklings: JRR Tolkien, CS Lewis, Owen Barfield, Charles Williams*: Farrar, Straus and Giroux.

Zenk, T. (2013). New atheism. In S. Bullivant & M. Ruse (Eds.), *The Oxford handbook of atheism* (pp. 245–260). Oxford University Press.

Zinnbauer, B. J., Pargament, K. I., Cole, B., Rye, M. S., Butfer, E. M., Belavich, T. G., & Kadar, J. L. (2015). Religion and spirituality: Unfuzzying the fuzzy. In W. Mirola, M. Emerson, & S. Monahan (Eds.), *Sociology of Religion* (pp. 29–34). Routledge.

Spiritual Hosting and Tourism: The Host-Guest Dimension

Yoel Mansfeld and Alison J. Mcintosh

ABSTRACT

This paper contributes to an understanding of how spirituality and faith can shape the nature of the host tourist encounter. Specifically, using the case of the New Zealand HIT (Hosting Israeli Travellers) network, this paper explores the phenomenon of 'spiritual hosting' and its influence on the host-guest interaction. By applying assumptions based on the 'Value-Stretch Model' and analysis of in-depth interviews, the paper characterizes the nature of the encounters between Israeli backpackers and their Messianic Christian hosts. The research analysis revealed that hosts fulfilled far more than their initial spiritual expectations by evangelizing Israeli travellers through hosting. These benefits were accompanied by strong socio-cultural experiences. Guests, who primarily sought low-cost accommodation, reported they also underwent profound spiritual and socio-cultural experiences.

Introduction and Literature Review

This paper provides an examination of the spiritual dimensions of tourism hosting. More specifically, the paper examines the extent to which hosts and guests expect, and then gain a shared sense of spirituality and/or religious ideology from their interaction. Within published research, scholarly discourse has given increased attention to the notion of spirituality, especially in disdplines such as management, psychology, theology, mental health counselling, and leisure. Spirituality is commonly regarded asanintenselyindividualphenomenoninvolvingpersonal transcendence, super-conscious sensitivity and meaningfulness (Heintzman 1999; Spilka and Mcintosh 1996); it is also often assodated with religion and theology and rituals. Many scholars define spirituality as multi dimensional (for example, Ashmos and Duchon 2000; Dean *et al.* 2003; Miner-Williams 2006). Often, the difference between ritualistic religion and spirituality is blurred. For others, the notion of spirituality is differentiated from ritualistic religion, as it is seen to be related to life's deeper motivations and an emotional connection to God; a 'personal religion', rather than having any institutional connotation (McCormick 1994). Generally, Mitroff and Denton (1999) argue that spirituality can best be summed up by the word 'inter-connectedness'; inter-connectedness with oneself, and/or with God, or with something else beyond. Spiritual experiences are generally referred to as transcendent, ineffable and an epiphany.

It has been argued that the increased emphasis on spirituality arose several centuries after Westerners had created a distinct separation between the outer world of common and routine extrinsic activities and the inner world of spirituality and religion (Neal and Biberman 2003). If this is so, we could be seeing a major reintegration of outer and inner life. Conversely, some argue that spirituality is a new form of consumerism (Belket *al.*1989; Lane 2001). Some recent studies have described the nature of what tourists consume or experience through travel to be quasi-spiritual in nature (for example, Bull2006; Collins-Kriener *et al.* 2004; Jamal *et al.* 2006; Noy 2004; Poria *et al.* 2003); while others have defined types of consumption of spiritually-based tourism products (for example, Attix 2002; Pernecky 2006).

Conceptually, modern tourism is seen as a contemporary manifestation of the traditional experience of pilgrimage (Cohen 1992a; Graburn 1989); a form of sacred or spiritual journey, a quest for meaning or spiritual refreshment with the spiritual experience of travel being found even beyond what was intended or proffered. In this way, personal spiritual experiences through travel can be gained that are neither collectively sought after nor expected (Little and Schmidt 2006; Zahra 2006). Arguably, most tourists actively construct their own spiritual experiences in authentic or subjective negotiation with the places they visit, transforming those places into spaces that are rendered personally meaningfuL or spiritual. As such, scholars argue that experiential evaluation of the 'journeying' is more important than the visit to the sacred site, or motive for pilgrimage (Carnegie and Devereux 2006; Crouch 2006). However, with the exception of the studies referred to here, the study of the relationship between spirituality and tourism generally remains an embryonic field of inquiry.

Historically, much previous research has focused on a particular approach to spirituality in tourism; that which focuses on the relationship between religion and tourism. Indeed, researchers have demonstrated a clear connection or relationship between tourism and religion (Lee 2006; Rinschede 1992; Timothy and Olsen 2006), as well as the influence of religion on tourists and hosts' behaviours (Din 1989; Fleischer 2000; Fleischer and Pizam 2002; Mattila *et al.* 2001; Poria *et al.* 2003). Previous research has predominantly dealt with the religious component of spiritual tourism relating to travelling pilgrims, tourist behaviour, and the comodification and/ or management of pilgrimage and religious sites. For instance, a wealth of previous research has sought to interpret the motivations of pilgrims visiting religious sites, or the outcome of pilgrimage travel (see for example, Cohen1992b; Coleman 2001; Collins-Kriener and Kliot 2000; Eade 1992; Hudman and Jackson 1992; Olsen 2006a; Olsen 2006b; Robichaud 1999; Sharpley and Sundaram 2005; Sizer 1999; Smith 1992; Vukonic 1996; Woodward 2004). Previous research has also examined the management of sacred sites for pilgrimage and tourism development (for example, Digance 2003; Din 1989; Henderson 2003; Nolan and Nolan 1992; Shackley 1999; 2001). Only limited research has examined the relationship between religion and tourism in relation to the experiences of tourism hosts, or the nature of host-guest encounter, and the wider spiritual (rather than merely religious) expectations and experiences which may result from a tourism encounter. Specifically, spiritual tourism experiences may involve unexpected as well as purposeful engagement; rather, they may be shaped or mediated by meaningful interaction and exchange, even between parties of differing faiths or ideologies.

A focus on the interactive nature of the host-guest relationship is important in an examination of the spiritual dimensions of tourism hosting, because previous studies have suggested that any posSible lack of mutual appreciation for the respective culture, faith or ideology of host or guest interactions can lead to unease or even conflict (Din1989; Mcintosh and Campbell2001; Mcintosh and Johnson 2004; Mcintosh and Bonneman 2006; Poria *et al.* 2003). As such, alternative forms of tourism, including cultural exchange programmes and volunteer tourism, have been associated with promoting the mutual appreciation and understanding of cultural differences between host and guest (Lyons 2003; Wearing 2001). Despite the obvious relationship between tourism hosting,religion and spirituality, the extent to which hosts and guests are motivated by, or gain a shared sense of spirituality or religion from their interaction, has not been considered in previous research. As such, it is important to understand what hosts and guests expect from their interaction and exchange, in particular the benefits that each party expects to gain from the encounter; their attitudes resulting from the interaction; and the likely impact of this interaction on both parties. Understanding these aspects better, will produce insights into what should be taken into account when such interaction is preplanned and promoted for similar spiritual tourism settings.

With a focus on understanding the consequences of host-tourist interaction, previous sodal exchange research has examined residents' attitudes towards community tourism impacts by evaluating how residents assess the expected costs and benefits of touriSln development in their community, and their reactions to tourism (see for example, Andereck *et al.* 2005; Hernandez *et al.* 1996; Jurowski and Gursoy 2004; Lee and Back2003; Waitt 2003). The underlying assumption of sodal exchange theory is that actors behave in a way that maximizes the rewards and minimizes the costs they experience; that is, people engage in an interaction process where they seek something of value (material, sodal or psychological) and, as long as the pattern of exchange is perceived as equitable, both parties will continue to engage in the exchange relationship (Madrigal1993). Social exchange theory generally suggests that actors evaluate tourism as either positive or negative in terms of the expected benefits or costs derived. A positive evaluation is suggested to occur only when both actors have high levels of sodal power or influence within the exchange relationship. Positive or negative evaluations are theorized in relation to the presence, or absence, of certain'antecedent conditions' or expected benefits. A positive evaluation may also be realized even if the pre-determined expectation is not met but something else offers satisfaction or unexpected benefits. As such, a positive evaluation will result if the expected (or unexpected) benefits meet the acceptable pre-determined level of satisfaction sought (Waitt 2003). As such, an examination of hosts' and guests' expectations, motivation, attitudes and benefits from the interaction and exchange is paramount. However, as mentioned above, the extent to which dimensions of religion and / or spirituality, feature specifically in the context and sodal exchange of the host tourist interaction remains unexplored. This provides a referent for the present paper.

The phenomenon of hosting networks that promote the sharing of religion and spirituality is an unusual but interesting platform where hosts and guests exchange spiritual values, as well as fulfilling travel and hosting motivations. Hence, the aim of this paper is to explore this possible exchange through a unique 'spiritual hosting' network. More specifically, the paper aims to unveil the benefits that are expected and gained by hosts and guests within the network, and to compare the benefits aspired by hosts and guests in order to find out if these encounters are mutually beneficial or conflicting. Anexamination of hosting in pursuit of spiritual benefits is, we argue, not a one dimensional phenomenon. In addition to the search for spiritual benefits, the encounter between host and guest may also facilitate sodo-cultural and/or economic benefits. The perceived importance of these three benefits may vary among hosts, among guests, and between hosts and guests in the hosting encounter. Both hosts and guests may be driven by a single leading expectatioflt or by combined expectations or situations. For example, guests or hosts may seek spiritual benefits. However, they can appreciate the other two benefits; or they can equally seek all three benefits, and thus be motivated by spiritual, economic and socio-cultural considerations.

'Spiritual hosting', therefore, may not simply represent a straightforward case of host-guest relations. At least one of the two parties'expected benefits are determined by their religious or spiritual faith and values. Thus, benefits from the hosting experience for at least one of the parties in the host-guest encounter do not merely result from their own individual motivations but by a set of norms and values defined by their faith. Hostinginitiated as a result of a given faith may cause hosts to engage themselves in hosting primarily for spiritual reasons. Consequently, economic and/or sodo-cultural expectations become either secondary or non-existent if host-guest relations are driven primarily by spiritual needs. Hosts, for example, may try to attract guests on the basis of efforts to fulfil their faith's evangelizing norms or values. Consequently, guests are aware that by being hosted they have to respond to this evident spiritual motivation. However, whilst hosts may be driven by spiritual needs, guests may be driven primarily by economic concerns. Thus, redprodty and the flow of benefits are dependent on the search for complementary benefits. Divergent benefits may create a conflict of benefits that impact on the sustainability of the hostingsystem. It is these issues, centred on the extent to which faith and spiritual ideology shape the nature of host-guest interaction, which provided the foundation for the present study.

The Host-Guest Dimension

Israeli Backpacking

Israeli backpackers are mainly (but not only) young budget travellers in their early to mid-twenties, who undertake an extended trip overseas after completing their military service, having saved enough money to cover a trip that lasts between several months to over a year (Noy 2005). Their most preferred destinations include South-East Asia, South and Central America, Australia and New Zealand. Previous research of Israeli backpackers has evaluated the operational, anthropological, cultural and psychological aspects of their travel behaviour, including their motivations, expectations and level of fulfillment of their travel needs (e.g., Uriely *et al.* 2002; Maoz 2005). Most have concluded that their travel motivations vary, but predominantly involve the urge to manifest their alienation from Israeli sodety (which imposes a compulsory military service on them) and seek a "full" sense of freedom and independence from strict military and other sodal frameworks (such as high schools and political parties) (Mevorach 1997; Maoz 2005). However, most of these studies have not considered the sodo-cultural imprint these backpackers leave as they visit their destinations. Thus, the question of sodal exchange and interrelations between Israeli backpackers and their hosts has yet to be explored. Since Israeli backpackers exploring New Zealand are the target guests of the HIT network, this paper will use this group in pursuit of understanding the sodal exchange characteristics of spiritual hosting.

HIT Characteristics and Background

Hosting Israeli Travellers (HIT) is a network of hosts located throughout New Zealand. Through paid membership of the HIT organizatiOflt the network provides almost 2,000 Israeli travellers each year with the opportunity, either free of charge or for a small nominalfee per night, 'to visit local people in their homes, experience the way they live, learn more about their faith and explore their culture first-hand' (http:/ /www.hit-nz.com). Most HIT hosts are New Zealanders who are Messianic Christians. As Messianic ("born again") Christians, they believe that the Jewish people are God's covenant people and want to meet and bless Israeli travellers who come to New Zealand. Messianic Christians place importance on the role of the Old Testament as the predominant source of Christian norms and values. They also believe in the centrality of Jesus Christ or 'Yeshua' (the Hebrew name of Jesus, meaning'salvation') as their Messiah and the guarantor of the continuity of the Jewish people and of those Christians who believe in the centrality of Judaism in the Christian faith. A Messianic Christian, according to their faith, is expected to convey this message to fellow Christians and to Jewish people through evangelism. It is not the intention of Messianic Christians to convert Jews to Christianity. On the contrary, according to their faith, the existence and preservation of theJewish people is the guarantor for future life for every believer. Evangelizing, therefore, aims to make people aware of the role of Jesus in pursuing these goals (http://www.homeworship101.com/fyi_Messianic_Judaism_ % 20Messianic_Orristianity.htm). The HIT hosting network was created as a medium to facilitate evangelizing by hosting that, through hosting, Messianic believers in New Zealand can get easy, direct and effective access to Jewish travellers (Ya'akobowich: interview 2005). Essentially, the network facilitated a meeting place between seekers of spiritual benefits and seekers of low-cost hosted accommodation.

HIT was established as a hosting network in 1999 by an Israeli backpacker who later took up residency in New Zealand and became a born-again (Messianic) Jew. Being a former backpacker in New Zealand and well aware of the Israeli backpackers' trend for budget accommodation and traveL he assumed that they could be an appropriate'captive market' for such an evangelistic encounter. Establishing this meeting place involved constructing a network of New Zealand Messianic hosts who were willing to open their homes to the Israeli backpackers, enabling them free or very low-cost touring of the country (Ya'akobowich: interview 2005). The idea of being hosted all over New Zealand-free, or almost free of charge-appealed to the psyche of the Israeli backpackers (Noy 2004), and thus, prospects of a 'win-win' situation jump-started the full operation of this hosting network Currently, trc nctworkis comprised of238 different households, scattered in both the north and south islands of New Zealand. Hosts were recruited predominantly through church meetings and bulletins in newsletters of Messianic Christian organizations.

Methodology

In order to understand the HIT network, specifically the expectations of HIT hosts and guests, and how this may influence the host-tourist relationship, exploratory qualitative research was undertaken, applying the assumptions of the 'Value-Stretch Model' (Mansfeld 1992, 1995, 2002; Mansfeld and Jonas 2006). The Value-Stretch Model allows a structured qualitative or quantitative detection of an individual's attitudes and perceptions of a given sodo-cultural situation or phenomenon (see Figure 1); in this instance, the phenomenon of'spiritual hosting'. It is based on the assumptions

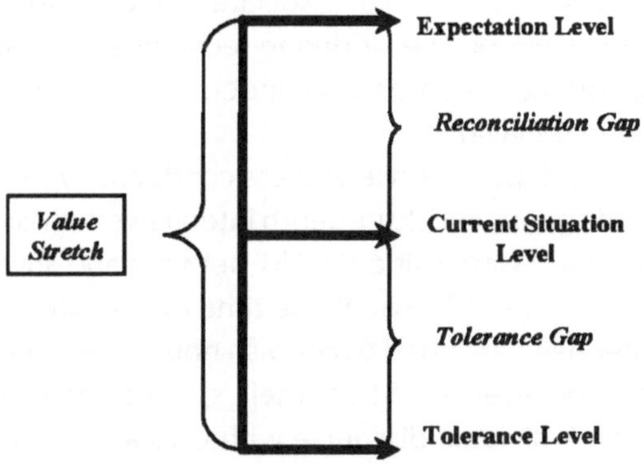

Figure 1. The Value-Stretch Model.
Based on Mansfeld (1992); Mansfeld and Jonas (2006)

that individuals approach sodocultural situations and phenomena on three levels of perception;a *tolerance level*, a *current situation level* and an *expectation level.* The attitudes detected in each leveL and the gaps or differences between these levels, expose individual/group multi-dimensional attitudes towards the phenomenon under consideration. The assumptions of the Value-Stretch Model were applied in the present study to design and structure the questions posed in in-depth interviews with HIT hosts and guests.

A semi-structured open-ended interview schedule was constructed, based on the primary assumptions of the Value Stretch Model outlined above. The interview schedule included five question themes covering the background characteristics of hosts and guests, their travel behaviour and patterns, hosts' and guests'motivations, the three levels of perception in the Value-Stretch Model, and host-guest relations. The Value-Stretchsectionformed the core of the data collection set. As determined by the three levels of perception in the Value-StretchModel(Figure 1), both hosts and guests were asked to express their minimum required benefits from partidpation in the HIT network ('Tolerance Level'), benefits that they currently experience ('Current Situation Level'), and future expected benefits ('Expectation LeveY) from the HIT network. At each of the three levels of perception, both HIT hosts and guests' initial responses to these questions were followed by the probing question, 'Why is that important to you?' As such, the interviews also employed the principles of the 'laddering' technique assodated with Means-End Chain theory (Reynolds and Gutman 1998; Mcintosh and Thyne 2005). The logic of the laddering technique is that follow-up probing questions elidt not just the salient but also the more abstract attributes that comprise an individual's (hierarchical) value system. Probing questions were used in this regard to elidt the underlying personal values and meanings that partidpants experienced or associate with participation in the HIT network. The laddering technique thus facilitates an inductive approach to understanding personal values and behaviour.

Thirty in-depth interviews were conducted with HIT hosts and agents, and 30 in-depth interviews with Israeli travellers who were using the HIT network and staying as guests with a HIT host at the time of the interview. Host interviews included twoJewish households and 28 Messianic believers who base their 'spiritual' practising mostly on individual discourse with God and with the Messiah, and disregard the institutionalized side of religious life. The other two are Jewish New Zealanders. The majority of hosts have been involved with the HIT network for the past three years. Most of them act as hosts only and others also function as regional agents and distributors of HIT membership cards. The majority of hosts are committed to hosting all-year round. However, some hosts prefer to be partially committed by hosting only occasionally and according to changing circumstances in their lives. Hosting conditions vary from very basic accommodation such as a tent outdoors, a mattress on the floor and shared facilities, to high standard self contained units.

Guest interviews were conducted at various HIT host locations around New Zealand. Thirty Israeli backpackers were interviewed using a pre-structured open-ended questionnaire. For the guests, becoming a HIT member involves purchasing a NZ$ 70 membership card allowing each backpacker to stay with HIT hosts either for free or for a minimal charge of NZ $5-10 per night. It also offers member discounts when booking various tourist services or attractions and when booking accommodations. Apart from the official role of HIT as a hosting network, members are often assisted by hosts in other ways such as in car purchasing, car maintenance, route planning, or solving health problems. The average length of stay of HIT members is 2-3 nights with each HIT host. The length of stay is determined both by the expressed wishes of hosts and the travel plans of each Israeli backpacker. Guests are expected to mingle with their hosts and to allocate time for discussion and exchange of ideas, beliefs, and experiences.

The interviews were tape recorded and then transcribed to ensure accuracy. Interviews with HIT hosts were conducted in English; the interviews with guests were conducted in Hebrew over a period of six months during the main tourist season in the north and south islands of New Zealand. Data collected were then manually coded into constructs based on the three levels of perception which enabled comparative analysis of hosts' and guests' required, experienced and expected benefits.

Hosts'and Guests'Perceptions and Attitudes

The research findings and the comparative analysis presented below follow the three levels of perception inherent in the Value-Stretch Model.

The Tolerance Level

Findings from the Value-Stretch analysis on the *Tolerance Level* revealed that both hosts and guests shared an array of requirementsin order to engage with the network. However, HIT hosts and their Israeli backpacker guests substantially differed in their requirements from the HIT hosting network and from this unique host-guest encounter. The findings suggest one important requirement and a variety of other secondary requirements among both hosts and guests. In the case of the HIT hosts, the most important requirement stemmed from the basic ideology of this hosting network- the spiritual dimension. This spiritual dimension was translated by hosts into two important activities: evangelizing and helping Israeli backpackers as part of their commitment to God and to Yeshua, the Messiah. As one host explained:' My most important benefit is being able to tell my guests about my faith, God, Yeshuaandmy love of Israel'

On the other hand, guests stressed that their most important requirement from HIT was to significantly lower their travel expenses in New Zealand by minimizing their accommodation costs.

In addition to their primary requirements, an array of additional secondary requirements was found among HIT hosts, including social, cultural and economic benefits. Hosts' social requirements included the ability to communicate with their guests, make friends with them and to understand the soda-politicalideologies of Israeli society. Quite surprisingly, cultural expectations as prerequisite requirements were elaborated only in terms of learning about Israeli culture and not about cross-cultural exchange (or, the sharing and exchange of culture; the primary motive here for hosts was to learn about Israeli culture first and foremost). Being a spiritual network, economic requirements were hardly expressed at this level, except for two cases where hosts admitted that for them the prerequisite for hosting is both spiritual and finandal. These two hosts take advantage of their geographical proximity to the Lake Taupo region one of New Zealand's main adventure tourism destinations. This region attracts most of the Israeli backpackers who come to experience those types of extreme tourist activities, yet the region has a major shortage in the availability of HIT hosts. Being the only two HIT hosts in the area, they realized the economic opportunity and charge a premium (more than double the average HIT) accommodation cost per night. As one of the hosts argued:

> I see nothing wrong in making some profits out of spiritual hosting. Being a single parent who raises four kids, one of whom is severely handicapped, I need to make some money out of HIT hosting. I'm confident that the Lord sends me these kids (the Israeli backpackers) because I am helping them and giving them a shelter. It's a win-win situation as I give them Spiritual Hosting and Tourism: Mansfeld and Mcintosh shelter and they provide me with much needed financial support.

The array of secondary requirements as expressed by the Israeli backpackers was not surprisingly centred on the quality of the expected low-cost accommodation. These requirements included the need for clean and tidy accommodation facilities as well as availability of a hot shower, a comfortable bed and hospitable hosts. These findings are interesting since they prove that guests' secondary requirements focus predominantly on the fadlity value of their hosting experience. Conversely, their hosts exposed a much more varied array of sodo-cultural and, to a lesser extent, economic requirements. Another interesting interpretation of the Value-Stretch data obtained for both hosts and guests at this *Tolerance Level* is that the relationships expected between the two sides on this level of the model do not reflect a reciprocal exchange of benefits. Unlike in social exchange theory, the 'give and take' relationship, in this case, forms a prerequisite requirement only by hosts. Hosts wish to benefit spiritually, sodally and economically, by giving the Israeli backpackers the opportunity to save money on their New Zealand tour. On the other hand, their guests on this Tolerance Level of the Value-Stretch Model expect only to gain a low-cost, convenient accommodation facility without intentionally benefiting the host in any possible way. As one of the backpackers reported:

> *When joining HIT, I did not really care what benefits were there for our hosts as long as I could save money on accommodation. At that time I was told by fellow backpackers that these people (HIT hosts) are a bit weird .... After all, in Israel nobody is ready to open his home and give up his privacy for nothing and just for the sake of serving God ...*

The Current Situation Level

Findings at the Current Situation level of the Value Stretch Model show further differences when perceived benefits by hosts and guests are compared. HIT hosts, unlike their guests, overwhelmingly expressed that all their primary (spiritual) and secondary (sodo-cultural) requirements were actually met. In the words of one of the hosts: 'God put His love in us like a mother hen, like a lioness'. In other words, hosts do believe that they are both blessing and being blessed by hosting Israeli backpackers. All of them expressed strong spiritual experiences and fulfillment of their prerequisite benefits. Although the main motivation for hosting was spiritual, respondents also currently experience significant sodal benefits based on their interaction and communication with guests, which often enriched their sodallife. For example, one host explained: 'Since my husband passed away, I am very lonely and my guests fillthis gap.'

Interestingly, HIT hosts expressed that their actual cultural experience reached a higher level than simply exposure to Israeli culture, which were regarded in the Tolerance Level asminimalrequirements. Hosts gladly termed their cultural experience as a process of cultural exchange. A quote from one host demonstrates this:

> *I really like the evenings with my guests. They wish to give me a good time by cooking a meal for us all. They are so proud of their Israeli food and also want to taste Kiwi dishes which I happily cook for them too. Our meals look like a feast of smells and cultures.*

The two hosts in the Lake Taupo region who indicated economic benefits as important requirements expressed their satisfaction in this respect since these basic expectations were met on the Current Situation level too. Although, being a marginal phenomenon among the HIT hosts, it is important to note that the few hosts who were concerned about making a living out of hosting felt that spiritual, cultural and economic requirement can be attained within the context of the spiritual hosting network.

In general, the findings on the Current Situation level revealed that, for the hosts, actual hosting experiences were overwhelmingly fulfilling. Moreover, if there were gaps between what was defined by them as the minimal required benefits and what they actually achieved, these were essentially positive gaps. Thus, their current cultural experiences were characterized by exchange relationships rather than simply a'one-way' cultural transfer from guests to hosts.

The analysis of guests' Current Situation experiences draws a more complicated picture than that of their hosts with respect to the differences between what was defined as prerequisite benefits on the Tolerance Level and what they had actually experienced while staying with HIT hosts. Findings of the Value-Stretch Gap analysis revealed that guests'requirements were only partially met. Many of them said that once they had stayed with several HIT hosts they realized that free or low-cost accommodation often comes with a'price'; as one of the guests explained:

> *My third host in this trip was a widow just north of Dunedin. She was a very nice and hospitable lady who charged only NZ$ 5 per night. However, the minute I arrived in her home I realized that I would have to sit with her in the living room all evening and listen to her efforts to convince me that Yeshua is God's only true representative on earth.... I was so tired and simply wanted to go to bed.*

The idea of hospitable hosts as a prerequisite benefit was, therefore, occasionally not met. Many Israeli backpackers interpreted their hosts'efforts to evangelize as "pushy" behaviour. This issue was raised by many guests not only during the interviews but also through partidpant observations. Quite frequently, one could hear guests recommending not staying with specific HIT hosts because they were perceived as being 'religiously too pushy'.

Furthermore, there was overwhelming agreement amongst guests that HIT accommodation fadlities were frequently unclean, untidy, occasionally lacked hot water when they needed to take a shower, sleeping conditions varied and, overall, the accommodation-fadlities ranged from basic and uncomfortable backpacker standards to highly comfortable, clean and very pleasant facilities on rare occasions. Most interviewed guests claimed that the upper end fadlities were greatly in demand, but scarce. Therefore, the chance to enjoy good quality accommodation fadlities required an early booking, which does not suit the modus vivendi of the average backpacker, who usually dedded on his/her next destination at the last minute.

Based on this evidence, guests expressed a negative Tolerance Gap as far as their basic required benefits from the HIT network were concerned. However, surprisingly, they reported other unexpected experiences which had not been antidpated by them prior to joining HIT. The most interesting finding in this respect was the strong spiritual experience that most of them shared as a result of being exposed to their hosts' beliefs and perceived religious behaviours. What shaped this unique spiritual experience was the understanding of the strong linkage in Messianic belief between being a Christian and being a religiously and spiritually uncompromising supporter of the Jewish people, because of the Jews' central and crudal role as God's covenant people. Almost all Israeli backpackers considered this unexpected experience as being mostly a cultural revelation, which they learned to appreciate more and more as they stayed with HIT hosts during their sojourn in New Zealand.

Only two backpackers reported no spiritual experiences.These two backpackers were orthodox Jews. For them, being hosted by Messianic Christians posed a major dilemma. Should they really trade off their religious Jewish commitment to stay away from Christianity and forego the low-cost accommodation? This dilemma was put to the test and after having experience of various HIT hosts, they dedded to stick to their Jewish values even if it meant spending more money on accommodation. As one of the orthodox backpackers described:

> *On the one hand, I knew that HIT can save me a lot of money, and can allow me to stay in New Zealand longer, which I would love to do. However, on the other hand, I could no longer stand the idea of sleeping with a Christian family, and with a cross hanging above my head.*

Whilst the majority of backpackers accepted the differentiation made by the host between evangelism and conversion as they became experienced HIT guests, these two orthodox backpackersrefused to accept these differences. They insisted that the hidden agenda behind evangelism is conversion. Therefore, they dedded to quit and no longer be part of the HIT network Once they had left the network, both felt a strong and positive spiritual experience. As one of them said: 'I have savedmyselffromfallinginto this trap.'

In addition to this unexpectedly strong spiritual experience, Israeli backpackers also reported significant cross-cultural benefits. Many of them said that they happily shared symbols of Israeli culture with their hosts and, in exchange, gained much exposure to 'kiwi' culture. As one of the guests noted:

> *Our HIT hosts are not only Messianic believers. They are at the same time ordinary Kiwis who share with us their way of living. I would never be able to expose myself to the Kiwis' way of life if I'd stayed in backpackers' or other low-cost accommodation facilities.*

Another interesting dimension of the unexpected benefits gained by guests in the Current Situation level was the feeling of safety that accompanied the HIT hosting experience. Backpackers reported on various safety and security incidents that proved that the commitment manifested by HIT hosts towards their guests goes far beyond merely accommodating them. For example, in cases where guests encountered car accidents, theft, loss of their belongings, medical problems, among other things, their HIT hosts were always there to help, support and negotiate on their behalf. This gave them a very strong feeling of safety. As one guest reported:

> *HIT is my insurance policy while travelling in NZ. I felt lost after colliding with another car and called my previous HIT host for help. He came immediately to the accident scene and helped me with all the formalities. Later, he towed my car to his home garage and fixed it. I don't know how my trip would have looked after this accident if he was not there to help me out.*

Another unexpected yet highly appredated benefit was the hospitable attitude most hosts demonstrated towards their guests. Interviewed guests perceived this attitude as being far beyond what you would expect to receive from a service provider in the hospitality industry. Guests often referred to this attitude as letting guests feel at "home away from home". Some hosts even called their guests "my kids", "my children" rather than simply "my guests" and, as indicated earlier, one of the Israeli backpackers called Trish, a HIT host from the-North Island, his"substitute mother".

Two more unexpected benefits were quoted by guests in the Current Situation level: first, enhancement of their self identity as proud Israelis as a result of their hosts' unconditional support of Israel and the Israeli people. Second, the fact that HIT has become an important networking platform, allowing guests to communicate effectively with other backpackers and coordinate with them parts of their travel route and trip arrangements. As one of the guests said:

> *I can use HIT hosts to inform their guests that I am looking for a partner to join me on certain segments of my trip. I can also meet other backpackers at HIT hosts and get some good advice on where to stay and what to do in places they have already visited.*

The Expectation Level

This level of the analysis looks at future antidpated benefits from the HIT network by both hosts and guests. Again, the data collected through the in-depth interviews showed that these two groups differ. As far as Future Expectations are concerned, HIT hosts seem to be generally satisfied with what they have already obtained through their membership. This conclusion stems from the fact that they hardly expressed any new benefits that had not been detected in the Current Situation level. Again, they expected to gain the same spiritual benefits in the future, and. perhaps enrich their sodal experiences by having the opportunity to spend more sodal time with their guests and by meeting them eventually also in Israel. In this respect, the only complaint some of the hosts made was that occasionally they did not have enough time to share their faith with guests and to evangelize. As one host commented:

> *They get back from their tracks too tired and what they really want to do is go to bed. Sometimes they tend to ignore the "deal" we struck with them when they became HIT network members ...*

Guests, on the other hand, expressed very openly their wish for hosts to somewhat ease their evangelizing pressure. In fact, what they wished for was more balanced spiritual activity by hosts- one which gave guests the prerogative to dedde if they wanted, or did not want, to become exposed to their hosts' need to evangelize. As one of the guests expressed it:

> *Although I know the difference between evangelizing and missionary efforts I did feel on some occasions that my hosts were trying to convert me. I wish this kind of activity would be mitigated and subject to my prior consent to listen and hear about Y eshua and Messianic beliefs.*

In terms of their future cultural expectations, most of the hosts hope to end up visiting the Holy Land to experience Israeli culture first-hand. Interestingly, when looking to the future, some of them also had economic expectations in the way of exchanging their willingness to host for free for some domestic and / or farm help. For example, one host reported:

> *I would have loved them to have cut the lawn; this would save me some money which otherwise goes to my gardener.*

Apart from the desire to ease hosts' evangelizing activity, guests' future expectations from HIT and HIT hosts were directed towards improving the tourism components of this network. Guests wanted a better, and more standardized, quality of accommodation. They also wanted to obtain more detailed information on the accommodation characteristics and fadlities offered by each host prior to their next HIT accommodation selection; they expected HIT hosts to be more proactive in guiding them to available tourism opportunities in their region and that each HIT host would become an effident and up-to-date source of travel information. They all reported that if HIT could improve in these ways, it would give HIT guest members much more added value than was currently being provided.

In sum, these findings show that despite the deep spiritual, cultural and sodal experiences that HIT purports to provide to Israeli backpackers, their attitude towards HIT, even after experiendng its unique qualities, were directed towards travel and accommodation benefits. Therefore, it might be concluded that HIT, as a spiritual and travel network, is mutually benefidal for both hosts and guests. However, each side perceived this situation quite differently. Hosts were overwhelmingly satisfied with what it had given to them and to their guests, while guests were generally satisfied but had their reservations and expectations for future improvements.

Discussion and Conclusion

This paper aimed to explore the spiritual dimensions of hosting and to evaluate to what extent spiritual hosting generated meaningful experiences for both hosts and guests. In this sense, the study contributed to the discourse beyond that of previous religious tourism research; it looked at spiritual tourism experiences not solely from the tourists' prism. Using a comparative approach, the study assumed that spiritual tourism experiences may be shared by both hosts and guests even though, initially, only hosts were motivated to undertake such spiritual tourism activity. In the case of the HIT network of New Zealand, the study revealed that the initial expectations of hosts and guests, and their actual exposure to each other, led to very interesting inter-relations which generally produced for both a benefidal and meaningful tourism experience.

Conclusions can be discussed on two levels. First, it is important to synthesize the research results in relation to the three levels of perception inherent in the Value-Stretch Model and the gaps between each level

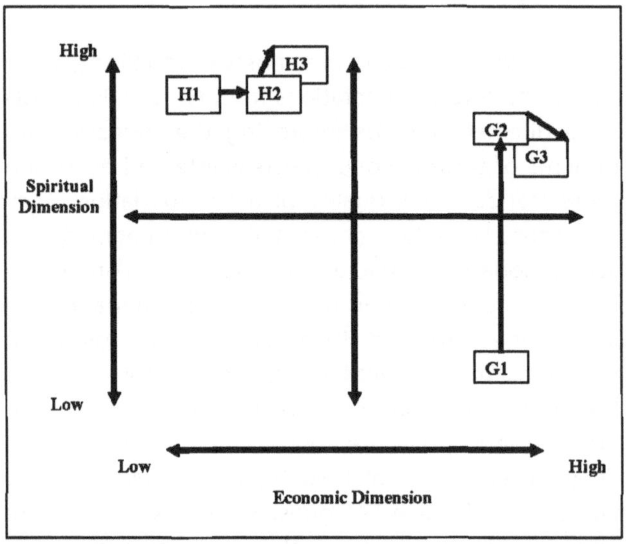

Figure 2. Hosts'[1] and Guests'2 Nonnative Change Along the Three "Value-Stretch" Levels.

1. Hl; H2; H3 denotes: Hosts on the three corresponding "value-stretch" model

2. Gl; G2; G3 denotes: Guests on the three corresponding "value-stretch" model

as well as between hosts and guests. Figure 2 illustrates how the two most
important dimensions to each party changed as hosts and guests expressed their perceived attitudes along the Value Stretch Model levels of perception.

Figure 2 shows that there were hardly any gaps between the Value-Stretch Model levels of perception obtained for the HIT hosts. Their primary benefit requirements (H1), their actual experience (H2) and future expectations from their encounters with the Israeli backpackers (H3), all centred around high levels of spiritual benefits; their wish to evangelize through hosting Israeli backpackers. Although remaining strongly inclined towards the high spiritual/low economic benefits, their actual experience slightly veered towards more economic benefits. This gap did not necessarily represent a major shift amongst hosts, but it reflected the difficulty oflong-term 'free of charge' hosting. Indeed, when interviewing hosts on this issue, those who used to accommodate between 100-500 Israeli backpackers each season indicated that they could not afford long-term free accommodation and had to charge a minimal fee of NZ$ 5 per person per night to cover their hosting expenses. As indicated above, a small minority of hosts even transformed their 'spiritual motivations to host' into a small backpacker accommodation service, charging market fees for their services. This situation has effectively created three types of hosts: a minority of non-charging spiritual hosts, a majority of spiritual hosts that charge minimal fees, and a small fraction of the HIT hosts' community that became fully commercial. As hosts expressed their future expectations from their HIT hosting, another economic dimension was revealed (H3). Many of them did not indicate that they want to gain more income out of hosting, but expressed their wish that their hosts give'in-kind' help instead. Although all hosts viewed such expected help as a token of gratitude for being hosted, it was clear that such help was needed in exchange for the cost involved in hosting them.

As for the guests, they exhibited a different process in the host-guest encounter. For them, the primary benefit requirements centred on the economic dimension, a more 'secular'interest, with no or hardly any spiritual expectations prior to joining HIT (G1). However, once moving to the current situation level, all guests underwent a meaningful and, in some cases, strong spiritual experience (G2). However, it must be noted that the gap between primary requirements (G1) and current situation (G3) was quite dramatic and highly rewarding; there were a few guests that could not cope with this unexpected and highly emotional situation. Some of them had to create spiritual walls between themselves and their guests, fearing that they might be carried away by their hosts'intention to convert them to Christianity. This spiritual'over dosage' created fear among some guests who expressed these worries when commenting on their expectations from their HIT encounter (G3). Many wished that their hosts be less 'pushy' and that they respect their wish to be left alone occasionally and simply rest rather than listen to their hosts' efforts to evangelize.

Secondly, the study findings can be discussed in relation to their wider theoretical contribution. The findings show that social exchange theory was an appropriate epistemologicalframework to analyse and explain the socio cultural exchange characteristics of host-guest encounters facilitated by the HIT network. Evidently, mostly positive exchange-based relations existed between the two groups. These relationships created an exchange of three types of benefits: namely, spiritual, social, and economic. Interestingly, in this case, the study revealed two more types of benefits: cultural benefits, and safety and security benefits. Both hosts and their guests underwent a more holistic, and, therefore, rewarding, experience than they initially expected. However, it seems that the perceived mutually beneficial nature of the HIT experience does not remain on the level of benefit exchange only. In some cases, the relationships that developed between guests and hosts became symbiotic; a mutual exchange of benefits that guaranteed the 'livelihood' of both. Based on the findings and analysis of the relationships between HIT hosts and guests, it became evident that the Messianic hosts willnever be able to undergo such a holistic spiritual experience without having Israeli backpackers staying with them and having Israeli sodety, culture, and faith exposed to them. By the same token, without having HIT as a spiritual hosting network, Israeli backpackers would have never had the chance to undergo such a multi-dimensional and profound tourism experience in New Zealand.

Asindicated in the theoretical background of this paper, sodal exchange theory is applicable only if both parties managed to attain their antidpated benefits from given interrelations in an equitable way (Madrigal1993; Waitt 2003). The question is whether this theory would have been useful to characterize exchange relationships if host/ guest relations had been less complementary? This depends on the expectations that form the relations between hosts and guest. Inthe HIT case, there was an evident complementary nature to host/ guests' expectations. It is true that the actual interaction between the HIT host and their guests brought about some 'conflicts'. However, the benefits to both sides outweighed the negatives. In any event, if such initial complementing and expected benefits do not exist, future exchange benefit studies will have to seek an alternative theoretical framework to characterize host/ guest expectations.

Finally, another theoretical concept that emerged from this study is the notion of 'spiritual hosting.' This hosting motivation is unique since it reflects not only a personal need to fulfil one's economic or other benefits, but also a normative behaviour that stems from being affiliated with a given sodal group that shares common beliefs, ideologies and/ or faith. Being such a unique phenomenon, it would be interesting to further explore whether holistic forms of hosting and tourist experiences, such as those attained by HIT hosts and guests, can be experienced in other locations, sodal, cultural, and spiritual tourism contexts.

References

ANDERECK, K. L., VALENTINE, K. M., KNOPF, R. C. and VOGT, C. A. (2005). Residents'Perceptions of Community Tourism Impacts. *Annals of Tourism Research* 32(4): 1056–1076.

ASHMOS, D. P. and DUCHON, D. (2000). Spirituality at Work. A Conceptualization and Measure. *Journal of Management Inquiry* 9(2): 134–145.

ATTIX, S. (2002). New Age-Oriented Special Interest Travel: An Exploratory Study. *Tourism Recreation Research* 27(2): 51–58.

BELK, R. W., WALLENDORF, M. and SHERRY, J. F. (1989). The Sacred and the Profane in Consumer Behavior: Theodicy on the Odyssey. *The Journal of Consumer Research* 16(1): 1–38.

BULL, A. (2006). Is a Trip to the Seaside a Spiritual Journey? Paper presented at 'Tourism: The Spiritual Dimension Conference', 5-8 April 2006. University of Lincoln. Lincoln. UK.

CARNEGIE, E. and DEVEREUX, C. (2006). Conversations with all Sorts: Self-reflection, Community and the Spiritual Journey. Paper presented at 'Tourism: The Spiritual Dimension' Conference, 5-8 April 2006. University of Lincoln. Lincoln. U.K.

COHEN, E. (1992a). Pilgrimage and Tourism: Convergence and Divergence. In Morinis, A. (Ed) *Sacred Journeys: The Anthropology of Pilgrimage.* Westport CO. Greenwood Press: 47–61.

COHEN, E. (1992b). Pilgrimage Centers: Concentric and Excentric. *Annals of Tourism Research* 19: 33-50. COLEMAN, S. (2001). Pilgrimage. *International Encyclopedia of the Social and Behavioral Sciences:* 11445–11448.

COLLINS-KRIENER, N. and KLIOT, N. (2000). Pilgrimage Tourism in the Holy Land: The Behavioural Characteristics of Christian Pilgrims. *GeoJournal* 50: 55–67.

COLLINS-KREINER, N., KLIOT, N. and MANSFELD, Y. (2004). *Between God and Guns: Christian Pilgrimage to the Holy Land during Security Crisis.* Aldershot, Hampshire. Ashgate.

CROUCH, D. (2006). Tourism and the Sacred. Paper presented at 'Tourism: The Spiritual Dimension' Conference, 5-8 April 2006. University of Lincoln. Lincoln. U.K.

DEAN, K. L., FORNACIARI, C. J. and MCGEE, J. J. (2003). Research in Spirituality, Religion and Work. Walking the Line between Relevance and Legitimacy. *Journal of Organizational Change Management* 16(4): 378–395.

DIGANCE, J. (2003). Pilgrimage at Contested Sites. *Annals of Tourism Research* 30(1): 143–159.

DIN, K. H. (1989). Islam and Tourism: Patterns, Issues and Options. *Annals of Tourism Research* 16: 542–563.

EADE, J. (1992). Pilgrimage and Tourism at Lourdes, France. *Annals of Tourism Research* 19: 18–32.

FLEISCHER, A. (2000). The Tourist Behind the Pilgrim in the Holy Land. *International Journal of Hospitality Management* 19: 311–326. FLEISCHER, A. and PIZAM, A. (2002). Tourism Constraints Among Israeli Seniors. *Annals of Tourism Research* 29(1): 106–123.

GRABURN, N. (1989). Tourism: The Sacred Journey. In Smith, V. (Ed) *Hosts and Guests: The Anthropology of Tourism* (Second edition). Philadelphia PA. University of Pennsylvania Press: 21–36.

HEINTZMAN, P. (1999). Spiritual Wellness: Theoretical Links with Leisure. *Journal of Leisurability* 26(2): 21–32.

HENDERSON, J. C. (2003). Managing Tourism and Islam in Peninsular Malaysia. *Tourism Management* 24(4): 447–456.

HERNANDEZ, S. A., COHEN, J. and GARCIA, H. L. (1996). Residents' Attitudes towards an Instant Resort Enclave. *Annals of Tourism Research* 23(4): 755–779.

HUDMAN, L. E. and JACKSON, R. H. (1992). Mormon Pilgrimage and Tourism. *Annals of Tourism Research* 19: 107–121.

JAMAL, T., HARTL, C. and LOHMER, R. (2006). Tourism, Pilgrimage and Planning Pedagogy: Cultural Heritage and Community Churches. Paper presented at 'Tourism: The Spiritual Dimension' Conference, 5-8 April 2006. University of Lincoln. Lincoln, U.K.

JUROWSKI, C. and GURSOY, D. (2004). Distance Effects on Residents' Attitudes Toward Tourism. *Annals of Tourism Research* 31(2): 296-312. LANE, B. C. (2001). Holy Silence: Invitation to Sabbath. *Christian Century* (October 24-31): 24–27.

LEE, C. and BACK, K. (2003). Pre- and Post-Casino Impact of Residents' Perception. *Annals of Tourism Research* 30(4): 868–885.

LEE, Y. (2006). Myth, Spirituality and Religion in Travel: Pre-Industrial Korea. *Tourism: An International Interdisciplinary Journal* 54(2): 97–106.

LITTLE, D. E. and SCHMIDT, C. (2006). Self, Wonder and God! The Spiritual Dimensions of Travel Experiences. *Tourism: An International Interdisciplinary Journal* 54(2): 107–116.

LYONS, K. D. (2003). Ambiguities in Volunteer Tourism: A Case Study of Australians Participating in a J-1 Visitor Exchange Programme. *Tourism Recreation Research* 28(3): 5–13.

MADRIGAL, R. (1993). A Tale of Tourism in Two Cities. *Annals of Tourism Research* 20: 336–353.

MANSFELD, Y. (1992). Group-Differentiated Perceptions of Social Impacts Related to Tourism Development. *Professional Geographer* 44(4): 377–392.

MANSFELD, Y. (1995). The Value Stretch Model and its Implementation in Detecting Tourists' Class-Differentiated Destination-Choices. *Journal of Travel and Tourism Marketing* 4(3): 71–92.

MANSFELD, Y. (2002). Reinventing a Tourism Destination Through a Network Designed Destination Management System. *Tourism: An International Interdisciplinary Journal* 50(4): 349–360.

MANSFELD, Y. and JONAS, A. (2006). Evaluating the Socio-cultural Carrying Capacity of Rural Tourism Communities: A 'Value Stretch" Approach. TESG- Tijdschrift voor Econornische en Sociale Geografie *(Journal of Economic and Social Geography)* 97(5): 583–601.

MATTILA, A. S., APOSTOLOPOULOS, Y., SONMEZ, S., YU, L. and SASISHARAN. V. (2001). The Impact of Gender and Religion on College Students' Spring Break Behavior. *Journal of Travel Research* 40(2): 193–200.

MCCORMICK, D. W. (1994). Spirituality and Management. *Journal of Managerial Psychology* 9(6): 5–9.

MCINTOSH, A. J. and BONNEMAN, S. M. (2006). Willing Workers on Organic Farms (WWOOF): The Alternative Farm Stay Experience? *Journal of Sustainable Tourism* 14(1): 82–99.

MCINTOSH, A. J. and CAMPBELL. T. (2001). Willing Workers on Organic Farms (WWOOF): A Neglected Aspect of Farm Tourism in New Zealand. *Journal of Sustainable Tourism* 9(2): 111–127.

MCINTOSH, A. J. and JOHNSON, H. (2004). Exploring the Nature of the Maori Experience in New Zealand: Views from Hosts and Tourists. *Tourism: An International Interdisciplinary Journal* 52(2): 117–129.

MCINTOSH, A. J. and THYNE, M. (2005). Understanding Tourist Behavior Using Means-End Chain Theory. *Annals of Tourism Research* 32(1): 259–262.

MAOZ, D. (2005). Young Adult Israeli Backpackers in India. In Nay, C. and Cohen, E. (Eds) *Israeli Backpackers: From Tourism to Rite of Passage.* New York. SUNY Press: 159–188.

MEVORACH, 0. (1997). *The Long Trip After the Military Service: Characteristics of the Travelers, the Influence and Significance of the Journey.* PhD. Dissertation submitted to the Department of Psychology, Hebrew University of Jerusalem (in Hebrew).

MINER-WILLIAMS, D. (2006). Putting a Puzzle Together: Making Spirituality Meaningful for Nursing Using an Evolving Theoretical Framework. *Journal of Clinical Nursing* 15: 811–821.

MITROFF, I. I. and DENTON, E. A. (1999). A Study of Spirituality in the Workplace. *Sloan Management Review* 40(4): 83–92.

NEAL, J. and BIBERMAN, J. (2003). Introduction: The Leading Edge in Research on Spirituality and Organisations. *Journal of Organizational Change Management* 16(4): 363–366.

NOLAN, M. L. and NOLAN, S. (1992). Religious Sites as Tourism Attractions in Europe. *Annals of Tourism Research* 19: 68–78.

NOY, C. (2004). This Trip Really Changed Me: Backpackers' Narratives of Self-Change. *Annals of Tourism Research* 31(1): 78–102.

NOY, C. (2005). Background: Backpacking in Contemporary Society - The Israeli Context *(Tarmilaut).* In Nay, C. and Cohen, E. (Eds) *Israeli Backpackers: From Tourism to Rite of Passage.* New York. SUNY Press: 1–44.

OLSEN, D. H. (2006a). Typologies of Spiritual Tourists. Paper presented at 'Tourism: The Spiritual Dimension' Conference, 5-8 April 2006. Lincoln, UK. University of Lincoln.

OLSEN, D. H. (2006b). Issues for Religious Heritage Attractions. In Timothy, D. J. and Olsen, D. H. (Eds) *Tourism, Religion and Spiritual Journeys.* London and New York. Routledge: 104–118.

PERNECKY, T. (2006). Co-Creating with the Universe: A Phenomenological Study of New Age Tourists. *Tourism: An International Interdisciplinary Journal* 54(2): 127–144.

PORIA, Y., BUTLER, R. and AIREY, D. (2003). Tourism, Religion and Religiosity: A Holy Mess. *Current Issues in Tourism* 6(4): 340–363.

REYNOLai, T. and GUTMAN, J. (1998). Laddering Theory, Method, Analysis and Interpretation. *Journal of Advertising Research* 28: 11–31.

RINSCHEDE, G. (1992). Forms of Religious Tourism. *Annals of Tourism Research* 19: 51–67.

ROBICHAUD, P. (1999). Tourist or Pilgrim? Rescuing the Jubilee. *America* 181(20): 10–14.

SHACKLEY, M. (1999). Tourism Development and Environmental Protection in Southern Sinai. *Tourism Management* 20: 543–548.

SHACKLEY, M. (2001). *Managing Sacred Sites: Service Provision and Visitor Experience.* New York. Continuum.

SHARPLEY, R. and SUNDARAM, P. (2005). Tourism: A Sacred Journey? The Case of Ashram Tourism, India. *International Journal of Tourism Research* 7: 161–171.

SIZER, S. R. (1999). The Ethical Challenges of Managing Pilgrimages to the Holy Land. *International Journal of Contemporary Hospitality Management* 11(2/3): 85.

SMITH, V. L. (1992). Introduction: The Quest in Guest. *Annals of Tourism Research* 19: 1–17.

SPILKA, B. and MCINTOSH, D. N. (1996). Religion and Spirituality: The Known and the Unknown. Paper presented at the American Psychological

Association Annual Conference, August. Toronto. Canada.

TIMOTHY, D. J and OLSEN, D. H. (Eds) (2006). *Tourism, Religion and Spiritual Journeys.* London and New York. Routledge.

URIELY, N., YONAY, Y. and SIMCHAI, D. (2002). Backpacking Experiences: A Type and Form Analysis. *Annals of Tourism Research* 29: 520-538.

VUKONIC, B. (1996). *Tourism and Religion.* Oxford. Pergamon.

WAITT, G. (2003). Social Impacts of the Sydney Olympics. *Annals of Tourism Research* 30(1): 194–215.

WEARING, S. (2001). *Volunteer Tourism: Experiences that Make a Difference.* Oxon. CABI Publishing.

WOODWARD, S.C. (2004). Faith and Tourism: Planning Tourism in Relation to Places of Worship. *Tourism and Hospitality Planning and Development* 1(2): 173–186.

ZAHRA, A. (2006). The Unexpected Road to Spirituality via Volunteer Tourism. *Tourism: An International Interdisciplinary Journal* 54(2): 93–196.

Website reference

Ya'akobowich, 0. (2005). interview available at http://www.homeworship10l.com/fyi_Messianic_Judaism_%20Messianic_Christianity.htm - Accessed on 15 March 2009.

Submitted: January 28, 2009
Accepted: May 27, 2009

Spirituality and Tourism: An Anthropologist's View

Sagar Singh

ABSTRACT

For many years, tourism has been studied as a ritual. Few scholars have sought to make an in-depth inquiry into the spiritual aspects or dimensions of tourism. This stalemate is a result of our thinking of tourism as essentially to do with physical behaviour, even though tourism marketers and some scholars study motivations and, hence, mental aspects. This paper seeks to explore and define spirituality in terms of its relationship with tourism anthropologically, and concludes that tourism is a spiritual activity that has to do with human sodety and its values as a whole. The paper synthesizes various anthropological concepts of tourism and arrives at a new definition of the social process, as well as explains why the spiritual aspects of tourism are more enduring than the ritual, since tourism is a holistic process and not mechanical.

Introduction

The relationship between spirituality and tourism has been less explored by scholars as opposed to rather direct inquiries regarding the relationship between tourism and religion (e.g., Vukonic1996; Cohen 2004; Raj and Morpeth 2007). The type of inquiry that has been employed appears to hinge upon the nexus and/or similarities and differences between pilgrimage and tourism, which has been the subject of much attention by tourism scholars. Is it spirituality or rituals that are more important in tourism and religion? Perhaps both. But which is more important in order to understand the true nature of tourism, in all its variety? According to many tourism scholars to date, rituals are more important since tourism and sightseeing appear to be 'modern rituals' (following MacCannell1976:13; Graburn 1983) that appear to have no direct relationship with spirituality. This is a result of our thinking about tourism as a special kind of circuitous movement, and movement implies ritual. Works on the social sdence aspects of pilgrimage and tourism have mostly chased the mirage of oneness between all religions as far as the similarities between various types of the ritual of pilgrimage - whether it is Hindu or Muslim, Christian or Sikh-are concerned (see, for example, Jha 1991; Vukonic1996; Timothy and Olsen 2006).Insuch treatments, pilgrimage and tourism are often equated and considered synonymous. Are they? This paper willshed some light on this assumption and attempt to clarify why this apparent fallacy is perpetuated.

A second line of thought has been to attribute sacredness to allforms of tourism, an opinion voiced early in the tourism literature by Graburn (1977,1989). Do rituals exist by themselves in a kind of vacuum divorced from their spiritual basis? What is the relationship between spirituality and rituals? What is the relationship between spiritualities of various kinds,i.e., among the great religions as well as the lesser tribal/peasant ones? What is the ontology of tourists and tourism with regard to spirituality? Can there be a metaphorical or metaphysical relationship between the spirit of travel and tourism and the philosophies of religions? Such questions, in the knowledge of this scholar, have not been asked, let alone answered. This paper is an attempt to shed light on some- but not all- of these important themes.

Methodology

The paper is the result of 22 years of episodic inquiry into various religions and their core, spirituality, through books, journals/ magazines, travel to places of pilgrimage and tourism, and discussion with religious functionaries over the finer aspects of religions (including Hinduism, Christianity, Islam, Sikhism, Jainism and Buddhism, and a few tribal religions), as well as interviews of common people who were either tourists or pilgrims. The inquiry covered analysis of books on tourism and religion, as well as journals that have published spedal issues on tourism and pilgrimage, including *Annals of Tourism Research* and *Tourism Recreation Research*. As a result, this research is a synthesis of both primary and secondary sources of information, as well as original thought. The textual sources of information are mentioned along with the discussion in this paper, in order to keep it short.

A brief overview of methodology, as well as the approach, is outlined here. First, the author started studying books on Hindu, Buddhist, and Christian thought at the age of 19. He also practised meditation, after reading books and visualizing (and some experimenting) at that age and thereafter. (Meditation continued, sporadically, in times of personal crises and otherwise.) After this, visitation of the western Indian Himalayas, which started at the age of 12, was undertaken again, although not with the explicit purpose of understanding pilgrimage and tourism, and the relationship between the natural and supernatural world. Though few structured interviews were taken, the diversity among people who were visitors at places of pilgrimage and tourist places was observed, as also the diversity among the local people. Unstructured interviews were often taken during later visits. Overall, conversations with some 100 persons were analysed over time, and the concepts evolved gradually. Tourism and pilgrimage were studied from a managerial view in1985, and from the academic as well as managerial point of view in several repeat visits in the1980s and 1990s, right up to 2001. Three excursions to the western Indian Himalayas were again undertakenin2004, 2007 and 2008. The author also visited holy places in India like Kurukshetra, Ayodhya, Orchha, Maheshwar, and Datia,Jain shrines, mosques, churches and *gurudwaras* (Sikh shrines) that are not in the Himalayas.

Friendship and acquaintanceship was made with people from all major religions (except Buddhism) and some tribal ones. An effort was made to understand religion and spirituality as understood by 'authorities' as well as laypersons. Tourism journals that were consulted included two special issues of *Annals of Tourism Research:* one on the anthropology of tourism [10(1), 1983] and the other on pilgrimage and tourism [19(1), 1992]; and one special issue of *Tourism Recreation Research* on 'Sacred Journeys'. Books published by the Association of Social Anthropologists of Great Britain and Ireland (ASA monographs) on anthropology of religion, social anthropology of complex societies, social anthropology and language, models in anthropology, and economic anthropology (as well as an anthropological book by economic historians: Polanyi *et al.*1957) were read and analysed.

Books on Karl Marx (e.g., McLellan 1975; Mills 1963; Howard and King 1976), by Marx (Grundrisse, tr. Nicolaus 1973) and on capitalism (Galbraith 1952, 1967), Max Weber (1958; Gerth and Mills 1958; Aron 1970), John Maynard Keynes (Moggridge 1976), as well as books and articles by economists like Paul Samuelson (1982) andJ.R. Hicks (1959) on economics, and on structural anthropologists like Oaude Levi-Strauss (Leach 1977), were studied in the 1980s, 1990s and later.Works of sociologists Emile Durkheim (Aron 1970), Talcott Parsons (1964), and Kingsley Davis (1981), among others, were also studied. This was done in order to understand the various viewpoints on religion and its relationship with other aspects of society, including economy. Overall, at least 50 books on anthropology and 40 books on tourism, and at least 60journal articles were studied. Books on geography and tourism (e.g., Bhardwaj 1973; Pearce 1987), and spirituality as opposed to religion (numerous books published by the Himalayan International Institute of Yoga Science and Philosophy, USA, and Radha Soami Satsang Beas, India), were studied in depth. The viewpoint that emerged was eclectic and hence anthropological in character. That is why sociologists and others have been referred to in this paper, but the article is essentially anthropological (or 'sociological' in its broadest sense).

The Meaning of Spirituality

The English term'spirit' or the Latin *anima* has shades of meaning that differ from culture to culture. The Persian and Urdu word for it, *rooh,*is derived from a Sanskrit-like language known as Proto Indo-European: *ru,* meaning 'being' or existence (as in *Rudra,*the Vedic god of the Hindus -later called *Shiva* -literally *ru,* 'one who was begot' and *dravya,* meaning 'from the primeval ocean'). But in all the major religions -in terms of numbers of followers, such as Christianity, Islam, Hinduism and many tribal religions, there is a similarity in meaning in terms of the spirit being the essence of life and the myriad colours of animal existence. Among the Native Americans or First Americans, as also some (East) Indian tribes, the prevalent form of religion used to be described by anthropologists as either animism or animatism (Madan and Majumdar 1966). In fact, some Native American religions also share a kind of mysticism (Brown 1980), although the mysticism among East Indian tribes has not been fully explored (see, e.g., Jha 1991; Timothy and Olsen 2006). Among these religions, thanatism (belief in death of the soul) has no place since there is a place for souls after death and God is said to do justice to all souls on the merit of their actions in this life.

In this way, Christianity and Islam, which believe in Judgement Day, are no different from Hinduism and Buddhism. In the last two religions as well, the cycle of birth and death is emphasized, since, ultimately, the belief rests on the merit of actions in *this* life. These merits are defined by the religion: good deeds or 'being a Good Samaritan' in Christianity, *sawaab* in Islam, and *punya* in Hinduism.

However, all these brief descriptions of earning 'merit' is based on materialism, which is the opposite of true spirituality, since the dominant view is that the gods or God can be pleased by certain actions, like donating money to the poor and needy (found in Hinduism, Islam, and Christianity). Thus, religions suggest ways of accumulating religious capital. Of course, religious capital can be earned both through rituals and by spiritual means, since both are part of religion. But spiritual merit can be earned only by spiritual behaviour, like meditation, which is practised in all the four major religions: Hinduism, Buddhism, Christianity and Islam (O'Brien 1978; Woods 1980; Nicholson 1980). However, it should be noted that there can be no religion without spirituality, whereas there may be spirituality without religion. One demonstration of this is yoga and its spiritual philosophy, which can be adapted to any religion (Swami Rama 1975; O'Brien 1978). This implies that yogic philosophy is not culturally restricted, and aspires to be a universal philosophy. Is it akin to an un-stated philosophy of the generalized tourist? Perhaps, but this can be ascertained only after looking at the meaning of spirituality and its relationship with tourism, and only if we agree that tourism is not only a universal phenomenon, but has existed (unstudied) since primordial times.

The Spiritual Side of Tourism: Preliminaries

How are actions weighed in religions? Where beliefs are magico-religious, as in many tribes across the world, the merit is confined to benefits and losses in this world (Beattie 1985; Mair 1984). But where religion comes in, the merit of actions is weighed both in this life and beyond, since all religions have a philosophy- even tribal ones, such as the religion of the Nuer tribe in Sudan (Evans-Pritchard1956) and all religious philosophies relate both to the material world and the spiritual world. And that is the difference between tourism as a ritual and the phenomenon seen in terms of spirituality.

MacCannell (1976: 13 and subsequent editions) had talked about sightseeing, and, by extension, tourism, as a ritual (Graburn1983). In other words, there appeared to be no spiritual side of tourism, or if there was, it remained less examined for reasons stated by this author (the emphasis on 'movement'). But forms like ecotourism, volunteer tourism, pro-poor tourism, VFR tourism, forest tourism, farm or rural tourism, educational tourism, and New Age tourism do not support claims that tourism is only a ritual, since in these forms of tourism, the 'ritualization of tourism' (Nash 1984) is not evident, since such tourists make a conscious effort to break free from the so-called'norm'. Here is one difference between usual pilgrimage and tourism: the former is most often undertaken by people who follow a norm, while the latter differs from person to person and from time to time: it is certainly illogical to presuppose, for example, that all ecotourists were and remain ecotourists, or that all ecotourists remain the same sort of ecotourists over time.

To come back to the distinction between the spiritual and ritual in tourism, it does not mean that the spiritual and the ritual are antipodal and binary opposites. There can be something spiritual about rituals - such as chanting the name of revered saints and gods and goddesses before sitting down for meditation *(jaap* in Hindi), or not treading with shoes on, on hallowed ground (common to Hindus and Muslims)- and there can be something ritualistic about spiritualistic actions, like sitting down for meditation at the same time every morning. All major religions enjoin upon the individual to devote some time to actions that are not concerned with the self and its longings. In other words, religions urge people to be sodal and that is why religions become the soul of a culture. Is this something that only modern humans have experienced? Probably not. Indeed, it may not be far-fetched to say that modernity has made humans less spiritualistic but there is a need for 'things spiritual'in all humans, hence the quest in Western societies to seek identity and belonging in Eastern cultures where religions still rule the roost. But the quest is not for religion *per se;* rather, the desire of tourists is to experience the different forms of spiritualism that binds peoples of different cultures together (such as Sufi'ism and yoga; the latter is fast becoming the mainstay for spiritualistic tours by backpackers: see, e.g., Maoz 2006; Noy 2006). This spiritualism common to all cultures does not have a name although it is an'attraction': tourists are attracted not only by exotic places but also by exotic cultures (which go together) (Boniface 1995); and many travel to places that are not-so exotic or not exotic at all, but still find the difference in culture refreshing and compare it with their own *in order to understand themselves;* what some scholars (see Gibson and Yiannakis 2002) call'anthropologist tourists'.

Defining Spirituality Anthropologically

The spirit, like the body, appears to be ephemeral and hence the popular (scientific?) but un-Christian Western belief in death as the be-ali and the end-all However, all the major religions stress that the soul does not die.

So spirituality cannot be defined with reference to the familiar dichotomy of life and death. Indeed, the soul, when separated from the body, becomes immortal, unless it takes the form of another human or animal or even a plant (Radha Soami Satsang Beas 2005: 35). Religions with animatism emphasize that after a human's death, the spirit or soul comes to reside in an animal, which is sometimes indicated before death by the dying persons themselves through dreams or self-induced hallucinations (Madan and Majumdar 1966; Mair 1984). This is found both in East Indian tribes and Native Americans.

Furthermore, the spirit is not the same as the soul. The spirit is both more and less than the soul. The Hindus differentiate between *atmaan* or *jiva* (soul) and *prana* (spirit). Explained from a sdentific view, we can look at the familiar example of organisms that are said to be in between the living and non-living: viruses. Viruses lie dormant for hundreds or thousands of years without life, but come alive or become active as soon as they find water and food (such as another organism). We may say that viruses perhaps have souls *(jiva)* but no spirit *(prana,* hence they remain motionless). The soul is said to reside in the body and transmigration of the soul, according to, say, Hindu and Buddhist belief, is a 'natural' process. The spirit, on the other hand, comes closer to the Latin *anima,*which is considered to be the animating element of life (such as'a spirited attack on communism' or when we say that a person is 'dispirited'; as also when we speak of liquor as 'spirit' that acts as a stimulant). Plants have souls, but they do not have a spirit similar to that found in animals and humans, which is why they are considered less evolved by religions (Radha Soami Satsang Beas 2005: 35), as also by biologists.

Animistic beliefs, on the other hand, lend an animation to natural features such as stones, sun, water and air (Madan and Majumdar 1966). Thus, natural features can also have something resembling spirit, but not the soul. As opposed to Christianity, Hinduism is an animistic religion as well. This does not mean that Hinduismis inferior. The other distinction usually made between Western sodeties and Eastern ones is that the former are said to be more outward looking and the latter more inward looking. These are attempts to label religions and sodeties as a whole, which is not a good way of making a comparison (Singh 2003, 2007b). One has only to remember litterateurs like T.S. Eliot and Rudyard Kipling, as also spiritualistic Americans like Emerson, Thoreau and Whitman (Arya 1978) to point out that spiritualistic beliefs are not the exclusive preserve of any one culture or sodety.

So how do we define spirituality? All humans seek play and recreation, and pursue leisure activities. Tourism, indeed, is a form ofleisure and recreation that only humans pursue (Singh 2007b). Recreation, like play, is an essential ingredient of tourism. And allcultures suggest ways in which individuals of the sodety can seek happiness. This usually takes the form of recreation like singing, dandng, playing musical instruments and games, telling anecdotes and stories, and so on (Singh 2007b). Tourism is distinct from other forms of recreation in that it involves a movement away from home and hearth and one of the widest recognized sodological 'type'of tourist is what Erik Cohen (1979) termed the 'existential tourist' (Maoz 2006).

'Existentialism' as a philosophy, however, is a cry of loneliness of the individual who does not believe in God and His Goodness (Bullock *et al.* 1988: 296-297). This philosophy, introduced by Kierkegaard (Harper 1965), and championed by Jean Paul Sartre and Albert Camus, emphasizes that humans are lonely strangers and that their life has no purpose-is meaningless. It should be noted that Kierkegaard speaks of universal love, but not God, whereas Nietzsche speaks of God but not love (Harper 1965). Camus does not speak oflove or God.Indeed,some of his philosophy appears to be an echo of the chapter in the Bible, 'Ecclesiastes' (1:2) where 'the Teacher' repeats throughout:

> 'Meaningless! Meaningless!' says the Teacher
> 'Utterly meaningless! Everything is meaningless!'
> (The Bible [New International Version] 1984: 469)

This is in contrast to spiritualism, where God's designs and purposes are known and acceptable - such as among the early American settlers and their descendants (Singh 2003). Moreover, Ecclesiastes is in contrast to the gospel and the perspective on life given by the New Testament whose teaching is that 'Life is meaningless without God, love, and Christ the Saviour, who was born so that we may live meaningful and joyful lives by God's Grace'. That is why most stimulating sermons are often delivered taking cues from the New Testament and not as often from the Old Testament. Ecclesiastes, in fact, goes out of the way to denounce useful things like 'work' as meaningless and in vain. Perhaps the author(s) of Ecclesiastes was/were trying to stress the importance of spiritualism and meditation, without directly saying so.

It appears that Maoz (2006) has not properly understood Cohen (1979) and is wide of the mark: tourists, including backpackers, are not existentialist humans, but beings seeking happiness in other places, other cultures. The ritual element of Christian and Jewish religion has diminished as a result of the impact of sdence on sodety but the spiritual element remains (Noy 2006). Religion helps institutionalize tourism through promoting discovery and self-realization (see Noy 2006, on Jewish religion's role in institutionalization of tourism in Israel; and Singh 2003, for a similar role of Christian and Hindu religions in America and India, respectively). In these cases, it is the spiritualistic element of these religions that is fuelling tourism, since the forms of religious tourism and pilgrimage, and notions of what is ritual, differ, but their effect on societies remains the same. Tourists are seeking what is missing in their everyday lives: the euphoria that spirituality/religion used to provide, but does not any more. And they are also seeking what fuels the knowledge economy: knowledge about the world. This is once more in contrast to the existentialist philosophy, perhaps again following (?) Ecclesiastes (1:18), where it is said that wisdom and knowledge are meaningless:

> For with much wisdom comes much sorrow;
> The more knowledge, the more grief.
> (The Bible [NN] 1984: 469)

Here again, perhaps, Ecclesiastes is trying to criticize knowledge about the *material* world, without saying it explidtly. Ritualism is about magical belief in the efficacy of some actions to undo what was or could be wrong, or do what is believed to be right in a *material* world. It is about behaviour of a kind where, by mere repetitive performance of certain actions, it is thought that the gods or God will be pleased. Spiritualism is about doing what is right in the ultimate test: death and beyond, the non-material world.It has less to do with the immediate world and more to do with the fourth dimension, which usually cannot be directly experienced by the senses, but can be experienced by meditation - but not by all. In ritual, performance is very important (Tambiah 1979), unlike in spiritualism where the mere thought or mental prayer is important. Why is performance important in ritual? It is important since performance appears to ensure that bodily movement or movement of the material and non-material elements that constitute the ritual-such as lighted earthen lamps in Hindu prayer or the light at the altar in Christianity or folding hands and chanting the 'right words' of prayer in Hinduism and Christianity -provide the link between this world and the other world (souls, gods and goddesses, Mother Mary and Christ, the Holy Ghost and, ultimately, God). In ritualism, this world is represented by humans and their culture (artefacts, language and meta-linguistic symbols), while the performance itself seems to provide a direct link to the supernatural world. In other words, the ritual has something superstitious or magical about it, as mentioned earlier.

Tourism does not appear to link us to the supernatural world through any movement or linguistic/meta-linguistic symbols. It cannot be called a 'secular ritual' either, since tourism does not appear to mediate between any two worlds - social and natural, or supernatural and natural – unlike pilgrimage (Singh 2004a). Secular rituals can mediate between two worlds, whereas sacred rituals may mediate between more than two worlds. For example, the secular ritual of national flag-hoisting on Independence Day in the US or in India mediates between the world of the person who hoists the flag and the sodal world; secular wedding ceremonies that take place, say, in court, mediate between the individual's (or the personal) and the sodal world. Death rites (sacred rituals) mediate between the material world (death of the body) and the sodal world (the end of sodal ties that existed with the dead person) as well as the supernatural world (rites are directed towards finding peace for the soul).Similarly, birth rites are also sacred rituals and mediate between the material world (birth of a body) and the sodal world (the beginning of sodal ties with the newborn) and the supernatural world (the birth is tagged as being either fortunate and lucky for the family and community, or unfortunate and boding ill). Does tourism display any characteristics of being a secular ritual? It seems not. However, it does seem to have the characteristics of a sacred act since, like the pilgrim, the tourist faces travails and the prospect of acddents and death is known and not beyond being envisaged.

Having considered these points, we can now define spirituality as 'therapeutic mental and corresponding physical behaviour that leads to lasting happiness and euphoria (including transcendental euphoria) for members of a particular quasi-group, group, community, sect or religion'. This working definition does not reek of 'behaviourism'for the simple reason that mental behavior is the opposite ofbehaviourism as understood in psychology and other sodal sdences. In fact, the term'activity' in place of 'behaviour' would have been inappropriate, since 'mental inactivity' is included in behaviour, but would have been the opposite of what was meant. So, meditation, which is an attempt to still the mind of everyday thoughts and concentrate on spiritual goals, is also behaviour. Euphoria can be defined as a form of long-lasting happiness that can be found in everyday waking life (simple euphoria, like recreation), sleep (silent euphoria, as in deep sleep), and during meditation (transcendental euphoria) (see Woods 1980 for a sdentific understanding of transcendental meditation). It must also be pointed out that all happiness is not euphoria, although all euphoria is happiness: passionate love is a form of happiness but not euphoria, since it does not last, and creates conflict, among other things; recreation is a form of happiness and also a form of 'simple euphoria' since the after-effects of recreation arelong-lasting. Unlike the typology provided by Cohen (1979), we can say that *all* tourism involves recreation and not just in the 'recreational tourist mode'. Hence we come to the logical conclusion that most forms of tourism have a spiritual side to them, since they provide recreation as well as leave a euphoria well after the event.

Defined in this way, spirituality becomes the basis of most forms of tourism such as cultural and ethnic tourism, ecotourism, adventure tourism, educational tourism, VFR tourism,farm tourism, volunteer tourism, pro-poor tourism, and, of course, religious tourism. The implications of this conclusion for tourism studies, if accepted, are vast. Theoretically, first, it dislodges MacCannell's (1976 and subsequent editions) stand that tourism is a 'modem ritual'. Second, it implies that world tourism is a phenomenon whose un-understood but purposive goal (or social function) is to unite humanity through exploring the dimensions of existence in various cultures. The practical implications of these two conclusions could be the subject of another paper, but will be briefly discussed in the end.

'Ritual'and 'Spiritual'Tourism

As mentioned earlier, human behaviour being what it is, it is frequently difficult and impractical to separate the ritual and spiritual aspects of life. But it has to be done if we are to understand the true nature of tourism. Shackley (2001), while concentrating on managing sacred sites, also speaks of managing secular sites. Shackley (2001: xv) also notes that 'operating sacred sites is an almost exclusively male prerogative'. This is not true of tourism. Women account for a substantial part of the tourism work force (Sinclair1999). Why males dominate management of sacred sites is simple to understand: most major religions - Christianity, Islam, Hinduism, Zionism, and Buddhism, as also Sikhism - are male dominated and centred. Even the yogic *Sankhya* philosophy, while giving almost equalimportance to males and females, speaks of *purusha* (the male element or superior consciousness) and *prakriti* (the feminine element, represented by Nature, something to be dominated). Inmany religions, including Islam and Hinduism, as well as many tribal ones, women are considered to be 'polluting', hence their absence from sacred sites. But this is important from the ritual point of view. Women are considered polluting because of menstrual blood (both among Hindus and Muslims), but post-menopausal women are also not allowed into the interiors *(sanctum sanctorum)* of temples. This is a ritualistic or fixed view. This is not true historically (or pre historically, to be precise), since female fertility cults are older than male ones, as evidenced by the excavation of female figurines or'venuses' in the Old Stone Age (Burkitt 1%3).

In Hinduism, one can see the truth of this statement when one recalls the oft-visited temples of *Kali* (a form of *Shakti*, the feminine force, which emanates from, but is stronger than, *Shiva)*. In India, there are 51 *Shakti peeths* ('elevated' shrines dedicated to *Durga* or *Kali)*, but only 12 *Jyotirlingas* ('effulgent' shrines dedicated to *Shiva)* (Jha 1991). Not surprisingly, though, even these *Shakti* shrines are managed by men. It is only with the rise in yogins (female yogis) in recent years that relative equality of religious status between men and women has started coming about in Hinduism.

To come back to the differentiation between 'ritual'and 'spiritual' tourism, or aspects of ritualism versus spiritualism in tourism, we may consider the following points as to why tourism is not so much of a ritual activity. One, tourism is often a repetitive activity; however, it is a continual not a continuous activity (Singh 2007a) and, unlike in a ritual, *there is no compulsion to perform*. Also, the places tourists go to are not always the same. Hence the difficulty, for most destinations, to retain attractiveness for repeat visitors, a problem that scholars utilizing the tourism area life cycle concept (following Plog 1972 and Butler 1980) sought to address.

Two, tourism is not always the same type of activity for the same group or individual, since many tourists evolve and the entertainment or recreational aspect for each individual changes over the course of his/her life.

Three, group tourism is increasingly being tailored to suit the demands of discerning cohorts (Middleton and Hawkins1998) and the ritual element is, thus, diminishing, and group tourists are becoming more like individual tourists.

Four, backpacker tourism is no more tourism on a shoestring budget and more like individual responsible tourism (Cohen 2006; Pearce 2006). More and more backpackers are emerging from enclaves and venturing out: hence they are becoming less ritualistic after gaining knowledge of the vast world without.

Five, codes of ethics are more and more being emphasized (Fennell and Malloy 2007), because many tourists are becoming aware of their responsibilities towards the environment and the host sodety, and are open to the idea of willingness to pay (WTP) (Fennell1999;Singh

2002a) for damage that they may be causing. This is in contrast to the apparently hedonistic or epicurean and ritualistic philosophy that was attnbuted to tourists (MacCannell1976; Turner and Ash1975).

Six, more and more ecotourists are apparently turning into mass tourists (Singh 2004b; Weaver 2007), and perhaps even vice-versa, but that is not necessarily bad (Buckley 2003): it merely means that the philosophy or educational aspect of ecotourism is spreading, which means that the ritualistic 'sun,sand, sea' or similar forms of tourism are decreasing in volume and the recreational-entertainment forms of tourism are increasing (see, for example, Pearce 2008).

Seven, some forms of tourism appear to be ritualistic, like visiting friends and relatives(VFR) tourism; but VFR is a broad category that subsumes differences such as between visiting relatives and visiting friends (Lockyer and Ryan 2007), and differences that can be understood with regard to rural and urban tourism. Hence, VFR tourism, as argued by Singh (2007a), is not 'ritual tourism' but keeps changing and hence not ritualistic. The timing of such tourism may be around the same time every year, since holidays in rural or urban areas for different sorts ofVFR tourists are subject to work contingencies like post-harvesting time in agrarian areas or national/regional cultures (or climate) that affect what is thought to be typical holiday time. But even this (culture) is changingand many people all over the world are taking short breaks increasingly due to their changing work culture and/or in response to efforts to stagger volume during peak seasons through marketing activities. This is equally true of marketing by tourist cities/regions/ nations and there is a conscious effort to direct tourist activity to these places in off seasons. Moreover, short breaks are not ritualistic since different destinations and different modes of transport may be chosen by the same tourists, so Singh (2007a) is wrong in placing'weekend tourism' in the ritual category. Since it not fixed, it is unlike a ritual, in which every action is more or less fixed.

Lastly, rituals are usually performed mechanically (Bharati 1991) while there is nothing mechanical about tourism. In a ritual, the mechanical element dominates since the decision to perform or not to perform has already (permanently) been taken. The decision to engage in tourism, on the other hand, has to be taken again and again and, sometimes, potential tourists decide to stay at home. Tourism is undertaken by individuals who have *not* taken a permanent dedsion to change place; if they do, they either become second-homers or migrants. Except for 'second-home tourists' (perhaps), tourism is not mechanical and, therefore, not a ritual.

Solidarity and Tourism

According to sodologists, there are two basic processes in sodety: cooperation and competition (Davis1981; Maciver and Page 1950). But this perspective misses out on what is more important for sodety out of these two processes. Competition leads to conflict, which may take the form of aggressive behaviour, such as physical violence and war. War has been known to sodety in historical times, so much so that many anthropologists and historian had come to view the history of humankind as a history laced with conflict (Montagu 1961a). Darwin's attributed statement 'survival of the fittest', which implied conflict to the uninitiated, was reviewed by physical anthropologists following neo-Darwinism, and its modified meaning was that this implies survival of individuals with the fittest genetic (and, by extension, physical) traits through greater procreation of those who were superior (Hammond 1978). Anthropologists differed from Thomas Hobbes *(Leviathan),* who viewed tribal and primordial humans (whom he described as humans in'a state of nature') as possessing 'no arts; no letters; no sodety; and which is worst of aR continual fear and danger of violent death; and the life of man solitary, poor, nasty, brutish, and short'. Archaeological anthropologists established that this was a condescending view of preliterate and primordial humans (see, e.g., Burkitt 1963) and that they lived almost as peaceful a life as modern humans, or perhaps had an even more peaceful life than some modern people (see also Hammond 1978).

From this we conclude that the most essential process for human society to survive is cooperation and not competition, since, if indeed competition and its attendant process of conflict were dominant, then sodety would have come to an end long ago. Tourism is also a cooperative process since, if it were a dominantly competitive or conflict process, international tourist arrivals would not have come dose to touching the 900-millionmarkin2007 (eTurbo News 2008). Domestic tourism in many countries has also risen phenomenally, so tourism does have to do with world sodety as a whole. According to the UNWTO's estimates (various years), domestic tourism all over the world is greater than international tourism in volume. In India, too, domestic tourism is far more voluminous than international tourism (Singh 2003). The logical conclusion is that if tourism is increasing by leaps and bounds, there must be a sodological side to tourism: it must be a cooperative process. Looking at it practically, if tourism suppliers and governments did not cooperate, there would be no tourism, *ceteris paribus.*

Societies sustain themselves through what classical sodologist Emile Durkheim called sodal solidarity. Sodeties survive through cooperation and the means of sustaining sodety by creating sodal organization or sodal structure is solidarity.Solidarity is the soul of sodety, just as religion is the soul of culture.Sodeties differ in the extent to which they have access to natural resources and may be rich or poor, technologically advanced or not, but they all share the process of social solidarity.In fact, the Latin phrase *vox populi* means that 'the voice of the people is the voice of God': this is not a mere metaphor, as so many thriving democrades show. Solidarity is of two types: simple and complex. Simple and complex solidarity operate at different levels: at the community, area or regional level (simple) or at national and internationallevels (complex). International tourism appears to be a form of complex solidarity, while domestic tourism appears to be a form of either simple or complex solidarity. These forms of tourism attempt to integrate human sodety at various levels and accompany globalization-through-other means, such as television, radio, motion pictures, transnational corporations, etcetera.

Tourism creates new sodal worlds in the minds of tourists or potential tourists and the hosts. When people meet, they exchange ideas, knowledge, and culture, which is why Nash (1996) called tourism a form of acculturation. But tourism has also been called a form of imperialism (Nash 1977) and neo-colonialism (Kobasic1996). This implies that tourism is a form of capitalism. However, monetary or 'personal capitalism' is divisive, whereas most forms of tourism do not divide sodeties but integrate them. This is the most compelling explanation of domestic tourism, which is thriving in most countries. International tourism, in fact, also being a cooperative process, seeks to unify human sodety. Monetary capitalism is a subset of sodal-cultural capitalism and the two go together.Tourism must, therefore, be a form of sodal and cultural capitalism as well Tourism educates and helps in sodalization and enculturation of dtizens of the world, and does this through providing recreation in new spatial settings in order to 're-create' the world. It creates new sub-cultures (also known as tourist cultures) as well. So we can define tourism afresh in the following manner.

> *Tourism is a primarily creative, cooperative process - a form of social, cultural, and monetary capitalism - that helps in socialization, enculturation, education, and recreation of people in a globalizing world. It is a driver of social, cultural, economic, and ecological change that leads to simple and complex social solidarity, domestically and internationally, through the process of continual mobility.*

Rituals help in sodalization and enculturation, too, but they do not educate - or, at least, do so in a minimal sense - and they do not provide recreation since they are done mechanically and in a fixed manner: tasks that have to be performed just because they ward off difficulties and ameliorate the condition concerned.When people traveL they are not warding off difficulties, but rather *find* many difficulties like renewing passports, getting a visa and life insurance, making reservations for transport and accommodation, reaching airports, railway or bus stations on time, carrying luggage or finding a porter, finding restaurants, and so on. It is important to recall the anthropological concepts of play, recreation, and leisure provided by Singh (2007b).'Play can be defined as a universal form of interaction between the individual and the sodal and the physicalenvironment in such a way that it promotes understanding of the sodal milieu and the physical world, while providing recreation to the partidpants'. On this count, tourism considered as play, is a form of sodalization and is not a ritual since secular rituals do not mediate between the individual and the sodal *and* the physical world.

Ritual is not play since it does not provide recreation, which latter 'is a universal form of mental and physical activity that helps in re-creating the world *as a particular culture sees it,* leading to happiness in ways that a society suggests' (Singh 2007b:72, emphasis in original). Rituals, surely, do not create happiness or euphoria: merely ringing bells in a temple does not provide happiness. It is the *ambience* of a temple that provides happiness, and that is a spiritual thing, a function of a mind that believes that one has arrived at a place of happiness and euphoria through ardent prayer. Of course, societies suggest rituals for happiness as well: pilgrimages are also rituals, often. But different sodeties suggest different types of rituals to find happiness, whereas tourism is a form of recreation that is common to all sodeties and cultures; peasants and tribal people also engage in VFR tourism and forest tourism (including huntingfor sport).So ritualism is bound to particular cultures, whereas tourism goes beyond sodal boundaries.

And finally, as far as tourism as a form of leisure is concerned: Singh (2007b: 72) defines leisure as 'free time where recreation becomes an end in itself', which is true of tourism as a 'spirited activity'. In a ritual, there is very little 'spirited activity', and it is usually mechanical; and recreation is not an end in itself, as far as leisure time taken up by rituals is concerned. So we conclude that tourism is not a ritual but a this-worldly activity that touches the souls of various sodeties: creating an integrated'global society with a glocal [global as well as local] culture' (Singh 2007a).

Discussion: Revisiting Some Concepts

Over the years, many scholars had got stuck in a vacuum of concepts regarding tourism and pilgrimage. The works of the Turners (e.g., Turner and Turner 1978) were no doubt seminal and important for the genesis of the anthropological or sociological perspective on these phenomena, but they should have been a starting point for analysis rather than being the end-all. Graburn (1983) followed this line of thought, thinking of tourism and pilgrimage as rites of inversion and rites of passage, as well as tre familiar concept of *communitas* and *societas* that scholars have resorted to, all too frequently. Rites of passage or sacred rituals are once-in-a-lifetime events and, as not discussed by these authors, mediate between three worlds at once. Most tourism, on the other hand, is not a once-in-a-lifetime event. Graburn (1983) spoke of sodal structure and'anti-structure' which has been critiqued by Singh (2003).It is an example of thinking in terms of binary opposites. Moreover, many authors have not fully explained the function of *communitas* and *societas*.It is plain to see that *communitas* and *societas* are nothing more than the kinds of euphoria that people of a community or sodety feel and their basis is spirituality, as induced by different religions at different points of time and at different scales.

Communitas and *societas* are instruments for social integration or for generating simple and complex sodal solidarity, respectively, between participants not only when the persons concerned are engaged in pilgrimage and tourism, but at other times, too, whenever people gather, such as during mass in a church among Christians, or during something akin to mass in Hindu sects, known as *satsang* (literally,'good and true company'), and during 'prayer time' or *namaaz* among Muslims. However, this does not always lead to euphoria during mass, *satsang* or *namaaz*.That is why Guru Nanak, the founder of Sikhism, criticized Muslims and Hindus who engaged in prayer but were not thinking of spiritual goals: the ritual of prayer became more important and was done mechanically, without elevating the act to its spiritual heights (Puri 2000). In fact, we can go one step further and theorize that all major religions seek to expand beyond any one community or society, but face the paradox of barriers like ethnicity and nationalism. That is why tourism becomes more important than pilgrimage in modem times, since it transcends boundaries and crosses barriers.

Singh (2004a) had concluded that the difference between tourism and pilgrimage is that the former usually involves mediation between the social and the natural worlds, while the latter involves a double mediation - between the social and the natural, and the natural and the supernatural. The 'supernatural'is not a monolithic whole, but is divided into ritual elements, the linguistic and emotive elements (mythologies), and the spiritual. This is as regards the spiritual aspect of pilgrimage as different from the ritual element, since it was also noted that merely going to a temple is not pilgrimage. Thus, the importance that society attaches to the spiritual and the ritual sides of life resolves into two dimensions: the sacred and the profane, which are not mutually exclusive, but mediated by the secular and the mundane. Profanity is sometimes used during pilgrimages (Singh2004a; see also Evans-Pritchard 1965; Jha 1991), since there is, in many societies, in the minds of pilgrims, a tension between the sacred and the profane. The pressure on individuals (ritual intensification: see Jha 1991) is so great that profanity is resorted to, to relieve that pressure (cf. Montagu 1961b). This does not happen in tourism.

As opposed to Singh (2004a), who said that only pilgrimage entails a double mediation between the sodal and the natural, and the natural and the supernatural, it is here suggested that tourism also involves a double mediation: between the social and the natural, and the natural and the spiritual.Sense of place has not merely to do with the sodal and spatial aspects of life, but also the spiritual: sometimes some places become sacred not because they were traditionally so, but because of some famous recent event or person that generations remember-such as Graceland (US rock star Elvis Presley's residence) or Woodstock (the site of one of the largest music concerts, also in USA) or Liverpool, UK (home of the British band The Beatles) (see *Tourism Recreation Research* 2003special issue on 'Sacred Journeys'). Music is a vehicle of spirituality (that is why Sufi saints frequently sing and play musical instruments) and a generator of tourism. Like spirituality and tourism, music is also a universal of culture and society. All societies have music and it is cultures that create music. Music is an attraction for most tourists, which is not surprising since it is only human societies (unlike ants and bees that also have societies (Davis 1981)) that have music. [It would be subjective to say which forms of music are more spiritualistic, but the author is of the opinion that'acid rock' (appreciated by drug addicts after taking drugs) is not spiritual at all.] As in music, which suggests ways of social integration, tourism is a means of constructing and maintaining social structure and it does that by creating sodal solidarity. A primarily creative, cooperative social process, tourism is a way that society remembers ideals that often get lost in ethnicity and nationalism: that societies have values that are paramount and far greater than mere ritual.

A tourist is a person who seeks new meanings in his / her own and other cultures and these meanings are not sought in order to please the gods and goddesses, but so that we understand what humanity means. That is why guides and interpreters are essential in tourism. And that is why the journey may not always be 'sacred' *ala* Grabum (1977,1989), because that would place the action in opposition to the profane, which is also this-worldly but relevant *only to particular cultures.*

The journey is the quest of the human spirit for oneness beyond the goals and purposes of communities and societies, what Nobel Laureate poet Rabindranath Tagore called beyond 'narrow domestic walls'. Tourism seeks new societies, new horizons, and the quest goes beyond the spirituality of any one culture. Tourists are un-realizing followers of a universal religion, what poet-mystic Tagore called 'The Religion of Man' (1931). This religion that Tagore speaks of was exemplified by the Bauls of Bengal in India, who were wandering musicians and spiritualistic, being neither Hindu nor Muslim nor Christian. Some are more steeped in such religion, some are less. Indeed, to recall the oft-used phrase, tourismis a force for peace, but rituals cannot be a force *qua* force, for peace. Tourism is, perhaps, a'spiritual force' for peace. It is a celebration of human society and human values in all their variety. Tourists are people 'romancing the world' (cf.Singh 2002b), its beautiful places and people. It is the same kind of romanticism or spiritualism thatisfoundin William Wordsworth's poems, such as the untitled one that follows.

'My heart leaps up when I behold
A rainbow in the sky;
So was it when my life began, So is it now I am a man,
So be it when I shall grow old
Or let me die!
The Child is father of the Man: And I could wish my days to be
Bound each to each with natural piety.'

Tourism: A Special Kind of Spirituality

Inthe end, it is necessary tore-understand spirituality and how it relates to tourism, in order to address the questions posed in the beginning.It has been clarified that both tourism and pilgrimage are similar in that they involve a double mediation.It was not clear how spirituality can be a part of the supernatural for both the phenomena of tourism and pilgrimage.In pilgrimage, the supernatural is represented by mythologies and the ritual is the journey and the activities undertaken at the centre of pilgrimage. The spiritual aspect of pilgrimage concerns the merit of undertaking the journey and the behaviour at that centre, but that is not necessary. For example, the Hindu concept of pilgrimage does not always stress the importance of undertaking the journey, but can be done sitting at home ('merely beholding the sight of the Himalayas is sufficient': see Kaur 1985; Singh 2004a), which probably implies meditation of a sort. On the other hand, in tourism the movement is essential, hence the confusion between tourism as a ritual and as a spiritual activity. This is why, despite so many advances in computer technology, cyber tourism is still not popular, since the sights and sounds of a destination and its local people can be recorded, but not the smells -like the scent of pine trees or local cooking or spice; or the feel of snow; or the genuine taste of local food; or the ambience of sitting in a restaurant in another location; or the contrast of lack of sunshine in the winters in your country and the warm or hot sun shining down upon you in a tropical or sub-tropical or equatorial country or region. Pilgrims are not motivated by such factors: they brave all sorts of difficulties to experience and live the journey and pray at the destination for material religious merit (Hindu *punya,* Islamic *sawaab)* and spiritual religious merit (e.g., Hindu *moksha,* freedom from the cycle of birth and death, and *trikuti;* spiritual salvation or God's grace among the Christians; attaining spiritual status *or Jannat,* or becoming a *Haji* among Muslims; *nibbana* and *parinibbana* among the Buddhists).

So it may be surmised that pilgrimage can be utilized to attain spiritual status, whereas tourism is an activity that does not result in spiritual status and is a never-ending activity: one may have travelled to 10 countries, but that does not mean that one cannot or should not travel to five more. A similar case can be made for domestic tourism: if you happen to live in a large country, a lifetime is not enough to discover the diversity. And that is one of the crucial differences: tourists seek unity in diversity and tourism is a celebration of unity in diversity of cultures, peoples, places - human values in all their variety, as pointed out earlier.

So what kind of spirituality is this? It certainly is not the kind of spirituality that religions teach or explain, except for the fact that all major religions promote brotherhood and peace, irrespective of how ritualistic followers interpret them and create conflict.Is tourism'a new kind of religion'? Inthe view of this author, tourism cannot become a religion and is not like religion. In religions, whatever may be argued, indoctrination is important. In tourism, there is no indoctrination, although there are codes of ethics.

Tourism is spirituality of a different kind, the kind that has to do with both the material world,including Nature ('a new kind of animism') (hence the confusion amongst many scholars about tourism as a ritual and a spiritual activity) and the non-material world (non-material culture, like values, beliefs, faiths, norms, knowledge, technology (as different from the products of technology), laws, languages, music, dance, drama, philosophies).In pilgrimage, the non material world relates only to the supernatural such as God and god-like personages, including saints, and a hope to achieve aims arrived at or understood by something akin to indoctrination. In tourism, the supernatural (more correctly, the spiritual) relates to God through representations like local people, their culture, and the un-understood - or perhaps partly understood- aim of uniting sodety. It is a spedal kind of spirituality: one that is directed by universal human feelings for (and of) solidarity in spite of cultural, ethnic, linguistic, religious, legal, political, geographical, environmental, psychological, monetary, and other obstacles.

In this way, the metaphorical and metaphysical relationship between tourist 'philosophies' and the philosophies of religions can be understood, since all major religions try to explain the same thing (that tourism frequently achieves, without an explicit philosophy): humanity is the same, and the apparent differences are, in a way,illusory. The ontology of tourists and tourism is derived from this: we are all the same presdent beings-some rich, some poor,some middling-and despite our differences, we all derive from the same background: the human race (what Jews, Christians and Muslims call Adam and Eve, and Hindus call *Manu)* and human culture in its totality.

Conclusion: The Economics of Spirituality and Tourism

It may well be asked: what are the practical implications of this study? How does it lead to better understanding of the sodal and tourist world, and hence, provide guidance regarding the economic aspects, which cannot be ruled out in any philosophy of tourism? While a detailed study is outside the scope of this paper, a few things may be noted as pointers towards the direction of future studies.

First, tourists are not ordinary consumers and do not just 'consume places' like any hedonistic group of sodety. They do not earn money from their wandering and do not set out to engage in tourism just because it may lead to better work later at the workplace or better income when they come back home. Their recreation and its economic aspects (such as increased productivity) is a corollary, not a causal factor leading towards a supposed end (better life or better income). When tourists buy local handicrafts (touristy or not), they do not just buy an indigenous product. The product, if original, may be costly and not all tourists can afford that. Most buy souvenirs or touristic art, which is often considered inferior (by those commenting on the nature of touristic transactions, such as anthropologists and sociologists). To restate the obvious, tourists buy souvenirs and take photographs in order to remember, re-create and relive the experience they had when away. They remember also the ambience and 'local colour' of the places visited and the people at the destination (mostly locals, but also outsiders), respectively. This is something that all humans share: remembering and reliving moments of happiness, or nostalgia. This has therapeutic value, and hence a form of spirituality. Its economic aspect is the increased happiness that leads to better health, what authors on 'wellness tourism' study ('Health is wealth'). Its economic value is recognized and it is thought of as something deserving targeted marketing ('nostalgic tourism', Russell 2008).

Second, tourists' recreation is unlike other forms of recreation: one can remember a motion picture but not enjoy it as much, seeing it again and again. In contrast, if a tourist enjoys a destination experience, he or she is likely to visit it again and again. This is so because a holiday and touristic experienceis more of a sodalexperience than going to a movie, or playing cricket or dominos or chess.Going to a restaurant in your own dty is not as enjoyable as going to a restaurant when on a tour. This also has an economic aspect: sodal interaction away from home leads to better social interaction when back home ('Sodal health is also wealth'). Thus, tourism often leads to better individual health and better social health and the two cannot be separated: sodal wellness is a reflection of wellness of the individuals who compose the sodety, just as sodal wellness leads to individual wellness. (Why is it that the USA and UK and Germany generate so much tourism and are among the leading economies of the world?) Both together lead to increased productivity and economic health: something that has not been given a name by economistsstudying tourism, but may well be called the 'economic health multiplier'.

Third, touristic spirituality leads to spiritual wealth: one reason why religious tourism and pilgrimage, as well as volunteer tourism and pro-poor tourism, are perhaps both increasing and being increasingly studied. This last leads to the concluding explanation of spirituality and tourism: tourism cannot always deal with loneliness (one can be on a group tour and lonely amongst a group or crowd), but can reduce the loneliness and create solitude (one can be alone on a tour and well-wrapped against the 'cold' by warm memories), like spirituality. This again leads to the conclusion that tourism involves spirituality of the kind mentioned earlier. Moreover, this has economic value since it may increase productivity, not only in terms of work, but also creativity. So, tourism can be a way of converting loneliness into solitude, something that has all of sodal, cultural, psychological, and economic ramifications. It is, thus, a complete sodal process and hence spiritual and holistic. As Wordsworth says in his poem, 'The Daffodils' (emphases added):

'I wander'd *lonely* as a cloud... When all at once I
saw a crowd: A host of golden daffodils...
I gazed - and *gazed* - but little thought What
wealth the show to me had brought... For oft,
when on my couch I lie...
They flash upon that inward eye
Which is the bliss of *solitude.*
And then my heart with *pleasure* fills...'

This explains the difference between loneliness and solitude, which tourism brings about; the tourist gaze; and the meaning of pleasure in tourism. Tourism, therefore, is human sodety discovering and rediscovering itself, and remembering places that may not exist in the future. To conclude with a metaphor, tourism is spiritual activity characterized by human mobility, that is 'Society Remembering Itself': something that economic historians should study, since times remembered are times recreated, and the future of tourism studies (and perhaps tourism) lies in understanding the past.

References

ARON, R. (1970). *Main Currents in Sociological Thought* (Part 2). Durkheim, Pareto, Weber. Harmondsworth, UK. Penguin.

ARYA, U. (1978). Introduction. In O'Brien, J. (Authored) *Yoga and Christianity*. Honesdale, Pennsylvania. Himalayan International Institute of Yoga Science and Philosophy.

BEAmE, J. (1985). *Other Cultures*. London. Cohen and West.

BHARATI, A. (1991). Grammatical and notational models of Indian pilgrimage. In Jha, M. (Ed) *Social Anthropology of Pilgrimage*. New Delhi. Inter- India Publications: 19–29.

BHARDWAJ, S. M. (1973). *Hindu Places of Pilgrimage in India*. Berkeley. University of California Press.

BONIFACE, P. (1995). *Managing Quality Cultural Tourism*. London. Routledge.

BROWN, J. (1980). The Question of 'Mysticism' Within Native American Traditions. In Woods, R. (Ed) *Understanding Mysticism*. New York. Doubleday: 203–211.

BUCKLEY, R. (2003). *Case Studies in Ecotourism*. Wallingford, UK. CABI Publishing.

BULLOCK. A., STALLYBRASS, 0. and TROMBLEY, S. (Eds) (1988). *The Fontana Dictionary of Modern Thought* (Revised edition). London. Fontana Press.

BURKITT, M.C. (1963). *The Old Stone Age: A Study of Palaeolithic Times* (Indian edition). New Delhi. Rupa and Co.

BUTLER, R. (1980). The Concept of a Tourism Area Cycle of Evolution: Implications for Management of Resources. *Canadian Geographer* 24: 5–12.

COHEN, E. (1979). A Phenomenology of Tourist Experiences. *Sociology* 13: 179-201.

COHEN, E. (2004). Tourism and Religion: A Comparative Perspective. In Cohen, E. (Authored) *Contemporary Tourism: Diversity and Change*. Oxford. Elsevier: 147–158.

COHEN, E. (2006). Pai - A Backpacker Enclave in Transition. *Tourism Recreation Research* 31(3): 11-27.

DAVIS, K. (1981). Human Society. New Delhi. Surjeet Publications in collaboration with the University of California Press.

eTURBO NEWS (2008). International Tourism Touches All-time High. Available at http://www.tourismexecutives.com Accessed on 28 May 2008.

EVANS-PRITCHARD, E.E. (1956). *Nuer Religion*. Oxford. Clarendon Press.

EVANS-PRITCHARD, E.E. (1965). Some Collective Expressions of Obscenity in Africa. In Evans-Pritchard, E.E. (Authored) *The Position of Women in Primitive Societies and Other Essays in Social Anthropology*. London. Faber and Faber: 76–101.

FENNELL, D. (1999). *Ecotourism: An Introduction*. London. Routledge.

FENNELL, D. and MALLOY, D. (2007). *Codes of Ethics in Tourism: Practice, Theory, Synthesis*. Clevedon, UK. Channel View Publications.

GALBRAITH, J.K. (1952). *American Capitalism: The Concept of Countervailing Power*. New York. The New American Library.

GALBRAITH, J.K. (1967). *The New Industrial State*. Boston. Houghton Mifflin.

GERTH, H.H. and MILLS, C.W. (1958). *From Max Weber: Essays in Sociology*. New York. Galaxy.

GIBSON, H. and YIANNAKIS, A. (2002). Tourist Roles: Needs and Life Course. *Annals of Tourism Research* 29(2): 358–383.

GRABURN, N.H.H. (1977). Tourism: The Sacred Journey. In Smith, V. (Ed) *Hosts and Guests: The Anthropology of Tourism*. Philadelphia. University of Pennsylvania Press.

GRABURN, N.H.H. (1983). Editorial: The Anthropology of Tourism. *Annals of Tourism Research* 10(1): 9–33.

GRABURN, N.H.H. (1989). Tourism: The Sacred Journey. In Smith, V. (Ed) *Hosts and Guests: The Anthropology of Tourism* (Second edition). Philadelphia. University of Pennsylvania Press.

HAMMOND, P. (1978). *The Human Species*. Englewood Cliffs, NJ. Prentice-Hall.

HARPER, R. (1965). *The Seventh Solitude: Man's Isolation in Kierkegaard, Dostoevsky and Nietzsche*. Baltimore, MD. John Hopkins Press.

HICKS, J.R. (1959). *Essays in World Economics*. Oxford. Oxford University Press.

HOWARD, M.C. and KING, J.E. (Eds) (1976). *The Economics of Marx: Selected Readings of Exposition and Criticism*. Harmondsworth, UK. Penguin.

JHA, M. (Ed) (1991). *Social Anthropology of Pilgrimage*. New Delhi. Inter-India Publications.

KAUR, J. (1985). *Himalayan Pilgrimages and the New Tourism*. New Delhi. Himalayan Books.

KOBASIC, A. (1996). Level and Dissemination of Academic Findings About Tourism. *Turizam* 44(7&8): 169–181.

LEACH, E.R. (1977). *Levi-Strauss. Fontana Modem Masters Series. Glasgow*. William Collins.

LOCKYER, T. and RYAN, C. (2007). Visiting Friends and Visiting Relatives - Distinguishing Between the Two Groups: The Case of Hamilton, New Zealand. *Tourism Recreation Research* 32(1): 59–68.

MACCANNELL, D. (1976). *The Tourist: A New Theory of the Leisure Class*. New York. Schocken Books.

MACIVER, R. and PAGE, C. (1950). *Society: An Introductory Analysis*. Macmillan. London.

MADAN, T. and MAJUMDAR, D.N. (1966). *Introduction to Social Anthropology*. New Delhi. Vikas Publications.

MAIR, L. (1984). *An Introduction to Social Anthropology*. London. Faber and Faber.

MAOZ, D. (2006). Erikson on the Tour. *Tourism Recreation Research* 31(3): 55–64.

MARX, K. (1973). *Grundrisse: Foundations of the Critique of Political Economy*. Nicolaus, M. (translator). The Pelican Marx Library. Harmondsworth, UK. Penguin.

MCLELLAN, D. (1975). *Marx. Fontana Modern Masters Series*. Glasgow. William Collins.

MIDDLETON, V.T.C. and HAWKINS, R. (1998). *Sustainable Tourism: A Marketing Perspective*. Oxford. Butterworth-Heinemann.

MILLS, C.W. (1963). *The Marxists*. Harmondsworth, UK. Penguin.

MOGGRIDGE, D.E. (1976). *Keynes. Fontana Modern Masters Series*. Glasgow. William Collins.

MONTAGU, A. (1961a). The Nature of War and the Myth of Nature. In Montagu, A. (Authored) *Man in Process*. New York. New American Library: 67–84.

MONTAGU, A. (1961b). On the Physiology and Psychology of Swearing. In Montagu, A. (Authored) *Man in Process*. New York. New American Library: 200–221.

NOY, C. (2006). Israeli Backpacking Since the 1%0s: Institutionalization and its Effects. *Tourism Recreation Research* 31(3): 39–53.

NASH, D. (1977). Tourism as a Form of Imperialism. In Smith, V. (Ed) *Hosts and Guests: The Anthropology of Tourism. Philadelphia*. University of Pennsylvania Press: 33–47.

NASH, D. (1984). The Ritualization of Tourism: Comment on Grabum's 'The Anthropology of Tourism'. *Annals of Tourism Research* 11(3): 503–505.

NASH, D. (1996). *Anthropology of Tourism*. Oxford. Pergamon/ Elsevier Science.

NICHOLSON, R. (1980). Sufis: The Mystics of Islam. In Woods, R. (Ed) *Understanding Mysticism*. New York. Doubleday: 179–191.

O'BRIEN, J. (1978). *Yoga and Christianity*. Honesdale, Pennsylvania. Himalayan International Institute of Yoga Science and Philosophy.

PARSONS, T. (1964). *The Structure of Social Action*. Glencoe, ILL. The Free Press.

PEARCE, D.G. (1987). *Tourism Today: A Geographical Analysis*. London. Pitman.

PEARCE, P. L. (2006). Backpacking and Backpackers - A Fresh Look. *Tourism Recreation Research* 31(3): 5–10.

PEARCE, P. L. (2008). Tourism and Entertainment: Boundaries and Connections (Editorial). *Tourism Recreation Research* 33(2): 125–130.

PLOG, S.C. (1972). Why Destination Areas Rise and Fall in Popularity. Paper presented to the Southern California chapter of the Travel Research Association, Los Angeles. Mimeographed.

POLANYI, K., ARENSBERG, C.M. and PEARSON, H.W. (1957). *Trade and Market in the Early Empires*. London. Routledge.

PURI, J.R. (2000). *Guru Nanak: His Mystic Teachings*. New Delhi. Radha Soami Satsang Beas.

RADHA SOAMI SAISANG BEAS (2005). *Spiritual Link* 1(7): 34–37. New Delhi Science of the Soul Research Centre.

RAJ, R. and MORPETH, N.D. (Eds) (2007). *Religious Tourism and Pilgrimage Management: An International Perspective*. Wallingford, Oxon. CAB International.

RUSSELL, D. (2008). Nostalgic Tourism. *Journal of Travel and Tourism Marketing* 25(2): 103–116.

SAMUELSON, P. (1982). *Economics: An Introductory Analysis* (12th edition). Princeton, NJ. Prentice-Hall.

SHACKLEY, M. (2001). *Managing Sacred Sites: Service Provision and Visitor Experience*. Continuum. London.

SINCLAIR, M.T. (Ed) (1999). *Gender, Work and Tourism*. London. Routledge.

SINGH, S. (2002a). Biodiversity and Ecotourism with Special Reference to India. *Tourism Recreation Research* 27(3): 92–95.

SINGH, S. (2002b). Love, Anthropology and Tourism. *Annals of Tourism Research* 29: 261–264.

SINGH, S. (2003). Travel and Aspects of Societal Structure: A Comparison of India and the United States. *Current Issues in Tourism* 6(3): 209–234.

SINGH, S. (2004a). Religion, Heritage and Travel: Case References from the Indian Himalayas. *Current Issues in Tourism* 7(1): 44–65.

SINGH, S. (2004b). *Shades of Green: Ecotourism for Sustainability*. New Delhi. TERI Press.

SINGH, S. (2007a). Is There Such a Thing as 'Mass Tourism'? *Tourism Recreation Research* 32(1): 107–111.

SINGH, S. (2007b). Conceptualizing Leisure, Recreation and Tourism from an Anthropological Perspective. *Tourism Recreation Research* 32(2): 67–74.

SWAMI RAMA (1975). *Meditation in Christianity*. Honesdale, Pennsylvania. Himalayan International Institute of Yoga Science and Philosophy.

TAGORE, R. (1931). *The Religion of Man*. London. George Allen and Unwin.

TAMBIAH, S.J. (1979). A Performance Approach to Ritual. *Man* (Journal of the Royal Anthropological Institute) 65: 113–169.

TIMOTHY, D. and OLSEN, D. (Eds) (2006). *Tourism, Religion and Spiritual Journeys*. London. Routledge.

TURNER, L. and ASH, J. (1975). *The Golden Hordes: International Tourism and the Pleasure Periphery*. London. Constable.

TURNER, V. and TURNER, E. (1978). *Image and Pilgrimage in Christian Culture*. New York. Columbia University Press.

VUKONIC, B. (1996). *Tourism and Religion*. Oxford. Pergamon / Elsevier Science.

WEAVER, D. (2007). Towards Sustainable Mass Tourism: Paradigm Shift or Paradigm Nudge? *Tourism Recreation Research* 32(3): 65–69.

WEBER, M. (1958). *The Protestant Ethic and the Spirit of Capitalism*. New York. Scribner.

WOODS, R. (Ed) (1980). *Understanding Mysticism*. New York. Doubleday.

Submitted: December 15, 2008

Accepted: May 15, 2009

Pilgrimage and Tourism Anthropological Perspectives

Sagar Singh and J. Anthony Paredes

Social Anthropology of Pilgrimage

Edited by Makhan Jha

The anthropological study of pilgrimage has taken on an ••c.1rncstnes·hitherto lacking, despite an overall decline in field-work in the past two decades. Coeval with this development has been an increasing interest in tourism among scholars previously indifferent to its study. The result has been an on- going discussion on the relation between the two.

But even as the study of pilgrimage has begun to carve out its own niche, to mark its own domain in anthropology, there are no indications as yet that it will form a sub-discipline or that it is in any position to. With input from different disciplines and various perspectives from within anthropology, the stage is all set for commensality, indeed a cross-fertilisation between the study of pilgrimage and tourism, on the one hand, and the two taken together with the rest of the discipline. It is a good augury that such a development has not spelt a severance of interests from the general body of anthropological thought. Where the study of tourism has gained a surer footing in an alliance with the study of pilgrimage, contrariwise, it is not surprising to find more than passing mention of tourism in books such as the *Social Anthmpology of Pilgrimage.*

Such a correspondence ofinterestand analysis was foreshadowed and given substance in the seminal work of the Turners. Their formulation, most quoted and utilised in subsequent analyses, as well as singly the most criticised, has not been fully comprehended even in otherwise sophisticated treatments. Eade (1991) and Cohen (1988) come closest to understanding it, but the former loses Turner in branding his discussion a "discourse *about* lsic] pilgrimage" (1991:5) and considering it, with Cohen, a contribution to the *sociology* of pilgrimage and tourism (!). This is no place to get involved in a deqatc on the 'Turnerian tradition' (for a fullerdiscussiorl see Singh, 1992).Suffice it is to say that *communitas* is just not a catchword; it is a concept and more-- an ideal typical construct *and* a value, to boot. It is a value that a Catholic Christian holds and an anthropologist, *pace* Nadel, who has 'intellectually assimilated' Catholic Christian culture upholds even in the profession (cf. Turner, 1977:61–62, 70). And while it is true that the concept at times hides more than it reveals, like 'peasantry', a better alternative is hardly forthcoming in the literature to suggest that the concept be retired. In any case, any contribution to the literature today is open to criticism as .regards its value, as a contribution to our understanding. Where landmark studies are few and far between, the subject is more averse to detractions and constantly seeks self-justification.

In looking at the book, the first question that confronts us is: Why the 'Social Anthropology of Pilgrimage'? It begs the question, to begin with, Can there be any Physical /Linguistic/ Archaeological anthropological perspective(s) on pilgrimage, and, if so, has there been any attempt to present it? The question is neither naive nor rhetorical. The significance of 'ascent' and the importance of mountains to various types of pilgrimage in diverse socio-cultural and temporal settings may be more than just symbolic (cf. Cohen, 1992:34, Hebrew *aliya* means both pilgrimage and ascent; the Mahabharata tale of the 'ascent' of the Panda vas, through long and arduous treks in the mountains, to heaven: *Swargarohana;* Evans-Pritchard, 1956:2, on Nuer religious concepts, '... for though the sky is not God, and though God is everywhere, he is thought of as being particularly in the sky...'). The physiological changes and their effect on mental states that physical ascent (say in the highlands of South America or in the Himalayas) involves is a subject not unworthy of attention. but little studied by physical/medical anthropology. A similar case may be made out for the archaeological and linguistic perspectives.

A second question one may ask relates to the use of the possessive preposition in the title. In the 'Social Anthropology of North American Tribes', 'The Anthropology of Religion' or 'The Anthropology of Evil', the reference as either to a physical I culturally demarcable area/people(s); that an adjectival usage could not be rendered without mishap; or that it presented a competi ive viewpoint to that of sociology. Sociology has been chastised (Sarana, 1983:26; Levi-Strauss, 1966: 113) for'creating' its own 'sub-disciplines' by conveniently adding its label through the use of the same, without the necessary withal of a perspective. The usage in this case raises no eyebrows if the emphasis is on the distinctive perspective that anthropology brings to the field; unfortunate if suggestive of the putative independent status of the study of pilgrimage.

The answer may simply lie in the fact that the book is a product of an international symposium held at Zagreb, Yugoslavia, under the aegis of the 12th International Congress of Anthropological and Ethnological Sciences in July, 1988, and hence takes its name after it.

The book does not raise, much less answer, the questions framed above. The editor, in his prefatory note, steers clear of making any statement as to direction, consensus or grey areas that the articles indicate. Articles, termed 'chapters', twenty in all, are somewhat arbitrarily divided into two distinct if continuous sections on the basis of whether they deal with Hinduism (in India and 'Hindu' North America) or Christianity-in such diverse forms as in Mexico, Peru, Costa Rica, Greece and the USA. No reason is forwarded for this division.

The task of summarising the diverse viewpoints presented or capable of articulation falls on Professor Agehanand Bharati of Syracuse University, New York. It is unsettling to note that Bharati's contribution, entitled 'Grammatical and Notational Models of Indian Pilgrimage', appears to be an unrevised version or draft. Aside from typographical errors ('pathological' repeatedly instead of 'patological'), the article is unnecessarilydismissive of competing interpretations and conceptions in the anthropology of religion. Scholars who have influ nced the course of such study are treated likewise, and the whole presentation shot through with references, allusions and comparisons that frequently seem neither here nor there.

To be fair to Bharati, he appreciates the viewpoints and analyses of Fritz Staal (1986) and S.J. Tambiah (1979), which, congruent with his own, are presented quite forcefully. But for the reader's ability to follow the rather difficult logic with which the conclusions are arrived at, everything appears hoity-toity. The conclusions lead us to consider the 'model' which "fits rather elegantly" (p.22) and reconsider the premises, something which he says he has been advocating all along these many years (p.20, 24). Bharati·s views can be summed up ina manner sympathetic to his own thus (where= means 'the same as or equals', → implication, ↔ mutual implication):

(i) idiography =(textual) study of particulars; Indology = textual study of Indian culture, society, language;
:. Indology = idiography;

[More precisely, in mathematical functional terms, where TS stands for Textual Study, Indology = TS (things Indian)]

(ii) nomothetics → (deductive) generalisations;
science → nomothetics;
Social Anthropology= science
Social Anthropology → (deductive) generalisations
(iii) Indian= Hindu;
Indian pilgrimage = Hindu pilgrimage ≠ any type of pilgrimage;
(iv) pilgrimage = circuitous locomotion of individuals, home base to destination(s) and back;
tourism= ditto;

Conclusions:

(a) pilgrimage = tourism;
(b) study of tourism-pilgrimage= study of such locomotion+ kinesics= patology
(c) patology → science → deductive generalisatiqns;
(d) deductive generalisations ↔ model(s)

Bharati would have it that in understanding pilgrimages as'ritual systems'we divorce all meanings from the actions, and study only the latter.

Despite the title, we are presented with a single model. And can a model substitute for theory, especially where it is known that whether it 'fits well' or not, it has limited explanatory value?

Kroeber, a twentieth century heavyweight in anthropology, once remarked that Indian anthropology has listened to British and American anthropology and, wisely, to herself. One is not sure whether such insulation has not resulted in paucity of concepts. In the study of pilgrimage at least the tradition of studying the 'sacred complex' of pilgrimage centres has held anthropologists in a vice. Once useful, the concept is now bedraggled, with no sign of a fresh impetus from within. Most of the contributions to the book by Indian anthropologists, including the editor's 'History, Patronage and Social Organisation of a Sakta Pilgrimage Centre' (Assam) and 'Studying the Sacred Complex of Kathmandu', fall in this tradition. The areas covered spread across

India, ranging from Arunachal Pradesh (Parsuram Kund, chapter 16) and Assam (Kamakhya temple, ch. 2) in the North-East to Puri (ch.12) in the East, from Badrinath and other centres in the North (Ch. 11, 15) to Chhotanagpur (ch. 6, 13) in the heart of India, and down south to Andhra Pradesh and Tamil Nadu (ch. 8 and 9 on Tirumala-Tirupati Devasthanam and Srisailam in Andhra, and Madurai/Sabarimalai in Tamil Nadu-Kerala, ch. 3).

Chapters three and five are distinctive contributions by Professors Lars Kjaerholm and Edward J. Jay on the cult of Aiyappan in Tamil Nadu and a *panch kosi yatm* in Chhattisgarh in Madhya Pradesh. Kjaerholm skilfullyusesdatagathered through fieldwork and from secondary sources to present his thesis that the cult of Aiyappan in Tamil Nadu, as different from Kerala and Karnataka, has grown out of the worship of the Tamil rural deity of Aiyanar. Illustrating the mythological linkages of Aiyappan diagrammatically, he presents pilgrimage as essential to but well integrated with the rest of religious beliefs surrounding the Aiyappan cult. Whereas the pilgrimage as described could have easily been analysed with Turnerian concepts, and certainly Van Gennep's, Kjaerhnlm eschews any such discussion.

Jay compares and contrasts Chhattisgarhi *panch kosi yatm* of Rajim with the well known *panch kosi* of Kashi, and in the process engages in a dialogue with Bharadwaj (1973). Analysing data obtained through fieldwork, he raises the important point whether visits to temples and their *parikrama* may not be considered comparable, If at different scales, with pilgrimages and the circumambulatory movement of pilgrims around the sacred zone. Prof. S.M. Bhardwaj discusses Hindu pilgrimage in North America and comes up with some interesting conclusions.

Contributions by Leslie Ellen Straub, Rene Gothoni, D.J. Davies and N. Ross Crumrine (ch. 17, 18, 19 and 20) also examine the definition of pilgrimage in the light of data from Costa Rica, Athos (Greece), Mormon culture (USA), and Mexico and Peru. Straub takes up pilgrimages at the national, regional and local level in Costa Rica. He finds the concept of 'replication' (of culture patterns), first suggested by Evon Vogt, to be singularly suitable to explaining the process of reproduction of the image of Our Lady of the Angels in Costa Rican pilgrimages. Straub holds that both "structural or conceptual replication" and such emic distinctions as *romeria/ perigrinacion* may prove heuristically productive and deserve attention (p. 289).

Gothoni has a close look at 'The Progress of Pilgrimage on the Holy Mountain of Athos'. Consulting field notes and some statistic he takes issue with one of the Eastern Churches' contentions that an increase in visitor influx necessarily means that there are more tourists than pilgrims coming to Athos (p.301). Stressing the effect of strenuous walking, the ambience and an ethos of discipline on the visitor's experience, Gothoni also gains an insight into youth tourism to Athas.

Davies's terse definition of pilgrimage as 'corporate soteriological movement' gains more in brevity and comprehensiveness than it loses in the usc of 'corporate'. Maintaining that emic analysis is specially fruitful when "there is a perceived distance between ernie categories and etic analysis" (p.311), Davies journeys into the world of Mormon pilgrimage. The question whether Mormon visitors from the U.S. to U.K. shrines will be considered pilgrims or tourists is resolved in the term "tourist pilgrims" (p.323). He prescribes a study of each "cultural case", giving it due "ideological, historical and social" weightage (p.324). A longish article by N. Ross Crumrine, taking up pilgrimage among the Mayos of North-West Mexico and in Catacaos, Peru, defines pilgrimage as the "dramatic mediation of two worlds or opposed poles throught the ritual process of reduction of the opposition"(p.365).

In sum: while two contributions (Sabita Acharya on Puri; S. Vijaya Kumar & M. Suryanarayana on Srisailam) indicate a slight break from the Indian tradition, is a mute but felt need today; it is apparent that the time is not yet ripe for any sharp break from the Turnerian tradition in studying Christian pilgrimage. Such is the strength of the concepts that Turner brought to the discussion that all such attempts risk throwing the baby out with the bathwater.Indeed, contributors to the book who offer a new approach and claim to disagree with Turner are cognizant of the fact. To be sure, new traditions must build upon earlier ones.Such a growing away only may, perhaps, help the subject and benefit scholars. While the get-up of the bonk needs to do just that, it presents a limited cross-section of work in progress across the world on pilgrimage and should prove a useful addition to the library.

References

BHARDWAJ, S.M. 1973. *Hindu Places of Pilgrimage in India*. Berkeley CA: University of California Press.

COHEN, E. 1988. Traditions in the Qualitative Sociology of Tourism. *Annals of Tourism Research* 15(1): 29–46.

COHEN, E. 1992. Pilgrimage Centres: Concentric and Excentric. *Annals of Tourism Research* 19(1): 33–50.

EADE, J. 1991. Introduction and Order and Power at Lourdes. In J. Eade & M.L. Sallnow, eds. *Contesting the Sacred: The Anthropology of Christian Pilgrimage*. London: Routledge.

EVANS-PRITCHARD, E.E. 1956. *Nun· Religion*. Oxford: Clarendon Press.

LÉVI-STRAUSS, C. 1966. The Scope of Anthropology. *Current Anthropology* 7:112–23.

SARAN A, G. 1983. Sociology and Anthropology. In *Sociology and Anthmpology and Other Essays.* Calcutta: Institute of Social Research and Applied Anthropology.

SINGH, SAGAR 1992. The Integrative Functions of Pilgrimage Traditions and Religious Tourism in India. Paper presented at the Seminar on Wildlife, Environment and Tourism, Ayodhya, October, 1992.

STAAL, F.J. 1986. The Sounds of Religion. *Numcn* XXXIII/1 /2:33–65, 185–244.

TAMBIAH, S.J. 1979. A Performance Approach to Ritual. *Man* 65:113–69.

TURNER, V.W.1977. Process, System and Symbol: A New Anthropological Synthesis. In *Discoveries and Interpretations: Studies in Contemporary Scholarship,* Vol. 1, *Daedalus,* Summer 1977:61–80.

In an introductory note, the editors, Julia Horne and David Walker, inform readers that the volume was based on a conference held at the University of New South Wales in June 1990. The work consists of twelve papers individually authored by seven historians, an art historian, a psychologist, an anthropologist, and two writers. Most of the articles are concerned specifically with Australia, but many have general relevance for understanding touristic phenomena.

The volume begins with a very brief but bitingly insightful essay by historian/ museologist Donald Horne. He assays various approaches to museums and historic preservation in Australia for creating tourist attractions and a sense of national past. His remarks, however, could well apply to other countries. To wit: "...any preservation of the material culture of the past is *never* authentic--because it lasted into a different age from the time when is was produced and it is being used for different purposes". Likewise, he sensitively comments on the problem of the presentation of aboriginal culture as both ethnography and art yet not neglecting the living native people themselves, while striving for the "best form of cultural tourism". Given current concerns with ethnic and cultural diversity in many places, Horne cautions against exhibitions that go in for "a sledgehammer didactic presentation, imposing a uniform idea of diversity".

Anthropologist Malcolm Crick likewise offers an evocative essay on "otherness" and tourism, which should be read by all who either promote or study tourism anywhere. No one is left unscathed in Crick's penetrating commentary. Considering even anthropologists as "a species of tourist," he notes that "it was colonial power that made the world safe for ethnography, just as nowadays entrepreneurs make the world safe for tourism". Rich with irony, Crick's analysis crackles with such observations as "most tourists cannot stand too much otherness", and "if tourism is about understanding between cultures, then tourists at the very least should endeavour to understand why so many locals are antagonistic towards them and how their 'cheap' holiday is only the reverse side of the poverty of their hosts".

Complementing Crick's essay is that by historian M.N. Pearson drawing on the works of Victor Turner and Edward Said. Pearson reminds us that such distinctions as traveJier, tourist, and pilgrim are not always cleanly made, that the pilgrim to Mecca will buy souvenirs and the country-and-western music tourist to Nashville "will have a lump in the throat". Weaving his analysis between the observations of an eighteenth-century Indian Muslim traveller to England, Pearson reveals some of the faults in Turner's sometimes too facile and too European theory of pilgrimage, even while making good use of such concepts as the liminality of travellers. His insights are often arresting, as in his observation that "the contemptuous attitude of the elite of mass tourism" is because "now the masses can do what their 'superiors' have been doing for a century or so". Finally, following Said, Pearson symbolically links tourist souvenir-buying to animai trophies and the looting of other countries' antiquities by imperial Britain and France.

The remainder of the essays are tied firmly to specifically Australian themes and vary considerably in their generalisability, though all are to some degree interesting in their own right. Meaghan Morris writes of the changing symbolism of Sydney Tower as national metaphor and offer the playful insight that in-country visitors to the Tower become part of the scene to be enjoyed by the foreign tourist. Psychologist Kevin Murray presents an informal report on the results of his conversations with 72 Melbourne citizens about their travel experiences and how these reia tc to one's sense of self and social position through what he calls "moral topography". In a nicely paired couple of articles, historians Roselyn Pesman Cooper and Alan Mayne write, respectively, about Australian travellers to Naples, Italy, in the period 1870-1930, and the account of an Italian missionary to Australia in the early 1870's; both reveal the smug superiority of the travellers' social and cultural perspectives of their time and place.

Art historian Sue Rowley offers a charming analysis of the way in which the structure of gender relations is modelled by metaphors of space, geography, and journey in nineteenth-century Australian literature and painting. Similarly, historians Julia Horne and David Goodman in their respective studies of portrayals of Australia's Blue Mountains and travellers' narratives of the Gold Rush plumb the intricacies of Australian national culture and character and their linkages to western civilisation. Along the way, both authors have things to say which have wider appeal, as when Goodman remarks that "Crocodile Dundee provides only one of lhe most recent proofs of the existence of a mass audience for stories of colonial divergence from metropolitan norms" or when Horne describes how with the beginnings of industrialism, "from being in- hospitable environments, many mountains became over-crowded tourist attractions".

Historian Ann McGrath's essay ostcnsibl y is about "the outback" as the "pivotal cultural symbol of Australia and Australianness" which "is central to national mythology". At another level, however, her analysis can be seen as almost an alle ory of western European colonialism, treatment of indigenous penples, and exploitation of natural resources everywhere, and of latter-day efforts to justify and make culturally palatable all that has gone before by touristic pilgrimages to the outback (or its metaphoric equivalent)--even by aboriginal peoples themselves. The parallels with the United States, its "old west", and American Indians are especially striking.

As with any collection of this sort, the writing and substance vary in quality and quantity, but on the whole all arc readable and informative. Nonetheless, the non Australian reader is sometimes left feeling somewhat like an outsider (or at least embarrassingly ignorant) by some of the references to Australian history and popular culture; some of the articles would have been enhanced by inclusion of a map showing places discussed. A number of the pieces are marred by self-conscious indulgence in being theoretically trendy (especially in the postmodernist mode) and not 'letting the data speak for themselves', as it were. Nothing so quickly dates otherwise useful information as dressing it up in the latest intellectual fashions of the moment.

The collection closes with Anna Gibbs' disquieting short-story end inin a poem.It's about airplanes, road travel, computer terminals, relationships, television, answering machines, tourists, cliches, urban de1nolition, going home, scandals, mass culture, and alternative life-styles. And it is very powerful. But rather obscure.

J. Anthony Paredes

Eudaimonic Benefits of Tourism: The Pilgrimage Experience

Marc van Iwaarden and Jeroen Nawijn

ABSTRACT

This paper proposes that tourism research needs to explore the potentially lasting psychological benefits of tourism by implementing a eudaimonic perspective of well-being. Consequently, this study adopted a eudaimonic approach to tourist well-being by investigating the reported psychological well-being of pilgrims and the concepts involved in the pilgrimage experience via the exploration of 74 personal narratives. Pilgrims' self-transcendent peak experiences encompassed spiritual moments of revelation, healing and epiphany. These experiences led to spiritual growth that seems to have impacted participants' religious spirituality, also long after their pilgrimage, which can be perceived as a potential long-term effect of the pilgrimage experience. The self-transcendent peak experience of the pilgrims can be explained via enhanced autonomy, the experience of positive relationships, achieved personal growth, a sense of purpose in and of life, their character development and a state of self-acceptance.

Introduction

During recent decades the relationship between tourism experiences and wellbeing has received a lot of attention, especially in recent years (e.g. Chen et al., 2013; Morgan et al., 2015). Whereas well-being is usually conceptualized as comprising both a hedonic and a eudaimonic perspective, tourism researchers have specifically directed significant interest toward hedonic and subjective well-being research (Lin & Nawijn, 2019), which is important for understanding tourist behaviour and destination marketing (Jabreel et al., 2017).

The study of hedonic subjective well-being in tourism experiences has taught us that hedonic happiness experienced during holiday trips does not last long after returning home (Nawijn, 2011). During holiday trips, tourists feel better than they do in their everyday lives (Nawijn et al., 2013). Novelty is an important determinant of these feelings (Mitas & Bastiaansen, 2018), as are social aspects (Mccabe, 2002) and a sense of autonomy (Cini et al., 2013). Furthermore, certain activities can influence feelings during tourism experiences, such as photography (Gillet et al., 2016). Tourists' personality explains some differences in how and when feelings are experienced during vacation (Lin et al., 2014).

Fewer studies have considered the eudaimonic aspect of well-being in a tourism context; a one-sided understanding of tourist well-being can be problematic (De Bloom et al., 2017; Filep & Laing, 2019). In Filep's words: 'there is room for less emotional and more eudaimonic perspectives in tourism research' (Filep, 2016, p. 3). This calls for re-evaluation and a swing back on the pendulum from hedonism to eudaimonia is slowly being answered in recent years. Matteucci and Filep (2017) identified eudaimonic characteristics within tourists' dance and music experiences and made a case for this tourism context as being powerful in bringing fulfilment and transformation to tourists. Li and Chan (2017) identified six eudaimonic themes within a diaspora tourism context, revealing that tourists' perceived benefits of home return travel are highly eudaimonic in nature. Kler and Tribe (2012) made a case for the flourishing of scuba divers by identifying themes of learning and personal growth.

Although this earlier work on eudaimonic well-being and tourism experiences has been insightful, it has failed to make a strong argument for the necessity of studying the lasting psychological impacts of travel in general. Most studies only assessed the effects of specific tourist activities and adopted a short-term perspective. This is surprising as, according to Kay Smith and Diekmann (2017), eudaimonic tourism experiences should allow tourists to experience more long-term, transformational benefits. This would mean that an understanding of the principles at work within eudaimonic experiences could further clarify tourism's role in the sustainable social and psychological development of society (Filep, 2014; Filep & Pearce, 2013).

This study contributes to the idea of positive tourism, postulated by Filep et al. (2016) and explores potential long-term eudaimonic effects in a pilgrimage context.

Pilgrimage is known for its meaning-making potential and transformational power (Schnell & Pali, 2013). One study found positive associations between the pilgrim experience and subjective well-being (Maheshwari & Singh, 2009), but eudaimonia is left undiscussed in academic work that considers these meaningful journeys. In this paper, pilgrimage is used as a context in which to assess the relevance of a eudaimonic approach to tourism well-being research. This study thus aims to explore how a eudaimonic approach to tourist well-being can aid our understanding of lasting psychological health benefits for tourists.

Literature review

Conceptualizing well-being

Well-being can be defined as optimal experience and psychological functioning (Diener et al., 2009; Ryan & Deci, 2001). Nowadays, most scholars agree that approaches to well-being can be grouped into two categories: hedonic well-being and eudaimonic well-being (Diener et al., 2009; Ryan & Deci, 2001). Hedonistic views regard well-being as being constituted by pleasure and happiness. The eudaimonic view, in turn, considers the actualization and optimization of human functioning and potentials (Ryan & Deci, 2001). This means that a person's well-being can be examined by assessing the extent to which they live in accordance with their true self. The process of reaching this true self (i.e. self-actualization), however, is difficult and, at times, at odds with short-term happiness (Waterman, 1984).

Well-being can be further conceptualized by looking at the concepts of subjective well-being and psychological well-being. Subjective well-being adopts the hedonist viewpoint by considering subjective pleasures and satisfaction (Diener et al., 2003), whereas psychological well-being reflects that of eudaimonia as it addresses the subject's human actualization (Ryan & Deci, 2001; Ryff, 1989).

Eudaimonia in a pilgrimage context

Pilgrimage is a journey that is both external as well as internal to the pilgrim. One travels physically but also experiences a shift in the self as they engage on a quest for spiritual meaning and truth (Courtney, 2015; Digance, 2006). Reader (2007) reasoned that pilgrimage might aid people in their quest for meaning. Schnell and Pali (2013) found that people experience increased meaning in life, both during and after their pilgrimage and get a taste of pilgrimage's transformative potential. Pilgrimage is looked at as a way of finding oneself, and especially in one's realization of the need for self-transformation, pilgrimage seems to be offering a helping hand (Coleman & Eade, 2004; Devereux & Carnegie, 2006; Schmidt, 2009).

Literature suggests that '[a]n attitude of openness to unusual experiences (spiritual, transcendental, peak, mystical) may be conducive to health and well-being' (Dierendonck & Mohan, 2006, p. 230). Pilgrimage can be considered as one such unusual experience that sparks meaning and transformation within the individual. Maheshwari and Singh (2009, p. 290) studied pilgrims in India and concluded that '[p]ilgrims seek happiness and religious blessings and separate themselves from everyday worldly concerns. They spend time in the presence of God in a place of special significance and meaning, which gives them the desired happiness and satisfaction'. Furthermore, Privette and Brundrick (1991) found that pilgrimage plays a positive role in decreasing anxiety and depression.

Religious spirituality. Pilgrimage is considered a form of religious tourism; many pilgrims journey each year because of their religiosity. Within the practice of religion, however, there is a distinction to be made between the act of going to a church, for example, and deciding to go on a pilgrimage. Pilgrims tend to intentionally seek a deeper personal and subjective experience of their relationship with God, making their journey relate to the spiritual also.

At times there seems to be a struggle between organized religion and personal spirituality (Zinnbauer et al., 1999). The first is structured, cold and mostly external. The second is fluid, experiential and personal. According to Reader (2007, p. 226),

> one of the appeals of pilgrimage is that it provides a means through which persons can engage in spiritual practices that help them deal with the modern world and find meaning in their lives without having to be committed to organized religious traditions.

As Reader (2007) suggests here, pilgrimage specifically seems to be an experience in which a more personal spirituality is treasured and sought after. Because the study context of this paper focuses on pilgrimage as a personal and transcendent (peak) experience, yet within the boundaries of what can be considered a religion, the more accurate way to define these pilgrims is as people on a journey of religious spirituality. Within tourism, examples of non-religious spirituality also exist, examples being wellness and meditation retreats (Norman & Pokorny, 2017). This distinction between religious and non-religious spirituality is important, as spiritual experiences during wellness or meditation activities will differ significantly from those of religious pilgrims.

Research is available on spirituality and its association with human well-being (Hill & Pargament, 2008; Unterrainer et al., 2010). Koenig (1998) viewed spirituality as a determining factor of better mental health. Unterrainer et al. (2010) also suggested that spirituality plays an important role in people's psychological well-being, just like Van Dierendonck (2004) did, who opted for integration of spirituality in the psychological well-being construct. Several studies furthermore suggest that spirituality is associated with subjective well-being for adults (Fave et al., 2013; Yoon & Lee, 2004).

In a pilgrimage context, Maheshwari and Singh (2009) identified positive associations between religiosity as a social construct and life satisfaction and happiness. More research is needed, however, that assesses pilgrims' religious spirituality and the role it plays in their psychological well-being specifically.

Closely related to meaning and pilgrimage is self-transcendence: the forgotten capstone of Maslow's well-known hierarchy of needs. Even though the concept has never gained popularity,

> [i]ncluding the step of self-transcendence allows us to apply a long-standing motivational theory to some aspects of religious experience and to understand how spiritual and even mystical experiences can be associated positively with mental health and other personality variables. (Koltko-Rivera, 2006, p. 312)

The motive towards self-transcendence implies that individuals 'seek a benefit beyond the purely personal and seek communion with the transcendent, perhaps through mystical or transpersonal experiences; they come to identify with something greater than the purely individual self, often engaging in service to others'. And '[a]t the level of self-transcendence, the individual's own needs are put aside, to a great extent, in favour of service to others and to some higher force or cause conceived as being outside the personal self' (Koltko-Rivera, 2006, p. 306). In his paper, Koltko-Rivera outlined that Maslow's motivational theory is not complete with the need for self-actualization at the top of the hierarchy. After writing his book 'A Theory of Human Motivation' (1943), Maslow identified self-transcendence as a construct that is strongly related to self-actualization, and even succeeds it on the hierarchy of needs, making it the capstone of the hierarchy. Frankl (1985) argued that self-actualization is an automatic result for people who pursue self-transcendence, and that self-actualization cannot be obtained when seen as a goal in and of itself. '[t]he more one forgets himself – by giving himself to a cause to serve or another person to love – the more human he is and the more he actualizes himself' (Frankl, 1985, p. 133).

The later insights that Maslow had regarding self-transcendence did not receive much attention in his time and the fact that textbooks today do not include self-transcendence in the hierarchy of human motivation proves that the concept hasn't gained much acknowledgment since. However, the importance of Maslow's revision to his motivational hierarchy should not be underestimated.

> The earlier model positions the highest form of motivational development at the level of the well-adjusted, differentiated, and fulfilled individual self or ego. The later model places the highest form of human development at a transpersonal level, where the self/ego and its needs are transcended. (Koltko-Rivera, 2006, p. 306)

Koltko-Rivera stated that adoption of the construct of self-transcendence by the field of positive psychology could facilitate its implementation in more disciplinary psychology. 'Spirituality and transcendent experience are domains addressed by positive psychology. The construct of self-transcendence as a motivational status can thus build a bridge between positive psychology and a major theory of motivation' (2006, p. 312).

The pilgrimage experience is a suitable candidate for the exploration of self-transcendence in a positive psychological context. Pilgrimage as a spiritual experience is often seen as an act of pursuit of or devotion to God or a higher power (Pandya, 2015). Pilgrims have escaped their usual environment and its problems in order to be able to focus on something that transcends these worldly yet personal issues (Maheshwari & Singh, 2009). The pursuit for meaning and purpose in life, seen in pilgrims, has also been mentioned as a major factor of motivation for the behaviour of self-transcendence (Koltko-Rivera, 2006; Nygren et al., 2005). Finally, apart from experience itself, pilgrimage seems to bring about self-transcendence in other life domains also. In a study by Schnell and Pali (2013, p. 900), pilgrims reported that after their journey, '[t]hey are more willing to take responsibility for affairs beyond their immediate concerns, keen on living in unison with nature and in loving relationships with others'. This hints at lasting personal transformation, and these characteristics, therefore, make self-transcendence a noteworthy concept to consider within the positive pilgrimage experience.

A third construct that is relevant in the exploration of eudaimonia in a pilgrimage context is that of peak experience (Ingersoll, 1994; Noble, 1987). First introduced by Maslow (1962), a peak experience implies a highly valued and fulfilling moment (Maslow, 1964; Privette, 1983), 'characterized as a transformational experience and one that surpasses the usual level of intensity, meaningfulness, and richness' (Dodson, 1996, p. 3). Peak experiences are extraordinary, and their impact on one's life goes beyond that of normal experiences that surround it in time and space (Laski, 1961; Leach, 1962; Noble, 1987).

Self-transcendence is fundamental in peak experiences, and spirituality cannot be understood without talking about peak experiences (Ingersoll, 1994). During a peak experience, there is, as Privette (1983, p. 1364) noted, a sense of 'fusion with the world or [an] object and sub-sequent loss of self'. A peak experience is an encounter with the transcendent (Noble, 1987); however, not only is self-transcendence part of the moment of a peak experience, those that have these experiences more regularly seem to carry a self-transcending attitude into other dealings of life as well (Maslow, 1964). The relationship between peak experiences, self-transcendence and spirituality is reciprocal; all 'nourish' each other and cannot be described exclusively (Wuthnow, 1978).

As one could expect, peak experiences thus play a role in psychological health also. 'Maslow [...] contends that these experiences are characteristic of psychological health, particularly in the direction of self-actualization. [...] Such experiences are typically followed by therapeutic aftereffects or dramatic personal growth' (Klavetter & Mogar, 1967, p. 171). Other literature supports this notion by reporting that 'peakers' (1) find greater meaning and purpose in their lives (Arnould & Price, 1993; Wuthnow, 1978); (2) score higher on personal talents and capabilities (Hayes, 2013; Wuthnow, 1978); (3) have high self-confidence, autonomy and internal strength (Arnould & Price, 1993; Maslow, 1962; Wuthnow, 1978); (4) have a high sense of personal growth and self-discovery (Arnould & Price, 1993); and (5) experience a change in self-concept and identity (Privette & Brundrick, 1991). It can thus be argued that peak experiences play a role in aspects of psychological well-being and in self-actualization (Noble, 1987; Wuthnow, 1978).

Literature that approaches pilgrimage as a potential peak experience is lacking. Studies that discuss pilgrimage as a mystical, spiritual or transcending phenomenon are more common, yet, surprisingly, these studies fail to make the connection to the topic of positive psychology (Cohen, 2006).

To enable us to explore how a eudaimonic approach to tourist well-being can aid our understanding of lasting psychological health benefits for tourists, we first need to understand the process of well-being in a pilgrimage experience in more detail and explain this state of well-being. Consequently, the research questions for the study are (1): how is well-being achieved, or not, via a pilgrimage? And (2): which characteristics of a pilgrimage experience suggest psychological health benefits?

Methodology

Population and sampling

The population was defined as contemporary religious-spiritual pilgrims motivated by self-transcendence and personal growth. The convenience sample was drawn from individuals that partook in a discipleship training school (DTS): an initiative from the non-denominational Christian organization 'Youth with a Mission' (YWAM). Most of the respondents were relatively young, which corresponds with the population of those generally participating in a DTS.

Each year around 25,000 young pilgrims travel to one of more than 1,200 locations worldwide with the objective to learn 'to know God and make Him known' (YWAM, 2017). 'Meaningful', 'intense' and 'transformational' are terms commonly used to describe this experience. Both the physical and inner journey is to ignite the (re)shaping of a personal and spiritual foundation on which participants can build the rest of their lives.

Data collection took place in 2017 by means of an online survey among those who completed a DTS. Online surveys are time efficient and cost-efficient, and they often allow better access to unique populations (Wright, 2005). The questionnaire was distributed with a link shared on social media (mainly Facebook), webpages and in groups that were likely to attract interest from pilgrims. As such, the survey made use of a convenience sample of individuals who had a recent DTS experience.

Personal narratives

As the aim was to explore, we used qualitative data to achieve this. These qualitative data were, however, collected via a questionnaire. This questionnaire contained an open response box, which allowed respondents to leave extensive narrative accounts of their experiences. As such, we used the more restrictive interpretation of personal narratives as discussed by Riessman (2002). This means that respondents were triggered to provide a discrete story related to a single event brought about by a single question. The question that was used was: 'is there anything else you'd like to share on how you experienced your DTS in the greater context of life?' These personal narratives were used as data for this paper. We regarded the responses as restrictive personal narratives due to the use of the aforementioned single question and the fact that the stories related to a single event (i.e. the DTS) and were thus clearly situated in time and space (cf. Riessman, 2002). The remainder of the survey contained items tapping various psychological constructs. These are, however, not discussed in this paper due to their cross-sectional nature, and as such, they do not add to the qualitative data. Seventy-four respondents provided personal narratives. Their narratives varied between several words and paragraphs of more than 200 words.

This study was executed from the Netherlands, and the data were collected via a non-interventional survey. Such studies do not require approval under Dutch national law. Regardless, the main author had the research proposal for this study reviewed by a team of Wageningen University professors, who gave their approval for the study.

Data analysis

The qualitative data were analysed using thematic analysis. 'Thematic analysis is a method for identifying, analysing, and reporting patterns (themes) within data. It minimally organizes and describes your data set in (rich) detail' (Braun & Clarke, 2006, p. 6). Personal narratives are suited for thematic analysis (Riessman, 2002).

Congruence in coding was achieved by having the main author analyse the data with a theoretical, deductive approach. The theories and concepts, as discussed in the literature review, served as a guideline for the coding and themes. The themes were identified at a semantic level, meaning that 'themes are identified within the explicit or surface meanings of the data' (Braun & Clarke, 2006, p. 13). To be able to gain a more detailed and nuanced understanding of the stories and concepts discussed, primarily those themes that are related to these concepts, were given attention and emphasized in further analysis and reporting. The main author and co-author actively discussed the coding while it took place. Next to these discussions, the quantitative survey data was helpful in aiding the process of researcher reflexivity. It enabled us to look at the qualitative data in different ways and, combined with the ongoing discussions between the researchers, this allowed for a situation in which we put the respondents' voices central instead of the researchers' possible pre-conceived notions on the nature of hedonia and eudaimonia. This thus required and forced both authors to look at the data and the concepts and engage in discussion on how to interpret the data in line with earlier work and their own views on the data. Data saturation was eventually reached after the congruence in coding had been achieved, and themes were derived via these codes. Combined with the notion that personal narratives rely on how stories are told more so than what is told (Riessman, 2002). Researcher reflexivity – in our case the ongoing discussions combined with comparisons to respondents' own quantitative data, allowed us to eventually reach data saturation.

In terms of validity (cf. Whittemore et al., 2001), we assured descriptive and theoretical validity by our use of personal narratives, combined with quantitative data and use of existing literature. Interpretative validity was safeguarded by our use of researcher reflexivity as well as presenting earlier work on the data to peers and receiving their views and ideas, forcing us to reinterpret data at times. Evaluative validity may affect our findings to some extent. We purposely discussed a very broad view of well-being concepts and theories. Yet, when interpreting the codes and themes in relation to these, we do engage in interpretation and provide potential explanations which may or may not be in line with the respondents' views. We thus acknowledge the possibility of a certain extent of bias in terms of evaluative validity. Generalizability is an issue for many qualitative studies as it is more typical for quantitative work. That also applies to this study, for which the sample is rather specific. We propose ideas for future work in the Discussion section of this paper.

Results

The first research question was: how is well-being achieved or not, via a pilgrimage? We discuss this process below, using three main concepts.

Peak experience. The data revealed that the reflections of pilgrims on their journey fit the characterization of a peak experience. Respondents described their experiences as transformational, intense, fulfilling and joyful. Pilgrims' journeys were commonly evaluated as being life-changing: '[i]t was totally God's guidance to lead me there and it gave fresh direction to my life and my calling got clearer. Life-changing experience' (male 1, 26). Respondents perceived themselves as coming out of the pilgrim experience differently. Most of the time these changes were reported as being positive in nature, referring to aspects of character, spirituality and/or outlooks on life. Participants thus seemed to consider their pilgrimage as a milestone in their journey of life. Yet, some also mentioned having more specific peak experiences within their pilgrimage. These peak experiences involved spiritual revelation, a sense of healing or epiphany, and are therefore, discussed as part of the religious spirituality concept.

Self-transcendence. The concept of self-transcendence plays a significant role in the journey that pilgrims experience. The focus of the pilgrim's journey is not on the actualization of the self, but rather meaning is found in the surrender and giving of the self to 'something' greater. Participants found that their pilgrimage gave them the opportunity to learn to fully devote themselves and their lives to the cause and beliefs of their religious spirituality. One of them evaluated his pilgrimage as being instrumental in the realization of self-transcendence as follows: 'I saw YWAM as an opportunity and something to get stuff out of myself. I realized that giving was the thing where I got most out of' (male 2, 27). Meaning and fulfilment thus were indirectly found by giving rather than receiving. This surrender of the self at the same time is not a destination or state that one reaches once one embarks on the pilgrimage; it is a disciplined path of learning and growing to put oneself out of the centre of attention, ideally with implications for the rest of one's life. Self-transcendence is thus a renewed way of thinking, and the pilgrim experience seems to be a well-suited place where one can learn to embrace and practice this. This progression of embracing through experience suggests that peak experiences facilitate and/or spark self-transcendence.

Religious spirituality. Religious spirituality was a prevalent theme in pilgrims' descriptions of their experiences. More specifically, spiritual growth was a frequently occurring theme. Spiritual growth in this context is an increasing awareness and experience of God. This increased awareness can come about in different ways, but the two most prevalent ways reported by this study's pilgrims seem to be a revelation (the receiving of spiritual insight and understanding) (also see Maslow, 1964) and epiphany (an appearance or manifestation especially of a divine being). As revelation and/or epiphany is experienced, spiritual growth is induced. 'I've seen His [i.e. God's] compassion and love and it's beautiful' (female 1, 23). The experience of seeing or perceiving the character of God sparks a new appreciation for it. This awareness is personal and is taken seriously. Because of its seemingly positive nature, it creates a sense of faith and trust in the unseen. 'It taught me more about having faith in God whether it's emotional, mental, physical, financial, and so much more' (female 2, 35).

As these pilgrims explored and discovered spirituality in their religious context and consequently experienced growth, their sense of connectedness with God increased through familiarity and trust. The growth of this trust is a personal spiritual experience, and it contradicts the more legalistic kind of religion that elevates law and formula above subjective experience (Reader, 2007). For this reason, the growth that is experienced also seems to reach into other aspects of the individual's psyche. For instance, significant mental and/or emotional healing was reported because of the spiritual experience.

> I was completely healed from clinical depression and eating disorder during DTS that I had struggled with for 7 years. I had seen multiple doctors and was put on several medications but nothing healed me quite like experiencing the Lord in such an impactful way. I also had a seizure disorder that doctors couldn't find the source of and therefore had a hard time treating, but I haven't had one since my DTS!! God is awesome!! (Female 3, 27)

Even though intentional efforts were put into getting better on different levels prior to the pilgrimage, eventual relief was found in engaging on a spiritual journey and consequently having peak experiences: 'experiencing the Lord in such an impactful way' and to 'encounter God'. As these peak experiences had reported physical, emotional and/or mental consequences, the impact of the peak experience on pilgrims' spiritual awareness and their connectedness seemed to have been reinforced. These positive experiences of healing can thus be considered a benefit of the peak experiences, but also experiences in themselves that are conducive to future spiritual growth and engagement. One's experience functions as a signpost pointing the way on the path of experiencing increasing spiritual connectedness with God. 'The experiences of my DTS were a constant reminder that God is real and He loves me, which is why I am now in a deep and intimate relationship with the living God' (female 4, 26).

Here, the interrelationship between peak experiences, self-transcendence and spirituality become clearer. Because of the self-transcendent peak experience with God, one's spirituality is awakened or brought to a new level to such an extent that the individual is inspired and determined to maintain and continue growing in the spiritual connection that one was introduced to because of the first experience. These changes are thus not ascribed to one's own efforts or skills, but rather to the increased awareness of and connectedness with God. The reader is reminded of the idea postulated earlier that self-actualization is truly cultivated as a by-product of self-transcendence.

Psychological well-being

The second research question was: which characteristics of a pilgrimage experience suggest psychological health benefits? The experiences that the participants went through led them to report experiencing greater magnitudes of psychological well-being, six components of psychological well-being were identified as sub-themes.

Firstly, autonomy was revealed mostly through respondents expressing growth in personal confidence and independence from others. One respondent said: '[i]t taught me a lot about being confident in your own personal beliefs and relationship with God; and trying to not compare your standing with others' (female 5, 20). Others indicated that their pilgrim journey helped them to be more independent Christians and become more confident in the identity they find in their religious spirituality.

Secondly, several respondents touched on the topic of positive relationships. They mention the formation of important relationships with friends: 'I still have relationships from my time there that are among the most important in my life' (female 6, 37), while others also mentioned laying the foundations for romantic relationships. In addition, pilgrims noted having gained new understandings of, and perspectives on, the nature of positive and meaningful relationships: 'discovering [...] the foundational value of authentic relationships' (male 4, 69).

Thirdly, personal growth was experienced in different ways. Some mentioned having learned things in general: 'I became a better me that is for sure' (male 5, 25), others mentioned having experienced growth in specific areas, such as language learning and social skills. An increase in self-knowledge was also mentioned regularly, but mostly in connection to the spiritual connectedness discussed in the previous section: 'I have learned more about myself and my gifts, but mostly about God and his provision for my life' (female 7, 40).

Fourthly, a sense of purpose in life was developed during the pilgrim experience. Participants seemed to find answers to the question 'why am I alive?', hence contributing to finding meaning. 'It [...] has made me see I'm made for so much more than what I could ever think. Bigger goals and dreams' (female 8, 22). Pilgrims experienced that they found a sense of direction and clarity as to what kind of person they want to be in life and what role they can play in society. Several respondents looked back on their pilgrimage as a launchpad into a meaningful life. 'It was a great foundation to my future perspective of my purpose here on earth' (male 6, 58).

Fifthly, respondents reported positive changes in their perception of self or self-acceptance. 'I discovered that I can do many more things than I thought I could. My picture of myself changed in a much more positive way' (male 7, 22). A common trend within this theme seemed to be that when pilgrims had spiritual experiences that involved receiving acceptance or love from God, this also helped them to establish and grow in self-acceptance. One respondent exemplified this, having experienced feelings of self-hatred. 'It made me see myself a little more how God sees me. Before [...] I hated myself; after I didn't and never went back to that way of thinking' (female 9, 23). For many, the pilgrim journey initiated a discovery of intrinsic values and their human potential.

Lastly, the thematic analysis revealed character development as an additional concept present in relation to pilgrims' eudaimonia. Pilgrims reported changes indicative of humanity: 'I love people so much more' or temperance: 'I am less apt to judge my fellow person' (female 4, 23). The virtues and character strengths that the data hinted at are open-mindednesses, love for learning, curiosity, bravery, zest, love, kindness, fairness, mercy, humility, hope and gratitude.

Discussion

Theoretically speaking, this paper affirms that the importance of studying the tourism experience and its potential contribution to optimal human functioning should not be underestimated. The identification of eudaimonic concepts will aid our understanding of what exactly is psychologically beneficial for an individual. Being aware of these benefits could positively influence the tourist as they plan and choose for travel experiences that are worth their money and will benefit them not only during their holidays but in the greater context of their lives. The tourism industry, on the other hand, will benefit from this better understanding in its efforts to develop and provide meaningful, authentic, and valuable tourism experiences (cf. Filep & Laing, 2019).

This study's attempt to identify and study some of these eudaimonic concepts demonstrated that the formation of self-transcendence and spirituality during pilgrimage was found to be a driving force behind better psychological well-being. The findings associated with the self-transcendence construct strengthen Koltko-Rivera's (2006) call for the significance of the construct for the field of positive psychology. There are some studies that discuss self-transcending types of tourism other than pilgrimage, such as volunteer tourism (Tukamushaba et al., 2017) or charity challenge tourism (Coghlan, 2015), but a greater understanding of the role of self-transcendence in tourist eudaimonia is still needed. Recent quantitative work suggests that eudaimonia is achieved in a physically challenging activity (Su et al., 2020). Our study suggests a challenging mental activity may have a similar effect.

Connections were observed between our qualitative data and the conceptualization of character by Peterson and Seligman (2004). Participants mentioned growth in different character strengths as a result of their pilgrimage experience, and all six virtues (humanity, wisdom and knowledge, courage, justice, temperance and transcendence) were represented in these personal narratives. Fowers (2012) reminds us of an interesting and important connection between eudaimonia and virtue, arguing that conceptualizations of the two should never be separated in the first place. 'From an Aristotelian perspective, virtues are simply the excellences that make it possible to live the good life, meaning that these two concepts mutually entail one another' (Fowers, 2012, p. 1). It also touches on the connection between hedonia and eudaimonia. Recent work suggests that some experience of hedonia is required to achieve eudaimonia (Lee & Jeong, 2020). Our study's findings seem to corroborate this link, as we observe that the peak experience includes the formation of positive relationships and the felt experience of several positive emotions, such as gratitude and hope.

Limitations and suggestions for future research

This study reported on qualitative findings. What is needed next are longitudinal quantitative data, to strengthen the theoretical contribution of future studies. Additional use of observations or data obtained from significant others (e.g. peers) would increase validity (Shoval et al., 2018). Furthermore, as this study was not longitudinal, and participants were asked to evaluate their pilgrim experiences post-pilgrimage – thus inquiring about a memory – the reported lasting effects are perceived as such by the pilgrim.

The nature of the sample is a last methodological consideration worth mentioning. The respondents in this study are assumed to score high on religious spirituality. The lack of non-spiritual respondents in this study thus limits the representativeness of the sample and consequently it is unclear to what extent the psychological concepts would act the same way in a less spiritual sample. Future research should thus address wider populations.

Furthermore, a pilgrimage can be viewed as a challenging mental activity. Recent work suggests physically challenging tourist activities are beneficial for eudaimonic well-being (Su et al., 2020). Future research should address mentally challenging tourism activities.

Finally, future studies could look at contexts other than pilgrimage, such as medical tourism (Heung et al., 2010), or volunteer tourism (Zahra & Mcintosh, 2007). Future work should also address contexts which contain religious elements or connotations, such as in the recent work on hotel wellness (Kim & Yang, 2021). This may allow us to gain a more in-depth understanding of all concepts at play in tourism experiences that potentially contribute to lasting psychological well-being. Having such an understanding is relevant to both the tourism industry and the tourist. The industry can benefit in terms of providing meaningful, authentic, and valuable tourism experiences, whereas the tourist may benefit in their journey of finding lasting psychological health and self-actualization.

Conclusion

Pilgrims' self-transcendent peak experiences encompassed spiritual moments of revelation, healing, and epiphany. These experiences led to spiritual growth that seems to have impacted participants' religious spirituality, also long after their pilgrimage, which can be perceived as a potentially long-term effect of the pilgrimage experience. The self-transcendent peak experience of the pilgrims can be explained via enhanced autonomy, experience of positive relationships, achieved personal growth, a sense of purpose in and of life, their character development, and a state of self-acceptance.

Disclosure statement

No potential conflict of interest was reported by the author(s).

References

Arnould, E. J., & Price, L. L. (1993). River magic: Extraordinary experience and the extended service encounter. *Journal of Consumer Research*, *20*(1), 24–45. https://doi.org/10.1086/209331

Braun, V., & Clarke, V. (2006). Using thematic analysis in psychology. *Qualitative Research in Psychology*, *3*(2), 77–101. https://doi.org/10.1191/1478088706qp063oa

Chen, Y., Lehto, X. Y., & Cai, L. (2013). Vacation and well-being: A study of Chinese tourists. *Annals of Tourism Research*, *42*, 284–310. https://doi.org/10.1016/j.annals.2013.02.003

Cini, F., Kruger, S., & Ellis, S. (2013). A model of intrinsic and extrinsic motivations on subjective well-being: The experience of overnight visitors to a national park. *Applied Research in Quality of Life*, *8*(1), 45–61. https://doi.org/10.1007/s11482-012-9173-y

Coghlan, A. (2015). Tourism and health: Using positive psychology principles to maximise participants' wellbeing outcomes – A design concept for charity challenge tourism. *Journal of Sustainable Tourism*, *23*(3), 382–400. https://doi.org/10.1080/09669582.2014.986489

Cohen, E. (2006). Religious tourism as an educational experience. In D. Timothy, & D. Olsen (Eds.), *Tourism, religion and spiritual journeys* (pp. 78–93). Routledge.

Coleman, S., & Eade, J. (2004). *Reframing pilgrimage: Cultures in motion*. Psychology Press.

Courtney, M. (2015). A review of the literature exploring the occupation of contemporary pilgrimage. *Journal of Occupational Science*, *22*(2), 170–182. https://doi.org/10.1080/14427591.2013.764816

De Bloom, J., Nawijn, J., Geurts, S., Kinnunen, U., & Korpela, K. (2017). Holiday travel staycations, and subjective well-being. *Journal of Sustainable Tourism*, *25*(4), 573–588. https://doi.org/10.1080/09669582.2016.1229323

Devereux, C. & Carnegie, E. (2006). Pilgrimage: Journeying Beyond Self. *Tourism Recreation Research*, *31*(1), 47–56. https://doi.org/10.1080/02508281.2006.11081246

Diener, E., Oishi, S., & Lucas, R. E. (2003). Personality, culture, and subjective well-being: Emotional and cognitive evaluations of life. *Annual Review of Psychology*, *54*(1), 403–425. https://doi.org/10.1146/annurev.psych.54.101601.145056

Diener, E., Wirtz, D., Biswas-Diener, R., Tov, W., Kim-Prieto, C., Choi, D.-W., & Oishi, S. (2009). Social indicators research series. *Assessing Well-Being*, *39*, 247–266. https://doi.org/10.1007/978-90-481-2354-4_12

Dierendonck, D. V., & Mohan, K. (2006). Some thoughts on spirituality and eudaimonic well-being. *Mental Health, Religion and Culture*, *9*(03), 227–238. https://doi.org/10.1080/13694670600615383

Digance, J. (2006). Religious and secular pilgrimage. In D. J. Timothy, & D. H. Olsen (Eds.), *Tourism, religion and spiritual journeys* (pp. 36–48). Routledge.

Dodson, K. J. (1996). Peak experiences and mountain biking: Incorporating the bike into the extended self. In K. P. Corfman, & J. G. Lynch (Eds.), *NA-Advances in consumer research* (pp. 317–322). Association for Consumer Research.

Fave, D. A., Brdar, I., Vella-Brodrick, D., & Wissing, M. P. (2013). Religion, spirituality, and well-being across nations: The eudaemonic and hedonic happiness investigation. In H. Knoop, & A. Delle Fave (Eds.), *Well-being and cultures* (pp. 117–134). Springer.

Filep, S. (2014). Moving beyond subjective well-being. *Journal of Hospitality & Tourism Research*, *38*(2), 266–274. https://doi.org/10.1177/1096348012436609

Filep, S. (2016). Tourism and positive psychology critique: Too emotional? *Annals of Tourism Research*, *59*, 113–115. https://doi.org/10.1016/j.annals.2016.04.004

Filep, S., & Laing, J. (2019). Trends and directions in tourism and positive psychology. *Journal of Travel Research*, *58*(3), 343–354. https://doi.org/10.1177/0047287518759227

Filep, S., Laing, J., & Csikszentmihalyi, M. (2016). What is positive tourism? Why do we need it? In S. Filep, J. Laing, & M. Csikszentmihalyi (Eds.), *Positive tourism* (pp. 17–30). Routledge.

Filep, S., & Pearce, P. (2013). *Tourist experience and fulfilment: Insights from positive Psychology (Vol. 31)*. Routledge.

Fowers, B. J. (2012). Placing virtue and the human good in psychology. *Journal of Theoretical and Philosophical Psychology*, *32*(1), 1–9. https://doi.org/10.1037/a0025819

Frankl, V. E. (1985). *Man's search for meaning*. Simon and Schuster.

Gillet, S., Schmitz, P., & Mitas, O. (2016). The snap-happy tourist. *Journal of Hospitality & Tourism Research*, *40*(1), 37–57. https://doi.org/10.1177/1096348013491606

Hayes, A. F. (2013). Model templates for process for SPSS and SAS. Retrieved December 12, 2013.

Heung, V. C. S., Kucukusta, D., & Song, H. (2010). A conceptual model of medical tourism: Implications for future research. *Journal of Travel & Tourism Marketing*, *27*(3), 236–251. https://doi.org/10.1080/10548401003744677

Hill, P. C., & Pargament, K. I. (2008). Advances in the conceptualization and measurement of religion and spirituality: Implications for physical and mental health research. *Psychology of Religion and Spirituality*, *S*(1), 3–17. https://doi.org/10.1037/1941-1022.S.1.3

Ingersoll, R. E. (1994). Spirituality, religion, and counseling: Dimensions and relationships. *Counseling and Values*, *38* (2), 98–111. https://doi.org/10.1002/j.2161-007X.1994.tb00827.x

Jabreel, M., Moreno, A., & Huertas, A. (2017). Semantic comparison of the emotional values communicated by destinations and tourists on social media. *Journal of Destination Marketing & Management*, *6*(3), 170–183. https://doi.org/10.1016/j.jdmm.2016.03.004

Kay Smith, M., & Diekmann, A. (2017). Tourism and wellbeing. *Annals of Tourism Research*, *66*, 1–13. https://doi.org/10.1016/j.annals.2017.05.006

Kim, B., & Yang, X. (2021). "I'm here for recovery": The eudaimonic wellness experiences at the Le Monastère Des Augustines wellness hotel. *Journal of Travel & Tourism Marketing*, 1–17. https://doi.org/10.1080/10548408.2021.1921093

Klavetter, R. E., & Mogar, R. E. (1967). Peak experiences: Investigation of their relationship to psychedelic therapy and self-actualization. *Journal of Humanistic Psychology*, *7* (2), 171–177. https://doi.org/10.1177/002216786700700206

Kler, B. K., & Tribe, J. (2012). Flourishing through scuba: Understanding the pursuit of dive experiences. *Tourism in Marine Environments*, *8*(1-2), 19–32. https://doi.org/10.3727/154427312X13262430524027

Koenig, H. G. (1998). *Handbook of religion and mental health*. Elsevier.

Koltko-Rivera, M. E. (2006). Rediscovering the later version of maslow's hierarchy of needs: Self-transcendence and opportunities for theory, research, and unification. *Review of General Psychology*, *10*(4), 302–317. https://doi.org/10.1037/1089-2680.10.4.302

Laski, M. (1961). *Ecstasy. A study of some secular and religious experiences*. Cresset Press.

Leach, D. (1962). *Meaning and correlates of peak-experience*. University Microfilms.

Lee, W., & Jeong, C. (2020). Beyond the correlation between tourist eudaimonic and hedonic experiences: Necessary condition analysis. *Current Issues in Tourism*, *23* (17), 2182–2194. https://doi.org/10.1080/13683500.2019.1611747

Li, T. E., & Chan, E. T. H. (2017). Diaspora tourism and well-being: A eudaimonic view. *Annals of Tourism Research*, *63*, 205–206. https://doi.org/10.1016/j.annals.2017.01.005

Lin, Y. K., & Nawijn, J. (2019). The impact of travel motivation on emotions: A longitudinal study. *Journal of Destination Marketing & Management*, *16*, 100363. http://doi.org/10.1016/j.jdmm.2019.05.006

Lin, Y., Kerstetter, D., Nawijn, J., & Mitas, O. (2014). Changes in emotions and their interactions with personality in a vacation context. *Tourism Management*, *40*, 416–424. https://doi.org/10.1016/j.tourman.2013.07.013

Maheshwari, S., & Singh, P. (2009). Psychological well-being and pilgrimage: Religiosity, happiness and life satisfaction of Ardh–Kumbh Mela pilgrims (Kalpvasis) at Prayag, India. *Asian Journal of Social Psychology*, *12*(4), 285–292. https://doi.org/10.1111/j.1467-839X.2009.01291.x

Maslow, A. H. (1943). A Theory of human motivation. *Psychological Review*, *50*(4), 370–396. https://doi.org/10.1037/h0054346

Maslow, A. H. (1962). *Toward a psychology of being*. Van Nostrand.

Maslow, A. H. (1964). *Religions, values, and peak-experiences (Vol. 35)*. Ohio State University Press.

Matteucci, X., & Filep, S. (2017). Eudaimonic tourist experiences: The case of flamenco. *Leisure Studies*, *36*(1), 39–52. https://doi.org/10.1080/02614367.2015.1085590

Mccabe, S. (2002). The tourist experience and everyday life. In G. Dann (Ed.), *The tourist as a metaphor of the social world* (pp. 61–75). CAB International.

Mitas, O., & Bastiaansen, M. (2018). Novelty: A mechanism of tourists' enjoyment. *Annals of Tourism Research*, *72*, 98–108. https://doi.org/10.1016/j.annals.2018.07.002

Morgan, N., Pritchard, A., & Sedgley, D. (2015). Social tourism and well-being in later life. *Annals of Tourism Research*, *52*, 1–15. https://doi.org/10.1016/j.annals.2015.02.015

Nawijn, J. (2011). Happiness through vacationing: Just a temporary boost or long-term benefits? *Journal of Happiness Studies*, *12*(4), 651–665. https://doi.org/10.1007/s10902-010-9221-y

Nawijn, J., Mitas, O., Lin, Y., & Kerstetter, D. (2013). How do we feel on vacation? A closer look at how emotions change over the course of a trip. *Journal of Travel Research*, *52*(2), 265–274. https://doi.org/10.1177/0047287512465961

Noble, K. D. (1987). Psychological health and the experience of transcendence. *The Counseling Psychologist*, *15*(4), 601–614. https://doi.org/10.1177/0011000087154004

Norman, A., & Pokorny, J. J. (2017). Meditation retreats: Spiritual tourism well-being interventions. *Tourism Management Perspectives*, *24*, 201–207. https://doi.org/10.1016/j.tmp.2017.07.012

Nygren, B., Aléx, L., Jonsén, E., Gustafson, Y., Norberg, A., & Lundman, B. (2005). Resilience, sense of coherence, purpose in life and self-transcendence in relation to perceived physical and mental health among the oldest Old. *Aging & Mental Health*, *9*(4), 354–362. https://doi.org/10.1080/1360500114415

Pandya, S. (2015). Pilgrimage and devotion to the divine mother: Mental well-being of devotees of mata vaishno devi. *Mental Health, Religion & Culture*, *18*(9), 726–737. https://doi.org/10.1080/13674676.2015.1112771

Peterson, C., & Seligman, M. E. (2004). *Character strengths and virtues: A handbook and classification (Vol. 1)*. Oxford University Press.

Privette, G. (1983). Peak experience, peak performance, and flow: A comparative analysis of positive human experiences. *Journal of Personality and Social Psychology*, *45*(6), 1361. https://doi.org/10.1037/0022-3514.45.6.1361

Privette, G., & Brundrick, C. M. (1991). Peak experience, peak performance, and flow: Correspondence of personal descriptions and theoretical constructs. *Journal of Social Behavior and Personality*, *6*(5), 169–188.

Reader, I. (2007). Pilgrimage growth in the modern world: Meanings and implications. *Religion*, *37*(3), 210–229. https://doi.org/10.1016/j.religion.2007.06.009

Riessman, C. (2002). Analysis of personal narratives. In J. F. Gubrium, & J. A. Holstein (Eds.), *Handbook of interview research: Context and method* (pp. 695–710). Sage Publications.

Ryan, R. M., & Deci, E. L. (2001). On happiness and human potentials: A review of research on hedonic and eudaimonic well-being. *Annual Review of Psychology*, *52*(1), 141–166. https://doi.org/10.1146/annurev.psych.52.1.141

Ryff, C. D. (1989). Happiness is everything, or is it? Explorations on the meaning of psychological well-being. *Journal of Personality and Social Psychology*, *57*(6), 1069–1081. https://doi.org/10.1037/0022-3514.57.6.1069

Schmidt, W. S. (2009). Transformative pilgrimage. *Journal of Spirituality in Mental Health*, *11*(1-2), 66–77. https://doi.org/10.1080/19349630902864242

Schnell, T., & Pali, S. (2013). Pilgrimage today: The meaning-making potential of ritual. *Mental Health, Religion & Culture*, *16*(9), 887–902. https://doi.org/10.1080/13674676.2013.766449

Shoval, N., Schvimer, Y., & Tamir, M. (2018). Real-time measurement of tourists' objective and subjective emotions in time and space. *Journal of Travel Research*, *57*(1), 3–16. https://doi.org/10.1177/0047287517691155

Su, L., Tang, B., & Nawijn, J. (2020). Eudaimonic and hedonic well-being pattern changes: Intensity and activity. *Annals of Tourism Research*, *84*, 1–14. https://doi.org/10.1016/j.annals.2020.103008

Tukamushaba, E., Xiao, H., & Ladkin, A. (2017). Self-transcendence: An alternative explanation for voluntourism in developing countries. *International Journal of Sustainable Society*, *9*(1), 61–75. https://doi.org/10.1504/IJSSOC.2017.085723

Unterrainer, H.-F., Ladenhauf, K. H., Moazedi, M., Wallner-Liebmann, S. J., & Fink, A. (2010). Dimensions of religious/spiritual well-being and their relation to personality and psychological well-being. *Personality and Individual Differences*, *49*(3), 192–197. https://doi.org/10.1016/j.paid.2010.03.032

Van Dierendonck, D. (2004). The construct validity of Ryff's scales of psychological well-being and its extension with spiritual well-being. *Personality and Individual Differences*, *36*(3), 629–643. https://doi.org/10.1016/S0191-8869(03)00122-3

Waterman, A. S. (1984). *The psychology of individualism*. Praeger Publishers.

Whittemore, R., Chase, S. K., & Mandle, C. L. (2001). Validity in qualitative research. *Qualitative Health Research*, *11* (4), 522–537. https://doi.org/10.1177/104973201129119299

Wright, K. B. (2005). Researching internet-based populations: Advantages and disadvantages of online survey research, online questionnaire authoring software packages, and Web survey services. *Journal of Computer-Mediated Communication*, *10*(3), 00–00. https://doi.org/10.1111/j.1083-6101.2005.tb00259.x

Wuthnow, R. (1978). Peak experiences: Some empirical tests. *Journal of Humanistic Psychology*, *18*(3), 59–76. https://doi.org/10.1177/002216787801800307

Yoon, D. P., & Lee, E.-K. O. (2004). Religiousness/spirituality and subjective well-being among rural elderly Whites, African Americans, and native Americans. *Journal of Human Behavior in the Social Environment*, *10*(1), 191–211. https://doi.org/10.1300/J137v10n01_05

Ywam. (2017). *Youth with a mission*. https://ywam.org/wp/

Zahra, A., & Mcintosh, A. J. (2007). Volunteer tourism: Evidence of cathartic tourist experiences. *Tourism Recreation Research*, *32* (1), 115–119. https://doi.org/10.1080/02508281.2007.11081530

Zinnbauer, B. J., Pargament, K. I., & Scott, A. B. (1999). The emerging meanings of religiousness and spirituality: Problems and prospects. *Journal of Personality*, *67*(6), 889–919. https://doi.org/10.1111/1467-6494.00077

Motivations of Urban Millennials for Spiritual Travel in India

Snigdha Kainthola, Nimit Chowdhary, Rahul Pratap Singh Kaurav and Pinaz Tiwari

ABSTRACT

The understanding of spirituality has come far from the traditional viewpoint of being a subset of religion. Consequently, contemporary spiritual tourism includes a vast realm of experiences to cater to heterogeneous spiritual tourists. Young travellers worldwide are increasingly attracted to spirituality. The research's principal objective is to determine the factors driving urban millennials in India to travel for spirituality. This objective was achieved through phenomenological analysis of their interviews which renders a meaningful understanding of their travel choices and behaviour. The push factors such as mental morbidities, corporate pressure, or curiosity shape the spiritual tourists' motivations, which are further manifested in the kind of activities they opt for during their journey. Therefore, the study identifies the prominent push factors and the motivations of the Indian Urban Millennial Spiritual Tourist (UMST). Further, the authors highlight the most desirable traits the spiritual tourists look for in a destination.

Introduction

The debate over spirituality and spiritual tourism is an old one. Traditionally, characterised as a phenomenon under leisure travel (Norman, 2012), spiritual tourism has expanded to include a range of experiences (Parsons et al., 2019). Several scholars have attempted to explain spirituality with precision (Norman, 2011; Reich, 2000; Roof, 1999; Schneiders, 1989; Sharpley, 2016; Willson et al., 2013; Zinnbauer et al., 1999) and in the process tried framing an intrinsic definition of spiritual tourism. However, the literary works on spiritual tourism sometimes lack the rigour in definitional austerity or are loosely framed to incorporate the residual traits to make it visibly inclusive. For several years, spirituality has been closely linked to religion, and the reflection is evident in the numerous works on the discussion (Benjamin & Looby, 1998; Kale, 2004; Pargament, 1999; Zinnbauer et al., 1997). However, contemporary work distinguishes the two. While spirituality is more focused on experiences, religion works to conceptualise that experience into a scheme systematically. Many people are increasingly being intrigued by relatively informal and less structured means of achieving spirituality (Kale, 2004; Zinnbauer et al., 1997). Tourism is identified as one such vehicle for spiritual fulfilment (Sharpley & Jepson, 2011).

More on the sacred side, Plante and Thoresen (2007) define spirituality as the quest for meaning in life, a relationship of an individual with God and nature (Plante & Thoresen, 2007). The foundation of the definitions advocated by several authors is primarily based on the tourists' motivations (Kale, 2004; Matheson et al., 2014; Moufakkir & Selmi, 2017; Norman, 2012; Roof, 1999; Sims, 1994; Singleton, 2017). The commonly accepted motivation for spiritual tourism is to escape the pattern of life, find meaning, come closer to oneself or higher authority, and foster spiritual betterment through personal experiences (Collins-Kreiner, 2010; Robledo & Batle, 2015). Several works investigated the travel motivations of the spiritual tourists to formulate a definition and provide a taxonomy (Moufakkir & Selmi, 2017; Norman, 2012; Sims, 1994; Singleton, 2017).

Research gap

Spirituality in India can be broadly classified into religious and nature-based spirituality. The religious spirituality seekers follow prescribed formal traditions of a religious institution (Haq & Jackson, 2009; Kale, 2004), whereas nature-based spiritual seekers are borderless (Roof, 1999). In the contemporary world of rising workload, pressure, stress, and depression (Bandyopadhyay & Nair, 2019; Parmar, 2016), the pursuit of life is happiness. An increasing number of youths are acknowledging it. Consequently, there is a rise in travel seeking wellness of mind, body, and soul through spiritual experiences (Ashton, 2018; Bone, 2013; Cohen et al., 2017; Sharma, 2020).

Though India is often associated with spiritual tourism, the existing literature does not comprehensively cover the motivations of India's millennials travelling domestically for spiritual pursuits. A large amount of work on spiritual tourism in India is focused on religious spirituality (e.g. Agrawal et al., 2010; Sharpley & Sundaram, 2005) with a scattered focus on its secular aspects. Also, there is little work on India's domestic spiritual tourists, especially with a scarce emphasis on urban and millennials. Through this work, the authors aim to understand Indian urban millennials' motivations to travel for spiritual reasons. The works investigating spiritual tourism domains are essential in building groundwork to refine the concept (Norman, 2012). This understanding, in turn, will help in more effective marketing and management of spiritual tourism. Also, the understanding of the tourists' motivations aids in the segmentation of the larger groups into smaller ones and further helps in illuminating the expectations of the travellers. With the understanding of these expectations and motivations, tourist satisfaction can be optimised.

The current research has three broad objectives. Primarily, the authors intend to decipher the motivations of the Urban Millennial Spiritual Tourists (UMSTs). The paper also investigates the nature of a UMST's activities during their trip. With the information from the first two objectives, the authors were able to identify the push and pull factors, especially for Indian domestic spiritual tourists. Consequently, the paper addresses three research questions:

RQ1 What are the travel motivations of UMSTs?

RQ2 What are the kind of activities that a UMST engages in at a destination?

RQ3 What are the push factors for the domestic UMST and pull factors of a spiritual destination?

A phenomenological approach was used to examine the lived experiences and subjective views of the spiritual tourists. Qualitative analysis is employed to provide a thorough understanding of the subject. A detailed, in-depth inquiry was conducted and this research identifies eight motivations of the urban millennials travelling for spirituality in India. The study on travel objectives contributes to understanding visitor intentions and consumption (Matheson et al., 2014). The research further categorises the activities undertaken by the spiritual tourists in a destination under seven broad themes that speaks volumes of the tourist motivations too. The research is a pioneer in identifying separate push and pull factors in the context of spirituality that paves the way for future research to identify and elucidate separate factors for different tourism types rather than generalisation. Understanding visitor motivation and behaviour is momentous for eloquent planning, managing, and marketing destinations.

Review of literature

The current research is conceptualised on the works of earlier scholars whose contributions have backed the research results. A thorough review of the literature has provided validation and reliability to the study results and provided the required theoretical foundation.

Spirituality

A multidimensional concept, spirituality, has been researched in varied genres such as management, psychology, administration, hospitality, and tourism (e.g. Bandyopadhyay & Nair, 2019; Carroll, 1997; De Souza et al., 2016; Fry, 2003). The initial discussion regarding spirituality is conspicuous in the scholarly works searching for the meaning of spirituality in religion. One of the oldest motivations of travel for humankind, the reminiscence of spirituality, was visualised in divinity by the authors, and often spirituality was considered a subset of religion. According to Pargament (1999), religion is an umbrella term in which spirituality plays an important role. In the same year, Stifoss-Hanssen (1999) argued that the analogy of religion and spirituality is different but has substantial overlapping. Religion is believed to be a road map to spirituality or a vehicle to aid the movement towards spiritual benefits (Elkins et al., 1988; Ingersoll, 1994; Kale, 2004).

However, studies often fail to acknowledge spirituality as separate from religion. Spirituality can be of two types- the one that is religiously achieved (Haq & Jackson, 2009; Roof, 1999) and the other that is independent of any religion empowered by the quest to seek meaning and happiness (Robledo, 2015; Stausberg, 2014). The contemporary literature recognises the two concepts as disparate and defines spirituality as a self-sufficient phenomenon to exist independently of religion (Hinterkopf, 1994). Authors like Benjamin and Looby (1998) propagate that spirituality can occur with or without the construct of religion, and not all religious activities are spiritual. As opposed to religion, spirituality is a broader concept centred on individuals' experiences (Kale, 2004; Zinnbauer et al., 1997). In contrast, religion aims to cage that encounter in an institution (Maher & Hunt, 1993).

Several authors have argued that spirituality is not easy to define. To have a definition unanimously accepted by all is a futile exercise (Kinjerski & Skrypnek, 2004; Koenig, 2012; Maher & Hunt, 1993). According to Abraham Maslow's Hierarchy of Human Needs, spirituality can be understood as the pyramid's highest rung. The tourists think beyond materialistic goals and strive towards self-actualisation, soul elevation, and self-development (Fathi et al., 2019). Whilst addressing this complex question, several works recognise the space for personal elucidation and self-transcendence facilitated by spirituality which empowers the essential attributes of spiritual growth, such as the search for the meaning of life (Carroll, 1997; Moal-Ulvoas, 2016; Vaughan, 1991), a quest for happiness and connection with oneself or the universe (Denton & Mitroff, 1999; Eckersley, 2002; Sharpley & Jepson, 2011).

Spirituality is a dynamic realm that can be perceived differently by different individuals. It is a subjective element that begins where religion ends (Sharpley, 2016) and can be perceived as a sacred personal experience (Vaughan, 1991; Zinnbauer et al., 1999). The authors propose an operational definition of spirituality for the paper as 'a dynamic idiosyncratic phenomenon where people are in a pursuit to attain the meaning of life, grasp the balance of mind and soul, and understand the purpose of being'. Spirituality is a broader domain than religion. However, it can also be found in religious practices and traditions, resulting in blurring the line between religion and spirituality. Spirituality can be understood as an individual quest for meaning rather than a collective experience where the concept is distinct for everyone, and the journey is the destination.

Spirituality and tourism

Spiritual tourism invariably attracts the attention of an increasing number of tourists and is believed to proliferate in the coming decades (Moufakkir & Selmi, 2017; Singleton, 2017). Spirituality is considered to be one of the most potent driving forces for travelling in the twenty-first Century as the majority of tourists is mindfully seeking spiritual satisfaction and are searching for meaning resulting in travel for self-actualisation- and wellness of the mind, body, and soul (Ambrož & Ovsenik, 2011). Various authors have tried defining spiritual tourism (Norman, 2011; Reich, 2000; Sharpley & Sundaram, 2005); however, a universally accepted definition is yet to be adhered to. Willson et al. (2013) describe spiritual tourism as undertaken by people searching for meaningfulness through travel.

Research helps acknowledge the role of tourism to provide patrons with the benefits of self-discovery, self-actualisation (Ee & Kahl, 2014). It enables purgative and promotes exorcising experiences among tourists (Caton, 2015; Zahra & McIntosh, 2007). The scholarly work bespeaks tourism's ability to facilitate identity formation (Desforges, 2000) and foster transformation (Filep, 2016; Noy, 2004; Voigt et al., 2011). Travel acts as a stimulus for the tourist to find meaning; hence, tourism can be understood as a phenomenon in which spirituality is rooted (Ambrož & Ovsenik, 2011). Norman (2012) proposed spiritual tourism as a quest, an experiment, healing, a retreat, and a collective.

The path of spirituality can be embarked upon through various encounters ranging from religious traditions (Agrawal et al., 2010; Sharpley & Sundaram, 2005), taking up pilgrimage (Haq & Jackson, 2009; Raj, 2012), and can even be observed in activities considered far away from religion or spirituality such as relaxing in the desert (Moufakkir & Selmi, 2017; Narayanan & Macbeth, 2009), attending a festival (Matheson et al., 2014), engaging in wellness retreats (Bandyopadhyay & Nair, 2019; Voigt et al., 2011), a mere four-wheel drive (Lehto et al., 2006; Snyder, 2007) or even through adventure sports (Bidder, 2018). Spiritual tourism, therefore, is any activity that benefits the spiritual desires of an individual. In modern times, people are less concerned about the rules and norms in their spiritual journey and seek less structured designs to achieve self-development (Sharpley & Jepson, 2011). The religious tourists are driven by specific norms affiliated to the God of the specific religion in contrast to the spiritual tourist who is motivated by chance to attain self-discovery and personal development through a realm of varied encounters (Griffin & Raj, 2017; Singleton, 2017).

Spiritual tourists and motivation

Any tourism has several objectives, such as leisure, adventure, and business, among others (Rinschede, 1992). The study of these motivations helps to place them in the appropriate type of tourism. A tourist travelling for spirituality consciously indulges in religious or non-religious traditions driven by the aim of spiritual growth and attaining the meaning of life (Collins-Kreiner, 2010; Norman, 2012). Spiritual tourists are motivated to achieve self-realisation and understand the more significant meaning of life (Hirschorn & Hefferon, 2013). The esoteric debate about the distinction between religious tourists and spiritual tourists is a continuous dialogue of juxtaposition. While there is little doubt that religion can lead to spiritual benefits for the participants, spiritual tourists' motivations are beyond just religion. Fedele (2012) coined the epithet 'new-pilgrims' to define the spiritual tourists who embark on pilgrimage for reasons contrary to the traditional objectives of a sacred pilgrim. The work of Tilson suggests that religious and spiritual tourism can co-exist (Tilson, 2005). He uses the phrase 'religious-spiritual tourism' to refer to the secular tourists motivated for spiritual growth away from religion.

Similar to the experimental type of tourist (Cohen, 1979), a spiritual journey of UMST may not necessarily be isolated from other tourism activities. A tourist may be an attendee of a festival (Matheson et al., 2014) or a visitor to the historical places, which may well fulfil their spiritual drive (Willson et al., 2013). A spiritual journey can have an assortment of activities that might not only be motivated by the ultimate goal of experiencing spirituality. A tourist can have multiple motivations like adventure, escape, or leisure from one trip. Still, the prominent of all the objectives should be for spirituality to be considered under spiritual tourism. A spiritual tourist seeks broad experiences that provides an authentic sense of discovery facilitated by the tourist's gaze irrespective of the external environment (Smith, 2003). The studies suggest that the ultimate goal of a tourist is to escape and seek meaningful experiences while travelling (Mannell & Iso-Ahola, 1987), which is central to the ideology of spirituality (Moufakkir & Selmi, 2017).

The study of tourist motivations has a long history in tourism, primarily based on the belief that it is a more significant driving force in tourist behaviour (Fodness, 1994). Crompton (1979) advocated that it is easier to understand the 'who', 'how', 'where', and 'when' of the tourist behaviour, but the 'why' is difficult to comprehend. Though other variables such as perception, culture, and social background influence tourist behaviour, motivation is believed to be the most compelling force behind all tourists' actions.

One of the most popular theory to explain the motivations of tourists is that of push and pull factors (Ashton, 2018; Crompton & McKay, 1997; Konu & Laukkanen, 2010). These factors lay light on the travel choices of the tourists and the selection criteria for choosing a destination. The theory of push and pull motivation, which provides an important structural foundation to the research, is derived from the work by Dann (1977) that evaluates the spiritual tourists' needs to retreat. The traditional meaning of push factors is those elements that drive an individual to travel. The pull factors are the attributes in a destination that attracts or suits best with the demand of the people (Chan & Baum, 2007).

While most of the studies using the theory are quantitative in nature (Ashton, 2018; Konu & Laukkanen, 2010), very few have used the qualitative approach (Chan & Baum, 2007; Prayag & Ryan, 2011; Said & Maryono, 2018). This study has adopted qualitative method along with push and pull factor theory largely because the research is exploratory in nature and the sample has not been researched thoroughly in the past. The inductive approach utilised in the study allowed a fresh data to be analysed in-depth to provide with a deeper understanding of the concepts which are often lost in numbers (quantitative research). The objectives of this study involve research in human emotions, their perception, and their behaviour, thus employing the qualitative method provides more comprehensive data.

While most authors consider the state of mind as the 'push factor' or the motivating factors behind tourist behaviour (e.g. Brown, 2005; Dann, 1981; Matheson et al., 2014), this study differentiates the two - the state of mind and the motivation. It implies that the motivating factors are different from the push factors. The 'push factors' induce motivation in a person and impact the objectives of an individual. For example, a person's state of mind can be 'stress' that 'pushes' the individual towards the motivation to escape. Another example is the person's religious beliefs (views) form the push factor, leading to the motivation to come closer to almighty, resulting in pilgrimage (Haq & Jackson, 2009; Raj, 2012).

The concept of spirituality is evident in the tourists' minds; the confusion is perhaps mainly an academic construct. Though the study on spiritual tourism is still in its early stages, scholars' focus on the discourse of the type of destinations that the tourists seek and their motivations are essential to study (Singleton, 2017). Spirituality is primarily personal; thus, their objectives, motivations, and attitudes are complex to interpret. A large number of authors have used empirical data from interviews and questionnaires to understand the motivation of the tourists and then advance categorisation theories of spiritual tourists or to deconstruct spiritual tourism (e.g. Ashton, 2018; Fry, 2003; Holmberg, 1993; Moal-Ulvoas, 2016; Moufakkir & Selmi, 2017; Norman, 2011). The motivations of different types of travellers can be different. The difference is apparent between the first-time and repeat tourists, foreign and domestic tourists, and the locals and the non-locals. The demography of the tourists plays a governing role in the formation of the motivations of the tourists. Therefore, it is appropriate to consider the tourist's demography, social, and cultural background before categorisation or advancing generalised theories for the entire market.

Spiritual tourism in India and millennials

India is renowned globally for its rich culture and diverse legacy (Singh & Shukla, 2016), which provides it with a competitive edge to become a world leader in spiritual tourism (James, 2000). In addition to the numerous religious circuits available, the Indian Ministry of Tourism has developed spiritual tourism circuits in adherence to the 12th Five Year Plan (an earlier national planning system). India's welcoming and largely tolerant demeanour has fostered many religious and spiritual ideologies (Medhekar & Haq, 2012). The birthplace of Yoga, Vedas, Buddhism, Jainism, Sikhism, and Hinduism, India is the world's spiritual capital (Aggarwal et al., 2008; Sharpley & Sundaram, 2005). With the copious natural beauty and easy availability of *sadhus* (hermits) and *gurus* in the country to guide towards the path of spirituality, India enables seekers' spiritual advancement. The arrival of *Beatles'* the lead guitarist, to India in 1966 and his revisit in 1969 with the entire band asserted its popularity for spirituality (Bandyopadhyay & Nair, 2019). The copious literature on spiritual tourism in India makes it a popular destination, especially for westerners (Norman, 2012). India is a destination with abundant religious and spiritual experiences that overwhelmingly attracts tourists for an authentic spiritual holiday (Aggarwal et al., 2008; Sharpley & Sundaram, 2005).

In today's fast-paced world, to have a balanced work-life, people are getting more attracted to health, wellness, and spiritual tourism – more so the millennials. Millennials, the people born between 1981 and 1996 (Dimock, 2019), have an abundance of knowledge, technology, and information at their disposal. They are also more sensitive towards their spiritual needs (Aboobaker et al., 2020). Victim of work-life stress and complex relationship scenarios, the millennials are increasingly seeking escape from the professional pressure and searching for serenity and ways to calm mind. The oldest form of travel for the millennia (Cohen, 1979; Rinschede, 1992), spiritual tourism has attracted the young urban adults of India. The growing interest of the westerners and the domestic demand for spiritual and religious tours influence this type of tourism. Subsequently, the rising global demand for spiritual and religious tourism has motivated the tourism sector to cater to their requirements. (Sharpley & Sundaram, 2005). The Westerners are attracted to Yoga and meditation. The adult domestic tourists are more inclined towards religion, and the millennials of India towards spirituality through solace, nature, peace, and serenity. This conjecture is aided through this research by the authors.

Methodology

This study deploys a qualitative approach to investigation. Qualitative analysis is best suited to understanding human behaviour, encounters, and psychology optimally and further studying the aspects hidden in numbers. As spirituality is subjective, its research involves the perusal of emotions, feelings, and experiences. Since there is no standardised instrument to study the phenomenon, a qualitative approach provides more in-depth knowledge and a thorough understanding. Nvivo 12 Plus software was employed to investigate the motivations of domestic UMSTs in India as the method provides extensive, socio-contextual, and elaborate descriptions along with insightful interpretations (Holloway & Galvin, 2010; Smith et al., 2011). A thorough, in-depth qualitative research approach is acceptable with a smaller sample size for investigating new research areas (Vaismoradi et al., 2016).

Data collection

The initial four participants were purposively selected whom the authors had identified to be spiritual travellers. Then snowball sampling was applied. The participants with the spiritual inclination and those who travel for spirituality were chosen. A total of twenty-two spiritual tourists between the age of 23 and 38 years were interviewed through purposive and snowball sampling (Denzin & Lincoln, 2017). The interviews were stopped when they sufficiently contributed to the research question, and a saturation level was achieved where additional interviews were not contributing any new information (Byrne, 2001; Guest et al., 2006; Kaurav et al., 2020). Twelve male and ten female respondents were in the sample, where 4 were students, 16 employed, and 2 were freelancers. The average number of the respondents' spiritual trips was 3, where the least number of trips in a year was two, and the maximum was 7.

The interview protocol was designed after a thorough examination of the existing research on tourists' motivations. As extensively discussed in the review of literature, several scholars have provided various motivating factors for spirituality and spiritual tourism through their research. However, very limited (in the light of spirituality) studies provide an extensive list of factors that illuminates the motivations of spiritual tourists of India. Also, lesser research work is available that talks about the motivating factors amongst *millennials*. Thus, with the help of the review of literature, the authors designed the interview questions that aid in answering the three research objectives set by the authors. Following which the Delphi method was used to develop the instrument with the help of the expert panel of academics (stage 1, Figure 1). Two professors and five research scholars (already researching similar themes), and one industry professional (involved in the selling spiritual tourism packages) were part of the panel discussion (stage 2 and 3, Figure 1). The anonymous, multiple-round survey procedure of Delphi was employed to produce an interview protocol that has been tested and unanimously accepted by the panel of experts after eradicating the original instrument's shortcomings in different rounds (Dalkey & Helmer, 1963) (stage 4, Figure 1). It effectively checks the instrument's reliability, validity, and trustworthiness (Markmann et al., 2020). Yeong et al. (2018) suggested the procedure utilised for developing the interview protocol of this study (refer to Figure 1).

Data collection process

Semi-structured interviews were carried out to allow meaningful discourse (Chima, 2020; Saunders et al., 2009) via video and phone calls with an average duration of 40 min. The interviews were conducted in the English language and were transcribed verbatim by the research team. No software was used for transcribing to reduce the errors. The final script was completed with a member checking by the respondents to affirm the transcripts' accuracy (Birt et al., 2016).

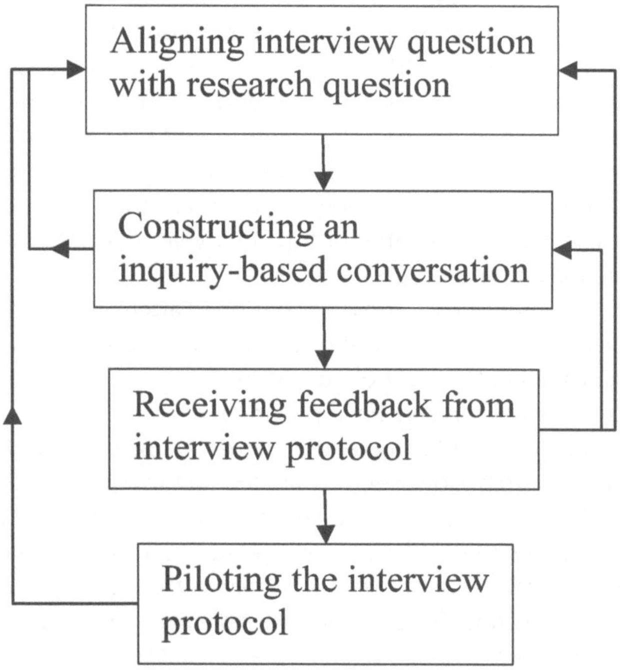

Figure 1. Process of development of interview protocol.

Questions such as 'How will you define spirituality?' 'How does travelling helps in attaining the spiritual goal set by you?' 'What motivates you to travel to a spiritual destination', 'What are factors you consider while choosing a spiritual destination?' along with others to generate answers aiding the research objectives. In total, twenty-one questions were asked to each respondent. The data were analysed using thematic analysis, a flexible tool that allows modifications as per the research's need to contribute detailed, trustworthy data (Braun & Clarke, 2006; Nowell et al., 2017). The thematic analysis method was adopted due to its innate ability to felicitate the human experiences' inquiry in objective settings (Percy et al., 2015). It is considered one of the most efficient ways of analysing written transcripts, especially interviews (Hannam & Knox, 2005; Walters, 2016), by providing rich cultural meanings from the textual data set. Thematic analysis is a method that helps in recognising, examining, and reporting a pattern in the detailed data (Alhojailan, 2012) and often helps in the interpretation of various aspects of the research topic.

Data analysis

The data was analysed in six steps as suggested by Braun and Clarke (2006), which began with familiarisation with the data, creating initial codes, and identifying the main themes and sub-themes. Qualitative data analysis software, Nvivo 12, was employed despite it being laborious since it is believed to provide meticulous efficacy to the research by adding rigour and veracity to the study (Alhojailan, 2012). The thematic maps worked as visual guides for easy understanding (Sharkey & Caska, 2019) as they eased the repeated reviewing and studying of the themes. The interview data's dominant themes were compared with the existing framework for motivations regarding spiritual tourists before proposing a construct (Tashakkori & Creswell, 2007). Direct quotes from the respondents support the findings from the data to support the results.

Findings

The data analysis process involved extensive use of Nvivo software to allow rigour and efficiency to the research findings. It is imperative to understand the answers to the rudimentary questions like 'Why, What, Where, When, and How' to deconstruct a millennial spiritual traveller's motivations. These fundamental questions form the backbone of the study. It is essential to understand the need to travel to a particular place (Ambrož & Ovsenik, 2011) undertaken at a specific time via a particular mode of travel and engaging in certain activities (Smith, 2003). The choice of the destination and mode of travel and the particular timing of the journey suggests plenty of the individuals' aims and objectives. It is a loop. The authors examine the activities undertaken by the spiritual tourists and their choice of a spiritual destination to effectively understand the spiritual tourists' motivations. The previous studies primarily focused on one of the two in their research (e.g. Matheson et al., 2014; Moufakkir & Selmi, 2017; Norman, 2012; Roof, 1999).

The current research focuses on India's millennials and their motivations to travel for spirituality domestically by evaluating their needs and choices of travel. People's motivations are transient and often influenced by the traditions and rituals around them, shaping contemporary trends (Holmberg, 1993). Considering the strong roots of religion, mythology, and traditions engraved with spiritual essence in India (Phukan et al., 2012; Sharma, 2016) and with the effect of globalisation on society (Kale, 2004), the information from the data analysis revealed that the millennials have a synthesis of western and Indian culture in their understanding of spirituality. The interview analysis revealed that the abundance of knowledge and easy availability of varied experiences to the youth has culminated into an evolved sense of spirituality which they increasingly view as a liberating force.

Several questions were directed to understand the definition of spirituality in the minds of the selected spiritually inclined millennials in India, such as, 'How do they define spirituality' and 'The top five words that they associate with spirituality'. The analysis resulted in an emphasis on the three fundamental characteristics of spirituality – firstly, it reaffirms the idiosyncratic nature of spirituality; secondly, the respondents had a confident clarity of the concept; and thirdly, they were aware that different people related differently to the notion of spirituality and the path varies from being religiously defined to being completely aimless loitering. To elaborate further, the most conspicuous finding from the replies of the respondents was that the notion of spirituality is entirely individualistic. Phrases like 'in my opinion', 'for me spirituality is', 'I think spirituality is', 'I can only speak for myself', or 'it might be different for others' were used before the respondents elaborated on the phenomenon. Their tone showed confidence and voiceled clarity of their thoughts while expounding on the experiences of spirituality. The confusion in defining spirituality instead seems more of a scholarly construct, and the interviews revealed that idea is quite evident in people's minds.

> I think spirituality is a deep source of nourishment for the soul, and for me, that is the most essential component of life. And as per my understanding, spiritual enrichment is the highest form of wellness which cannot be achieved without efforts and asking questions. For me, spirituality is an everlasting journey which I myself do not want to end (Male, Age 36 years, CEO).

Dynamics of spirituality

The respondents were asked the top five words that they instantly and deeply identify with spirituality (Figure 2). Approximately two-thirds of the respondents associated spirituality with words like 'peace' and 'tranquillity' - which were considered synonyms due to their semantics. Whereas half of the interviewees linked spirituality with a 'meaning-making project', and almost half of them have defined it as a means of connecting with self. Spirituality was also connected to travel and nature. Through the course of the interview, the respondents emphasised the importance of the spiritual destination to be in a natural setup. Also, the efficacy of travel in spiritual development was strongly mentioned.

Additionally, words like isolation, quietness, solitude, *nirvana*, sacred, understanding of the world, meditation, and unanimity were used mainly. These top words that the respondents identified with spirituality can be understood as their underlined motivations and aim for this spiritual journey. There was no mention of religion in the five words associated with spirituality; however, approximately one-fifth of the interviewees used the words sacred and almighty. Therefore, the current study supports the argument that religion and spirituality are two distinct concepts with overlaps in their paths, and spirituality can exist without religion.

> My ideal year includes a trip to Varanasi as I feel close to a higher power there, with the chants of mantras and the sound of the bells creating a divine aura. Further, a weekend trip to Kasol after a hectic week also provides me the harmony between my mind and my body. Staying close to nature is therapeutic on its own (Male, Age 24 years, Trainee).

Activities of the travellers

While studying activities the spiritual tourists indulge in, the challenge is that the term 'spirituality' is too loosely defined, comprising a broad spectrum of experiences. Undoubtedly, there can be endless activities that are spiritually inclined (Robledo, 2015). The thematic map of the interview data revealed that the travellers' activities consisted of seven broader themes with multiple subthemes (Figure 3). The map demonstrates the comprehensive activities that a millennial spiritual tourist

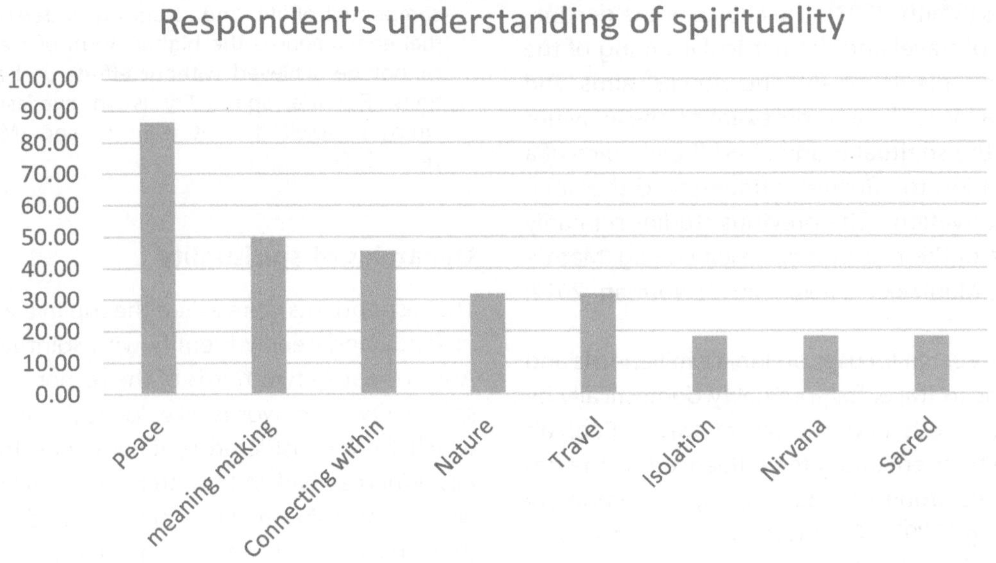

Figure 2. Word association with spirituality.

undertakes during their journey. The evaluated data and the spiritual activity that a person indulges in their surroundings might differ from those adopted during the escapade. The broader themes are as follows:

1. Nature Related activities;
2. Self-indulgent;
3. Experience;
4. Helping others;
5. Wellness;
6. Assisted activities; and
7. Religiously motivated spiritual activities.

The interview data revealed that about nine-tenths of the respondents preferred to enjoy nature in their trip in ways such as listening to the flowing river's sound, watching the sunsets, long walks in a beautiful landscape, and experiencing nature up close. Approximately four-fifths of the interviewees revealed that they engaged in self-indulgent activities during their spiritual travel through activities like spending time alone, aimless loitering, and preferred to do things as per their terms without any plan. The data further revealed that about three-fifths of the respondents chose to indulge in experiential activities such as adventure sports, connecting with the locals, and trying the local cuisine to experience cultural inversion. Such a result also validates the ideology that spirituality is idiosyncratic and resides in various experiences that are not conspicuously spiritual (Bidder, 2018; Kale, 2004) but hold experiences that target the perceived spiritual side of an individual.

Around three-fifths of the respondents said that they like to help the locals with their services. The antecedents of spirituality in volunteering are supported by the research, which is increasingly being recognised in the contemporary scholarly discourse (Shalbafian & Zarandian, 2019). The data further illustrated that about half of the interviewees engaged in meditation, breathing exercises, aromatherapy, and Yoga during their trip for spiritual, mental, and physical wellness. These tourists adhered to holistic living even in their daily lives. Around one-third of the interviewees indulged in the assisted form of activities conducted with the help of gurus, sages, and organisations like the Art of Living. These tourists were found to be already associated with those organisations of interest. They believed that gaining knowledge from an already well-read spiritual person will help them in their spiritual journey. The vibes of the *Ashram* (Hermitage) positively contribute to their spiritual development. Further, the presence of like-minded people with a similar goal works as an added incentive.

> The good part about Ashrams is that you meet people with similar goals and relatable mindsets. You can learn from their experience and have the privilege of being in their storeys. The dialogue of knowledge is an essential part of spirituality which can only constructively help in your journey. The people create the vibe of the place, and the energy of the Ashram is filled with positivity and life (Female, Age 29 years, Team Leader).

The responses revealed that only one-third of the respondents found spirituality in religiously driven activities such as pilgrimage, visiting temples, and chanting mantras. The experience of the journey to the well-renowned temple was of utmost importance to them as they felt much closer to divinity.

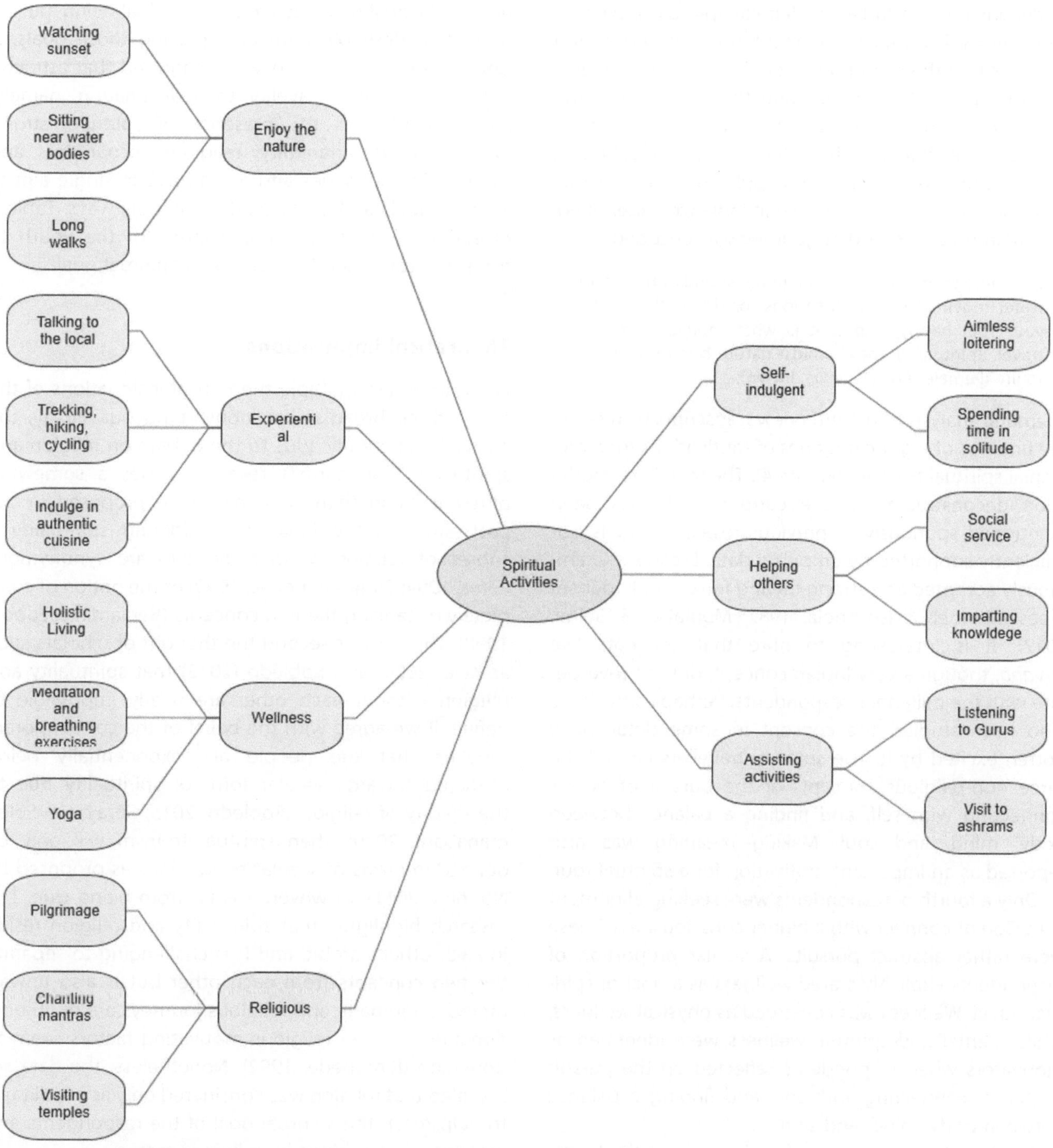

Figure 3. Thematic map of the spiritual activities mentioned by respondents.

> I am a religious person but taking the pilgrimage for me has a whole new reason. I enjoy the experiential journey to the temple and value the energetic, lovely people you meet in this escapade. My quest for spirituality starts from my doorstep till I reach the desired temple, mostly in the mountains like Kedarnath or Badrinath. It is more than a religious journey or an adventure for me; it is just spiritual. The encounters during my pilgrimage provide me with answers to several questions I have in mind and bring me closer to my true self (Male, Age 27 years, Executive).

Push and pull factors of spiritual tourism

The questionnaire was divided into four themes as per the objectives of the study. The first theme was aimed to understand the perspective of the spiritual tourist towards spirituality. The second theme envisaged to enlist the activities they indulged in during spiritual tourism. The third theme was centred around the purpose and goals of a spiritual traveller, which helped to deconstruct an Indian millennial spiritual traveller's motivations. The fourth theme intended to understand the traits that the respondents desired in a spiritual destination. It aided in expounding the various determinants that attract tourists for spiritual travel to a particular destination. The interview data resulted in three broad findings where the 'pull' and 'push' factors for a spiritual millennial traveller in India were identified. The main objective of their journey was decoded.

> Can there be more peace than in the sound of birds and water moving? You tend to look for things that make your soul happy, and that is what motivates me to travel. Spiritual journeys amidst nature bring me closer to life (Female, Age 24 years, Intern).

Thematic analysis of the interview transcripts resulted in the findings of eight categories of motivations for a millennial spiritual traveller (Figure 4). The travellers' motivation deconstructed in the current study has been mentioned spuriously in previous research that is not efficiently supported by empirical data. Escape is unanimously accepted as a strong driving force for all tourism types (Mannell & Iso-Ahola, 1987; Moufakkir & Selmi, 2017). It is interesting to note that concepts like *nirvana*, though a very Indian concept, did not reverberate with the millennial respondents. Perhaps only those who have studied the concept in some detail have gotten excited by it. The spiritual travellers chased the three non-religious concepts of the pursuit of peace, connecting with self, and finding a balance between body, mind, and soul. Making meaning was also reported as an important motivation for a spiritual tour.

Only a fourth of respondents were seeking alignment with God or connect with a higher consciousness. These were rather abstract pursuits. A similar proportion of respondents explicitly stated wellness as a goal of spiritual travel. Wellness was construed as physical wellness, while mental and spiritual wellness were identified as motivators when respondents reflected on the pursuit of peace, connecting with self, and finding a balance between body, mind, and soul.

The travellers' push was often because of the hectic lifestyle or religious beliefs and curiosity or need to break the monotony. The driving force towards the formulation of tourists' motivations is considered the 'push factors'. Frequently, a journey has several motivations and subordinate aims (Rinschede, 1992). Therefore, it was visible in the analysis that the same person was interested in travelling for wellness and attaining peace. Broadly seven push factors were recognised: work pressure, mental morbidities, curiosity, Weltschmerz, imbalance, religious beliefs, and traditions. These 'push factors' or the 'state of mind' form the motivations of an individual that leads to an individual being 'pulled' towards a desirable path of achieving those goals. A total of seven pull factors are recognised that attract a millennial spiritual traveller to a destination, namely, natural landscape, the presence of solitude, strong local aesthetic, unanimity, religious recognition, and the presence of sages and ashrams. One single aim of an individual can be achieved in multiple ways. Table 1 presents the push and pull factors for the youth to travel to attain their long-standing spiritual goal.

Theoretical implications

While there are multiple theoretical insinuations of the study, three broad implications stand out. Firstly, the research coherently adds to the debate on religion and spirituality. The current research poses a somewhat different stand than what has been proposed in the past, such as the idea of considering spirituality a subset of religion or that the two are synonymous (Sims, 1994; Zinnbauer et al., 1999) or the notion of completely distancing the two concepts (Benjamin & Looby, 1998). The authors second the thought of scholars such as Kale (2004) and Robledo (2015) that spirituality and religion without each other are nearly impossible to define. If we agree with the belief of the contemporary scholars that the people are exponentially being attracted towards secular form of spirituality due to the rigidity of religion (Robledo, 2015; Sorayya & Soleimanabadi, 2019), then spiritual tourism will only be defined in terms of secular expressions as proposed by Norman (2011). However, it is far from being true. The research highlights that spirituality and religion reside in each other's ambit, and it is challenging to separate the two concepts from each other but is also unwarranted. Pilgrimage and religious journeys are multifunctional even when religious motivating factors seem to dominate (Rinschede, 1992). Nonetheless, the data set revealed that religion was considered only as a facilitator to help reach the spiritual goal of the respondents and was not wholly driven by religious motives.

The second significant theoretical implication of the paper is in the study about millennials in India. A topic which is increasingly being researched in the past decade under different genres (Myers & Sadaghiani, 2010; Aboobaker et al., 2020). The current study contributes to understanding the travel choices of millennials, especially in a spiritual tourism context. The millennials that form a dominant part of today's workforce is called the 'hoping generation' (Ertas, 2015) as they are unable to stick to one company unless they feel their material, as well as spiritual needs, are realised in the

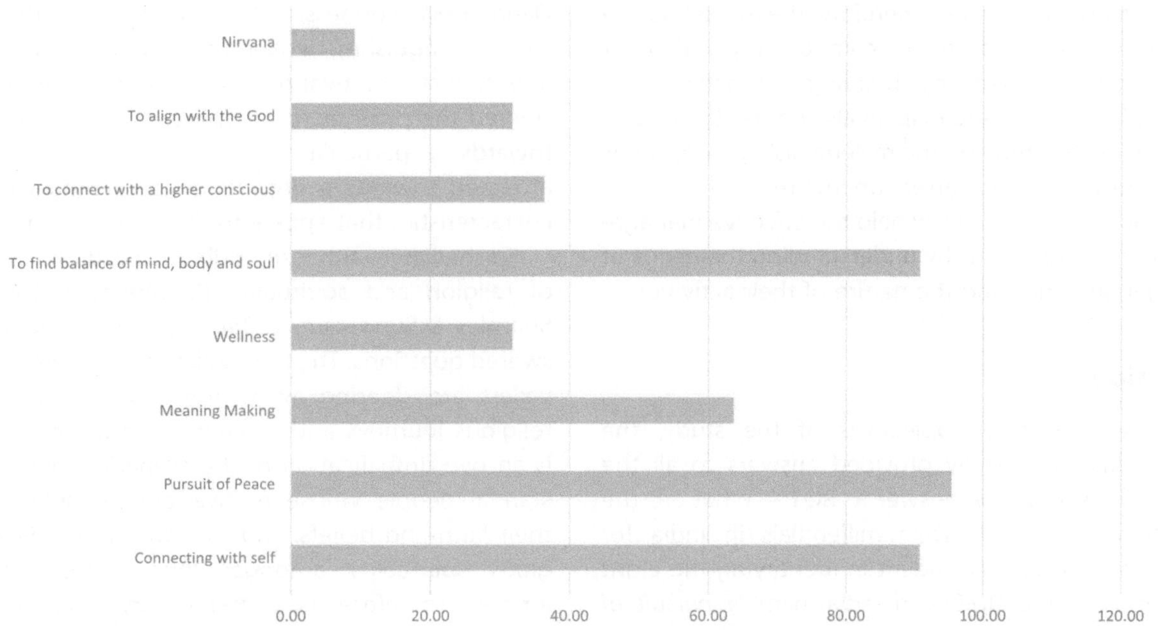

Figure 4. Aims of a spiritual traveller.

organisation (Aboobaker et al., 2020). India's labour force has a significant proportion of millennials, which will only increase by the day. This component of the workforce should be mentally and emotionally well off. Spiritual tourism being a critical intervention, suppliers need to understand the UMST market. This understanding is vital in organising and catering to the sector effectively. This research considerably adds value to the current body of knowledge on the domestic spiritual traveller, especially in the Indian context, which has received little attention from academics.

Thirdly, the research identifies and establish separate push and pull factors in the context of spirituality and spiritual tourism. This prominent implication of the study allows way for future researchers to examine and propose separate factors for different tourism types rather than a mere generalisation. Since all forms of tourism are different from each other and the tourist destination have their own characteristics, it is time to acknowledge that there are different push and pull factors surrounding them. The current study provides specialised push and pull factors for the urban millennial spiritual travellers in India which is important information for future researches on the topic and provides insights on their consumption behaviour.

Table 1. Push and pull factors.

Push Factor	The aim of the spiritual journey	Pull Factors/desired traits of a destination
Corporate Pressure, Hectic Life	Escape	Natural Landscape
Mental morbidities, stress	Transcendence	Isolation, solitude, and calmness
Curiosity, Experience	Meaning Making	Local aesthetic
Weltschmerz	The pursuit of Peace, Connecting with self	Away from city, solitude, and harmony
Imbalance	Wellness	Calmness and Unanimity
Religious Beliefs	To align with the sacred, *nirvana*	The religious quotient of the destination/religious site
Pilgrimage, Traditions	To connect with a higher power/ consciousness	Spiritual/religious recognition of the destination. Presence of gurus and ashrams

Practical implications

The research provides insights into the motivations of the urban millennials in India during their domestic travel, which can help organisations formulate marketing schemes and strategies to efficiently cater to their desired segment. For example, to target the urban millennials, the destinations can create budget-friendly experiences for tourists closer to nature. Further, the tourists' description of the kind of activities that they would like to engage in at a destination provides insight into the type of destination they will choose. Therefore, this allows the supply side to comprehend the psyche of their customers. Understanding tourist motivation can result in designing and marketing tourism products as solutions to consumer's needs. For instance, multiple activities such as camping, hiking, or bungee jumping can be added to the itinerary to attract a larger audience striving for spirituality. The paper highlights the growing market of spiritual tourism among millennials. Therefore, the marketers can easily showcase the product's compatibility with their target audience's aims by building campaigns that broadly connect with the millennials and use their vocabulary. Since the study on the millennials' behaviour has recently attracted much attention, the research on their behaviour can constructively help the effective management of the destinations by understanding the needs of the target audience and the nature of their activities.

Conclusion

Considering the three objectives of the study, the research has successfully provided answers to all the research questions. The answer to RQ1 – 'What are the motivating factors of urban millennials in India for their spiritual travel?' is evident in identifying the eight motivations of the UMSTs in India, namely *pursuit of happiness, desire to find balance, connect with self, wellness (physical and mental), to find meaning in life, to align with God, to connect with higher consciousness, and nirvana*. These motivations have scattered mentions in the previous studies on spiritual tourism as well. The data reflected that the pursuit of happiness, desire to find balance, connect with self, wellness, and find meaning in life were the dominant objectives of the travellers. Also, to align with God, connect with higher consciousness, and *nirvana* were echoed in the results.

In response to RQ2 – What are the types of activities that a millennial urban spiritual tourist engages in a destination? – this study has categorized activities under seven themes that speak profusely of the tourist motivations. These themes include nature-related activities, self-indulgent, experience-based, to help others, wellness-oriented, religiously inclined, and some which are performed with others' help. Also, the activities that a spiritual tourist indulges in are varied and often reflect on a spiritual traveller's motivations. A single tourist can have multiple motivations for travel which results in overlapping. Therefore, when placing a tourist in a particular typology, it is essential to acknowledge the grey area. For instance, the motivation of a tourist can be for spiritual development through adventure sports. Further, a tourist embarking on a pilgrimage can visit architectural sites en route. Therefore, the commonality must not be ignored, and the segmentation cannot be rigid.

The third research question (RQ3), 'What is the push for the domestic UMST and pull factors of a spiritual destination?' was also coherently answered through the study. The authors proposed seven specific 'push' and 'pull' factors concerning a spiritual tourist. In contrast to the previous studies on motivations (Brown, 2005; Dann, 1981; Fodness, 1994; Unger, 1991), the current work distinguishes between the tourists' state of mind and tourists' motivations. The state of mind is considered the push factor, shaping a person's motivation towards a particular act. The spiritual tourists get attracted towards a destination pertaining to suitable characteristics that appeal to the tourists' motivations.

Furthermore, although India is considered a rich land of religion and spirituality (Medhekar & Haq, 2012; Sharpley & Sundaram, 2005), there are plenty of unanswered questions. The firm belief that contemporary travellers are steering away from rigid institutionalised religious journeys and inclined towards spiritual travel is an oversimplification of the scenario. There are substantial people whose motivations are influenced by their faith and beliefs, and the strong presence of religious spirituality amongst the people cannot be ignored. Therefore, the contemporary definition of spirituality or spiritual tourism must incorporate and acknowledge the phenomenon's religious and secular aspects.

Limitations and future directions

While this research lays down the motivations and travel behaviour of the UMSTs in India, it is not without its limitations. Firstly, the current research is based on qualitative analysis, which has its limitations. Thus, the future studies can apply a mixed methodology keeping the outcomes from this study as their foundation. Secondly, with the use of both qualitative and quantitative methodology, the sample size can be increased and the results can achieve more reliability. Thirdly, the study is limited to the urban millennials, whereas a larger rural population of India has not been explored. So a comparative study can also be conducted between the urban and the rural millennials regarding their understanding of spirituality, their motivation for the spiritual journey, and the role that religion plays in their lives. It is critical to understand travellers' motivation both from scholarly as well as market perspectives. As a result, destinations would be able to offer better tourist experiences. The current research is primarily focused on UMSTs. The perspective of the millennials is much different from the baby boomers. Therefore, the study on understanding spirituality by the baby boomers and their motivations needs to be researched. This study takes further the long-standing debate of religion versus spirituality since the older generation in India overwhelmingly considers the two concepts as closely overlapping.

Further, the 'push and pull' factors of spirituality identified in this study can be re-deliberated as per different age groups and places as each place has its own demography, culture, and history, along with several other influencing factors.

Disclosure statement

No potential conflict of interest was reported by the author(s).

ORCID

Snigdha Kainthola http://orcid.org/0000-0002-0472-532X
Nimit Chowdhary http://orcid.org/0000-0002-7335-2379
Rahul Pratap Singh Kaurav http://orcid.org/0000-0001-9851-6854
Pinaz Tiwari http://orcid.org/0000-0003-0631-3377

References

Aboobaker, N., Edward, M., & Zakkariya, K. A. (2020). Workplace spirituality and employee loyalty: An empirical investigation among millennials in India. *Journal of Asia Business Studies*, *14*(2), 211–225. https://doi.org/10.1108/JABS-03-2018-0089

Aggarwal, A., Guglani, M., & Goel, R. (2008). *Spiritual & yoga tourism: A case study on experience of foreign tourists visiting Rishikesh, India*. Conference in Tourism in India-Challenges Ahead. http://dspace.iimk.ac.in/handle/2259/588.

Agrawal, M., Choudhary, H., & Tripathi, G. (2010). Enhancing Buddhist tourism in India: An exploratory study. *Worldwide Hospitality and Tourism Themes*, *2*(5), 477–493. https://doi.org/10.1108/17554211011090102

Alhojailan, M. I. (2012). Thematic analysis: A critical review of its process and evaluation. *West East Journal of Social Sciences*, *1*(1), 39–47.

Ambrož, M., & Ovsenik, R. (2011). Tourist origin and spiritual motives. *Management*, *16*(2), 71–86.

Ashton, A. S. (2018). Spiritual retreat tourism development in the Asia Pacific region: Investigating the impact of tourist satisfaction and intention to revisit: A chiang Mai, Thailand case study. *Asia Pacific Journal of Tourism Research*, *23*(11), 1098–1114. https://doi.org/10.1080/10941665.2018.1526198

Bandyopadhyay, R., & Nair, B. B. (2019). Marketing kerala in India as god's own country! For tourists' spiritual transformation, rejuvenation and well-being. *Journal of Destination Marketing and Management*, *14*(December), 1–8. https://doi.org/10.1016/j.jdmm.2019.100369

Benjamin, P., & Looby, J. (1998). Defining the nature of spirituality in the context of Maslow's and rogers's theories. *Counseling and Values*, *42*(2), 92–100. https://doi.org/10.1002/j.2161-007X.1998.tb00414.x

Bidder, C. (2018). Outdoor adventure tourism: Exploring the spiritual dimension of wellness. *International Journal of Academic Research in Business & Social Sciences*, *8*(16), 199–217. https://doi.org/10.6007/IJARBSS/v8-i16/5128

Birt, L., Scott, S., Cavers, D., Campbell, C., & Walter, F. (2016). Member checking: A tool to enhance trustworthiness or merely a nod to validation? *Qualitative Health Research*, *26*(13), 1802–1811. https://doi.org/10.1177/1049732316654870

Bone, K. (2013). Spiritual retreat tourism in New Zealand. *Tourism Recreation Research*, *38*(3), 295–309. https://doi.org/10.1080/02508281.2013.11081755

Braun, V., & Clarke, V. (2006). Using thematic analysis in psychology. *Applied Qualitative Research in Psychology*, *3*(2), 77–101. https://doi.org/10.1191/1478088706qp063oa

Brown, S. (2005). Travelling with a purpose: Understanding the motives and benefits of volunteer vacationers. *Current Issues in Tourism*, *8*(6), 479–496. https://doi.org/10.1080/13683500508668232

Byrne, M. (2001). Sampling for qualitative research. *AORN Journal*, *73*(2), 522–525. https://doi.org/10.1016/S0001-2092(06)61990-X

Carroll, M. M. (1997). Spirituality, alcoholism, and recovery. *Alcoholism Treatment Quarterly*, *15*(4), 89–100. https://doi.org/10.1300/J020v15n04_07

Caton, K. (2015). Critical commentary: Growing on the go? Moral development and tourism. *Annals of Leisure Research*, *18*(1), 1–8. https://doi.org/10.1080/11745398.2015.1004253

Chan, J. K. L., & Baum, T. (2007). Motivation factors of ecotourists in ecolodge accommodation: The push and pull factors. *Asia Pacific Journal of Tourism Research*, *12*(4), 349–364. https://doi.org/10.1080/10941660701761027

Chima, J. S. (2020). Are participants in ethnonationalist movements rational-choosers, socially-embedded actors, or psychologically-instinctive beings?: Motivations and worldviews of Sikh separatists from India. *Studies in Conflict and Terrorism*, *0*(0), 1–22. https://doi.org/10.1080/1057610X.2020.1711604

Cohen, E. (1979). A phenomenology of tourist experiences. *Sociology*, *13*(2), 179–201. https://doi.org/10.1177/003803857901300203

Cohen, M. M., Elliott, F., Oates, L., Schembri, A., & Mantri, N. (2017). Do wellness tourists get well? An observational study of multiple dimensions of health and well-being after a week-long retreat. *Journal of Alternative and Complementary Medicine*, *32*(2), 1–9. https://doi.org/10.1089/acm.2016.0268

Collins-Kreiner, N. (2010). Researching pilgrimage continuity and transformations. *Annals of Tourism Research, 37*(2), 440–456. https://doi.org/10.1016/j.annals.2009.10.016

Crompton, J. L. (1979). Motivations for pleasure vacation. *Annals of Tourism Research, 6*(4), 408–424. https://doi.org/10.1016/0160-7383(79)90004-5

Crompton, J. L., & McKay, S. L. (1997). Motives of visitors attending festival events. *Annals of Tourism Research, 24*(2), 425–439. https://doi.org/10.1016/S0160-7383(97)80010-2

Dann, G. M. (1981). Tourist motivation an appraisal. *Annals of Tourism Research, 8*(2), 187–219. https://doi.org/10.1016/0160-7383(81)90082-7

Dann, G. M. S. (1977). Anomie, ego-enhancement and tourism. *Annals of Tourism Research, 4*(4), 184–194. https://doi.org/10.1016/0160-7383(77)90037-8

Dalkey, N., & Helmer, O. (1963). An experimental apllication of the Delphi method to the use of experts. *Management Science, 9*, 458–467.

De Souza, M., Bone, J., & Watson, J. (2016). *Spirituality across disciplines: Research and practice*. Springer. https://doi.org/10.1007/978-3-319-31380-1.

Denton, E. A., & Mitroff, I. I. (1999). A study of spirituality in the workplace. *Sloan Management Review, 40*(4), 83–92. http://strandtheory.org/images/Spirituality_in_the_workplace-Mitroff_Denton.pdf.

Denzin, N. K., & Lincoln, Y. S. (2017). *The SAGE Handbook of Qualitative Research* (5th ed.). (eds). Sage.

Desforges, L. (2000). Traveling the world. *Annals of Tourism Research, 27*(4), 926–945. https://doi.org/10.1016/S0160-7383(99)00125-5

Dimock, M. (2019). Defining generations: Where millennials end and generation Z begins. *Pew Research Center, 17*(1), 1–7.

Eckersley, R. (2002). Spirituality, progress, meaning and values: Implications for leadership and management. *Journal of Spirituality, Leadership and Management, 1*, 15–28.

Ee, A. E. T. Z., & Kahl, C. (2014). My mind and soul suspended on a ticket to self-discovery. Now boarding, Koh Phi Phi!. *Procedia – Social and Behavioral Sciences, 144*, 5–14. https://doi.org/10.1016/j.sbspro.2014.07.268

Elkins, D. N., Hedstrom, L. J., Hughes, L. L., Leaf, J. A., & Saunders, C. (1988). Toward a humanistic-phenomenological spirituality: Definition, description, and measurement. *Journal of Humanistic Psychology, 28*(5), 5–18. https://doi.org/10.1177/0022167888284002

Ertas, N. (2015). Turnover intentions and work motivations of millennial employees in federal service. *Public Personnel Management, 44*(3), 401–423. https://doi.org/10.1177/0091026015588193

Fathi, M. R., Maleki, M. H., Koksal, C. D., Yuzbaşıoğlu, N., & Ahmadi, V. (2019). Future study of spiritual tourism based on cross impact matrix and soft systems methodology. *International Journal of Tourism & Spirituality, 3*(2), 19–41.

Fedele, A. (2012). *Looking for Mary magdalene: Alternative pilgrimage and ritual creativity at catholic shrines in France*. Oxford University Press. https://doi.org/10.1093/acprof.

Filep, S. (2016). Tourism and positive psychology critique: Too emotional? *Annals of Tourism Research, 59*, 113–115. https://doi.org/10.1016/j.annals.2016.04.004

Fodness, D. (1994). Measuring tourist motivation. *Annals of Tourism Research, 21*(3), 555–581. https://doi.org/10.1016/0160-7383(94)90120-1

Fry, L. W. (2003). Toward a theory of spiritual leadership. *Leadership Quarterly, 14*(6), 693–727. https://doi.org/10.1016/j.leaqua.2003.09.001

Griffin, K., & Raj, R. (2017). The importance of religious tourism and pilgrimage: Reflecting on definitions, motives and data. *International Journal of Religious Tourism and Pilgrimage, 5*(3), 2–9. https://doi.org/10.21427/D7242Z

Guest, G., Bunce, A., & Johnson, L. (2006). How many interviews are enough? An experiment with data saturation and variability. *Field Methods, 18*(1), 59–82. https://doi.org/10.1177/1525822X05279903

Hannam, K., & Knox, D. (2005). Discourse analysis in tourism research a critical perspective. *Tourism Recreation Research, 30*(2), 23–30. https://doi.org/10.1080/02508281.2005.11081470

Haq, F., & Jackson, J. (2009). Spiritual journey to hajj: Australian and Pakistani experience and expectations. *Journal of Management, Spirituality and Religion, 6*(2), 141–156. https://doi.org/10.1080/14766080902815155

Hinterkopf, E. (1994). Integrating spiritual experiences in counseling. *Counseling and Values, 38*(3), 165–175. https://doi.org/10.1002/j.2161-007X.1994.tb00834.x

Hirschorn, S., & Hefferon, K. (2013). Leaving it all behind to travel: Venturing uncertainty as a means to personal growth and authenticity. *Journal of Humanistic Psychology, 53*(3), 283–306. https://doi.org/10.1177/0022167813483007

Holloway, I., & Galvin, K. (2010). *Qualitative research in Nursing and healthcare* (3rd ed.). Wiley Blackwell.

Holmberg, C. B. (1993). Spiritual pilgrimages: Traditional and hyperreal motivations for travel and tourism. *Visions in Leisure and Business, 12*(2), 19–27.

Ingersoll, R. E. (1994). Spirituality, religion, and counseling: Dimensions and relationships. *Counseling and Values, 38*(2), 98–111. https://doi.org/10.1002/j.2161-007X.1994.tb00827.x

James, N. (2000). Double identity in orissa's golden triangle. *Antiquity, 74*(285), 682–687. https://doi.org/10.1017/S0003598X00060063

Kale, S. H. (2004). Spirituality, religion, and globalisation. *Journal of Macromarketing, 24*(2), 92–107. https://doi.org/10.1177/0276146704269296

Kaurav, R. P. S., Suresh, K. G., Narula, S., & Baber, R. (2020). New education policy: Qualitative (contents) analysis and twitter mining (sentiment analysis). *Journal of Content, Community and Communication, 12*(1), 4–13. https://doi.org/10.31620/JCCC.12.20/02

Kinjerski, V. M., & Skrypnek, B. J. (2004). Defining spirit at work: Finding common ground. *Journal of Organizational Change Management, 17*(1), 26–42. https://doi.org/10.1108/09534810410511288

Koenig, H. G. (2012). Religion, spirituality, and health: The research and clinical implications. *ISRN Psychiatry, 1*, 1–33. https://doi.org/10.5402/2012/278730

Konu, H., & Laukkanen, T. (2010). Predictors of tourists' wellbeing holiday intentions in Finland. *Journal of Hospitality and Tourism Management, 17*(1), 144–149. https://doi.org/10.1375/jhtm.17.1.144

Lehto, X. Y., Brown, S., Chen, Y., & Morrison, A. M. (2006). Yoga tourism as a niche within the wellness tourism market. *Tourism Recreation Research, 31*(1), 25–35. https://doi.org/10.1080/02508281.2006.11081244

Maher, M. F., & Hunt, T. K. (1993). Spirituality reconsidered. *Counseling and Values*, *38*(1), 21–28. https://doi.org/10.1002/j.2161-007X.1993.tb00817.x

Mannell, R. C., & Iso-Ahola, S. E. (1987). Psychological nature of leisure and tourism experience. *Annals of Tourism Research*, *14*(3), 314–331. https://doi.org/10.1016/0160-7383(87)90105-8

Markmann, C., Spickermann, A., von der Gracht, H. A., & Brem, A. (2020). Improving the question formulation in Delphi-like surveys: Analysis of the effects of abstract language and amount of information on response behavior. *Futures & Foresight Science*, (July), 1–20. https://doi.org/10.1002/ffo2.56

Matheson, C. M., Rimmer, R., & Tinsley, R. (2014). Spiritual attitudes and visitor motivations at the beltane fire festival, Edinburgh. *Tourism Management*, *44*(2014), 16–33. https://doi.org/10.1016/j.tourman.2014.01.023

Medhekar, A., & Haq, F. (2012). Development of spiritual tourism circuits: The case of India. *GSTF Journal on Business Review*, *2*(2), 212–218. https://doi.org/10.5176/2010-4804

Moal-Ulvoas, G. (2016). Spirituality and the travel motivations of older adults. *International Journal of Religious Tourism and Pilgrimage*, *4*(4), 79–85. https://doi.org/10.21427/D7BB14

Moufakkir, O., & Selmi, N. (2017). Examining the spirituality of spiritual tourists: A sahara desert experience. *Annals of Tourism Research*, *70*, 1–12. https://doi.org/10.1016/j.annals.2017.09.003

Myers, K. K., & Sadaghiani, K. (2010). Millennials in the workplace: A communication perspective on millennials' organizational relationships and performance. *Journal of Business and Psychology*, *25*(2), 225–238. https://doi.org/10.1007/s10869-010-9172-7

Narayanan, Y., & Macbeth, J. (2009). Deep in the desert: Merging the desert and the spiritual through 4WD. *Tourism Geographies*, *11*(3), 369–389. https://doi.org/10.1080/14616680903032783

Norman, A. (2011). Approaching spiritual tourism. In *Spiritual tourism: Travel and religious practice in western society*. https://doi.org/10.1017/CBO9781107415324.004.

Norman, A. (2012). The varieties of the spiritual tourist experience. *Literature & Aesthetics*, *22*(1), 20–37.

Nowell, L. S., Norris, J. M., White, D. E., & Moules, N. J. (2017). Thematic analysis: Striving to meet the trustworthiness criteria. *International Journal of Qualitative Methods*, *16*(1), 1–13. https://doi.org/10.1177/1609406917733847

Noy, C. (2004). This trip really changed me: Backpackers' narratives of self-change. *Annals of Tourism Research*, *31*(1), 78–102. https://doi.org/10.1016/j.annals.2003.08.004

Pargament, K. I. (1999). The psychology of religion and spirituality? Yes and no. *International Journal for the Psychology of Religion*, *9*(1), 3–16. https://doi.org/10.1207/s15327582ijpr0901_2

Parmar, S. M. (2016). Wellness tourism: Review and growth. *Paripex Indian Journal of Research*, *5*(9), 33–35.

Parsons, H., Houge Mackenzie, S., & Filep, S. (2019). Facilitating self-development: How tour guides broker spiritual tourist experiences. *Tourism Recreation Research*, *44*(2), 141–152. https://doi.org/10.1080/02508281.2019.1582159

Percy, W. H., Kostere, K., & Kostere, S. (2015). The qualitative report generic qualitative research in psychology recommended APA citation generic qualitative research in psychology. *The Qualitative Report*, *20*(2), 76–85.

Phukan, H., Rahman, Z., & Devdutt, P. (2012). Emergence of spiritual tourism in India. *International Journal of Marketing and Technology*, *2*(4), 300–317.

Plante, T. G., & Thoresen, C. E. (Eds.) (2007). *Spirit, Science, and Health: How the Spiritual mind Fuels Physical Wellness*. Praeger Publishers.

Prayag, G., & Ryan, C. (2011). The relationship between the "push" and "pull" factors of a tourist destination: The role of nationality – An analytical qualitative research approach. *Current Issues in Tourism*, *14*(2), 121–143. https://doi.org/10.1080/13683501003623802

Raj, R. (2012). Religious tourist's motivation for visiting religious sites. *International Journal of Tourism Policy*, *4*(2), 95–105. https://doi.org/10.1504/IJTP.2012.048997

Reich, K. H. (2000). What characterises spirituality? A comment on pargament, emmons and crumpler, and stifoss-hansen. *International Journal for the Psychology of Religion*, *10*(2), 125–128. https://doi.org/10.1207/S15327582IJPR1002_05

Rinschede, G. (1992). Forms of religious tourism. *Annals of Tourism Research*, *19*(1), 51–67. https://doi.org/10.1016/0160-7383(92)90106-Y

Robledo, M. A. (2015). Tourism of spiritual growth as a voyage of discovery. In *Tourism Research frontiers: Beyond the boundaries of knowledge* (pp. 71–86). Emerald. https://doi.org/10.1108/S1571-504320150000020009.

Robledo, M. A., & Batle, J. (2015). Transformational tourism as a hero's journey. *Current Issues in Tourism*, *20*(16), 1736–1748. https://doi.org/10.1080/13683500.2015.1054270

Roof, W. C. (1999). Spiritual marketplace: Baby boomers and the remaking of American religion. *Princeton University Press*, *62*(1), 140. https://doi.org/10.2307/3712241

Said, J., & Maryono, M. (2018). Motivation and perception of tourists as push and pull factors to visit national park. *E3s Web of conferences*, 31, 1–5. https://doi.org/10.1051/e3sconf/20183108022.

Saunders, M., Lewis, P., & Thornhill, A. (2009). *Research Methods for Business students*. Pearson Education.

Schneiders, S. M. (1989). Spirituality in the academy. *Theological Studies*, *50*(4), 676–697. https://doi.org/10.1177/004056398905000403

Shalbafian, A. A., & Zarandian, N. (2019). Volunteer tourism; An approach to realise spirituality (A new look of Iranian pro-poor tourism). *International Journal of Tourism & Spirituality*, *4*(1), 9–31. https://doi.org/10.22133/ijts.2019.95828

Sharkey, J., & Caska, B. (2019). Work-life balance versus work-life merge: A comparative and thematic analysis of workplace well-being. *DBS Business Review*, *3*, 49–72. https://doi.org/10.22375/dbr.v3i0.59

Sharma, N. P. (2020). Meditation tourism, Buddhist philosophy in practice: A case study of lumbini, Nepal. *The Gaze Journal of Tourism and Hospitality*, *11*(1), 37–59. https://doi.org/10.3126/gaze.v11i1.26615

Sharma, S. K. (2016). Spiritual tourism for economic growth and sustainable development by rediscovering values and linking cultures. *International Journal of Transformation in English & Education*, *1*(1), 47–50.

Sharpley, R. (2016). Tourism and spirituality: An evolving relationship. *International Journal of Tourism and Spirituality*, *1*(1), 8–24. https://doi.org/10.22133/IJTS.2016.43074

Sharpley, R., & Jepson, D. (2011). Rural tourism: A spiritual experience? *Annals of Tourism Research*, *38*(1), 52–71. https://doi.org/10.1016/j.annals.2010.05.002

Sharpley, R., & Sundaram, P. (2005). Tourism: A sacred journey? The case of ashram tourism. *India. International Journal of Tourism Research*, *7*(3), 161–171. https://doi.org/10.1002/jtr.522

Sims, A. (1994). "Psyche" - spirit as well as mind? *British Journal of Psychiatry*, *165*(4), 441–446. https://doi.org/10.1192/bjp.165.4.441

Singh, N., & Shukla, N. K. (2016). Spiritual tourism: A socio-cultural approach for holistic development of society (a feasibility analysis with kanpur city, UP). *IJHPD*, *5*(2), 43–50.

Singleton, A. (2017). The summer of the spirits: Spiritual tourism to america's foremost village of spirit mediums. *Annals of Tourism Research*, *67*(C), 48–57. https://doi.org/10.1016/j.annals.2017.08.002

Smith, J., & H, B., & Cheater, F. (2011). Theoretical versus pragmatic design challenges in qualitative research. *Nurse Researcher*, *18*(2), 39–51. https://doi.org/10.7748/nr2011.01.18.2.39.c8283

Smith, M. (2003). Holistic holidays: Tourism and the reconciliation of body, mind and spirit. *Tourism Recreation Research*, *28*(1), 103–108. https://doi.org/10.1080/02508281.2003.11081392

Snyder, S. (2007). New streams of religion: Fly fishing as a lived, religion of nature. *Journal of the American Academy of Religion*, *75*(4), 896–922. https://doi.org/10.1093/jaarel/lfm063

Sorayya, A. K., & Soleimanabadi, F. (2019). A methodology study of the concept of spirituality in tourism according to fuzzy logic. *International Journal of Tourism & Spirituality*, *3*(2), 85–105. https://doi.org/10.22133/IJTS.2019.84370

Stausberg, M. (2014). Religion and spirituality in tourism. In A. A. Lew, C. M. Hall, & A. M. Williams (Eds.), *The Wiley Blackwell companion to tourism* (pp. 349–360). Wiley.

Stifoss-Hanssen, H. (1999). Religion and spirituality: What a european ear hears. *International Journal for the Psychology of Religion*, *9*(1), 25–33. https://doi.org/10.1207/s15327582ijpr0901_4

Tashakkori, A., & Creswell, J. W. (2007). Editorial: The new era of mixed methods. *Journal of Mixed Methods Research*, https://doi.org/10.1177/2345678906293042

Tilson, D. J. (2005). Religious-spiritual tourism and promotional campaigning: A church-state partnership for St. James and Spain. *Journal of Hospitality & Leisure Marketing*, *12*(1–2), 9–40. https://doi.org/10.1300/J150v12n01_03

Unger, L. S. (1991). Altruism as a motivation to volunteer. *Journal of Economic Psychology*, *12*(1), 71–100. https://doi.org/10.1016/0167-4870(91)90044-T

Vaismoradi, M., Jones, J., Turunen, H., & Snelgrove, S. (2016). Theme development in qualitative content analysis and thematic analysis. *Journal of Nursing Education and Practice*, *6*(5), 100–110. https://doi.org/10.5430/jnep.v6n5p100

Vaughan, F. (1991). Spiritual issues in psychotherapy. *Journal of Transpersonal Psychology*, *23*(2), 105–119. http://www.atpweb.org/jtparchive/trps-23-91-02-105.pdf.

Voigt, C., Brown, G., & Howat, G. (2011). Wellness tourists: In search of transformation. *Tourism Review*, *66*(1/2), 16–30. https://doi.org/10.1108/16605371111127206

Walters, T. (2016). Using thematic analysis in tourism research. *Tourism Analysis*, *21*(1), 107–116. https://doi.org/10.3727/108354216X14537459509017

Willson, G. B., McIntosh, A. J., & Zahra, A. L. (2013). Tourism and spirituality: A phenomenological analysis. *Annals of Tourism Research*, *42*(2013), 150–168. https://doi.org/10.1016/j.annals.2013.01.016

Yeong, M. L., Ismail, R., Ismail, N. H., & Hamzah, M. I. (2018). Interview protocol refinement: Fine-tuning qualitative research interview questions for multi-racial populations in Malaysia. *The Qualitative Report*, *23*(11), 2700–2713.

Zahra, A., & McIntosh, A. J. (2007). Volunteer tourism: Evidence of cathartic tourist experiences. *Tourism Recreation Research*, *32*(1), 115–119. https://doi.org/10.1080/02508281.2007.11081530

Zinnbauer, B. J., Pargament, K. I., Cole, B., Rye, M. S., Butter, E. M., Belavich, T. G., Hipp, K. M., Scott, A. B., & Kadar, J. L. (1997). Religion and spirituality: Unfuzzying the fuzzy. *Journal for the Scientific Study of Religion*, *36*(4), 549. https://doi.org/10.2307/1387689

Zinnbauer, B. J., Pargament, K. I., & Scott, A. B. (1999). The emerging meanings of religiousness and spirituality: Problems and prospects. *Journal of Personality*, *67*(6), 889–919. https://doi.org/10.1111/1467-6494.00077

Index

Note: **Bold** page numbers refer to tables and *italic* page numbers refer to figures.

One night in '98: This guy I know invites me uptown to smoke a blunt. As we are puffing away, his itty-bitty boombox catches the munchies and starts snacking on his mixtape. While the guy tries to salvage his shit with the precision of a brain surgeon, a white chick rolling a shopping cart interrupts. With a voice faltering like pantyhose scraped across sandpaper, she scuffs along, "...Uhhh...Could yous watch my cart...for a minute...I gotta go upstairs...to....uhhhh... take a shower?"

The guy answers, "Yeah. Go ahead."

"I just gotta clean up real quick."

"No problem," he assures her, but she doesn't move. Again, he promises, "Don't worry about it. We'll be here."

"...Uhhh...Yeah...Thanks...." Her eyes switch from us to the building, fingers twitching for a fix. She darts past her cart, which happens to be stocked with—among other things—a set of striped bath towels.

While lighting a cigarette, I hear a door slam.

"Did you see that?" the guy asks, his voice edgy with excitement.

"See what?"

"You didn't see that?"

"Okay," I sigh, speaking slowly. "Tell me what you saw, and I'll tell you if I saw it."

The guy pauses and shakes his head in frustration, then continues. "Right after she went inside, some guy jumped out that door over there and grabbed something out her cart." Watching my face intently, he adds, "It's like he was watching through the window."

My eyes widen and furtively glance at the door in question. "Do you think he's watching now?"

This guy hadn't considered that. "Let's go over to Broadway. I gotta stop by this restaurant for a minute."

The restaurant is packed. It's happening. It's the place to be, what with it being Tuesday and everything. We squeeze past the customers perched like birds at the counter. An odd sense of anticipation energizes the room, though everyone plays it cool. We are seated up front, inside the picture window, underneath a neon sign that shines a warm light on our table. The walls are lined with mirrors, the seats are stuffed with people, and the air conditioning cannot compete with the resounding body heat. A waitress brings us a couple of cups, but no menus. In my reflection, I notice no one is eating.

A stunning Dominican woman, barely covered by swaths of the softest pink suede, saunters towards us, speaking *espanol muy rapido*, all smiles as she takes a seat. *"Hola Papi...Qué necessitas?"* Turning towards her, this guy places his order while I eye myself from a multitude of angles, trying to determine whether my hair is glorious golden brown or simply succulent caramel. So involved am I in me that I barely notice when Pink Suede Sweetie leaves.

Ten, fifteen minutes later, she returns with an even bigger smile and sits down at the table, where Styrofoam cups of murky water masquerading as coffee are quietly ignored. Exercising discretion, she passes this guy a package under the table. Distracted, he holds the plastic bag up to the light to get a look. Taken aback, she gasps while I burst into a laugh. He shrugs and gets up. "Let's go. I want to take you somewhere."

We head west, down a side street to Riverside Park with an expansive view of the West Side Highway, and just beyond that, the Hudson River. "Fuck!" the guy mutters to himself.

"What?" I wonder.

"The shopping cart," he remembers, stopping mid-step and turning east. "We just left it."

I start laughing. "Shit!"

"I know. I feel bad."

"Nahh. Not that. I mean, I gotta piss."

"Why didn't you go in the restaurant?"

"'Cause that place was nasty!"

"But pissing in the park? That's classy."

"Ohhhhh! I'm just gonna take a little walk." I pivot quickly, swish down the ramp, and nip around the corner to a secluded spot at the bottom of a steep bank where a storm drain beckons. With a quick zip and tug at the hips, my jeans are around my ankles and I aim carefully to avoid the dreaded backsplash all females know well. Amazed at the volume now traveling into the city sewer, I have plenty of time to check out my surroundings and almost fall over when I look up into the sky. There, just above me, are cars cruising the Drive, headlights igniting their rides, spotlights zooming through the night.

I press on, then zip up.

At the top of the ramp, I find this guy with his nose in a dollar bill but decline his offer, having sworn off ski trips after an eight ball and a date with Oscar Wilde. "We should go." We stroll along a path at the edge of the park illuminated by old streetlamps once filled with gas but now lit courtesy of Con Ed. The breeze through the trees conspires with the cobblestone scene, making Harlem on the Hudson something like Paris on the Seine.

When we reach Broadway, we grab a gypsy cab. What starts off as a pleasant ride turns into a hideous wreck when cops stop the car before we get on the West Side at One Two Five. An undercover wearing a Mets cap asks the driver our destination, but the cabbie barely speaks English and stutters incoherently. Doom seems imminent. We are finished like Orwell's Burmese elephant.

Looking the cop dead in the eye, clasping a purse littered with dimes (while this guy slinks further and further into the backseat in a sad attempt to blend into the upholstery), I tell Five-O that we're going to Twelfth and Sixth, *Twelfth and Sixth*. The cop nods in agreement and lets us go.

Close, but not quite. I promise I'll try harder next time...

—Miss Rosen

Photograph © **Mia Petzäll**

powerHouse magazine

Issue 5

Staff
Retired Detective First Grade: Daniel Power
Idealist: Craig Cohen
Dance Fever Finalist '81: Miss Rosen
Aerial Camera Operator: Mine Suda
Grifter: Ariana Barry
Fifth Beatle: Daoud Tyler-Ameen
NEET: Riyo Mochizuki

Contributing Artists
Keiji Ando, Basty, Orkan Benli, Nora Bibel, Boogie, NOV, Nicola Cairns, COPE2, Matthew Charles Crabe, Raphaël Dallaporta, DJ Disco Wiz, Deborah Dragon, Ron English, Derek Erdman, Larry Fink, Yoav Galai, Jesse Gammage, Arlene Gottfried, Roc Herms Pont, Steven Hirsch, James Huges, JR, Alex Lukas, John Lurie, Narayan Mahon, Sarah Anita McNeill, Caleb Neelon, ONE9, Mark Peterson, Mia Petzäll, Joseph Rodríguez, Joseph Rivera, Ivan Sanchez, Colin Simmons, Lilla Szasz, Nathaniel Welch, Kurt Wilberding, Hank Willis Thomas, Geordie Wood, Wooz, Meiko Xavier, Joy Yoon

Contributing Writers
Patti Astor, Basty, Nora Bibel, Loucious Broadway, Nathan Brown, Negri Cabrini, COPE2, Matthew Charles Crabe, DJ Disco Wiz, Michael Ellsberg, Ron English, Derek Erdman, Samantha Gainsborough, Yoav Galai, Michael A. Gonzales, Griddy Grimes, Steven Hirch, Pinch Hudson, Idris Intifada, Todd Kelley, Narayan Mahon, Brantly Martin, Craig Mathis, MISTA, Caleb Neelon, Joseph Rodríguez, Miss Rosen, Christophe Salet, Ivan Sanchez, Lilla Szasz, Maureen Valdes, Nathaniel Welch, Geordie Wood, Joy Yoon

For advertising inquiries please contact Miss Rosen
tel: 212 604 9074 x105, email: sara@powerhousebooks.com

For distribution inquiries please contact Wes Del Val
tel: 212 604 9074 x103, email: wes@powerhousebooks.com

powerHouse Magazine
37 Main Street, Brooklyn NY 11201
Tel: 212.604.9074
Fax: 212.366.5247
Email: magazine@powerhousebooks.com
ThepowerHousemagazine.com

Cover art © **Derek Erdman**
Photograph © **Mia Petzäll**

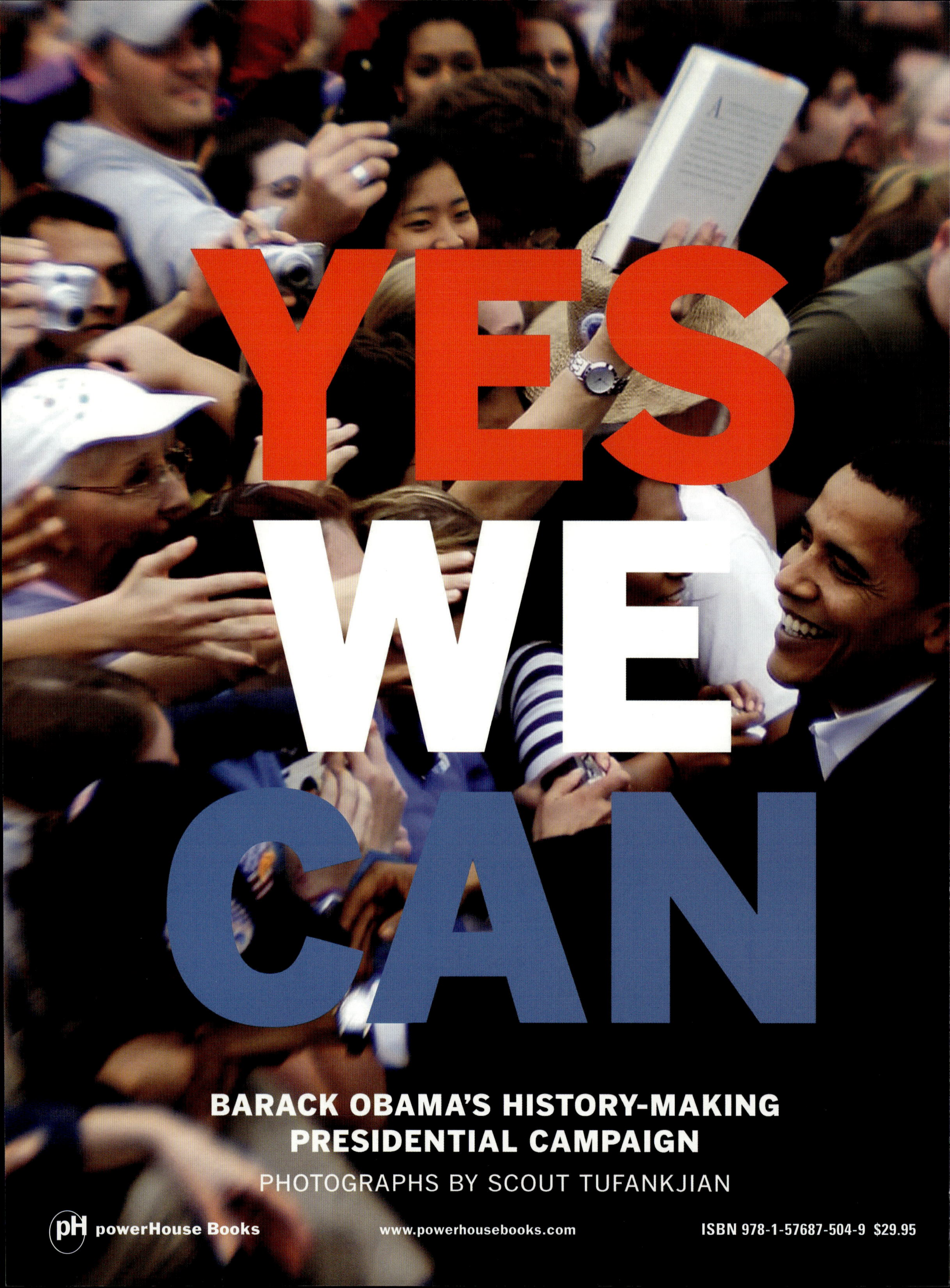
YES
WE
CAN
BARACK OBAMA'S HISTORY-MAKING
PRESIDENTIAL CAMPAIGN
PHOTOGRAPHS BY SCOUT TUFANKJIAN
pH powerHouse Books
www.powerhousebooks.com
ISBN 978-1-57687-504-9 $29.95

CONTRIBUTORS

Keiji Ando was born in NYC and raised one block from the powerHouse Arena, at 31 Washington Street. The village that raised him included sculptors, painters, writers, teachers, and troublemakers. He currently resides in Brooklyn.

Founder of the FUN Gallery, **Patti Astor** has appeared in a dozen films, including *Wild Style, Rome '78, The Long Island Four, Snakewoman,* and *Underground U.S.A. FUN! The True Story of Patti Astor* (Miss Rosen Editions/powerHouse Books) will be published in conjunction with the documentary film *Patti Astor's FUN Gallery* and a feature film with Roberts/David Films.

Basty is a neuroscientist and photographer whose work moves around the nebulous border of brain science, visual arts, and psychologically charged narratives. He lives and works in Manhattan and his photographs and slide presentations have appeared in various shows on the East Coast and in his native Berlin.

Orkan Benli was sleeping in his bathtub on New Year's Eve 2000. He did not believe the hype. Having finished his studies at the Berliner Technische Kunstschule, he came straight to NYC to first intern and then work for powerHouse Books as a graphic designer.

Nora Bibel, born in Munich in 1971, has been living and working in Berlin as a freelance photographer since 1999. She shoots commercially in addition to her own fine art photography, and her work has received several awards and been exhibited internationally. www.nora-bibel.de

Boogie was born and raised in Belgrade, Serbia, and now lives in Brooklyn. He is the author of *It's All Good, Boogie*, and *Belgrade Belongs to Me* (powerHouse Books, 2006, 2007, and 2008). www.artcoup.com.

Loucious Broadway a.k.a. Dumar Brown a.k.a. Nov York a.k.a. Mokie G a.k.a. BK Joe a.k.a. Babylove Johnson a.k.a. Allah Sun Jackson a.k.a. Genghis Khan a.k.a. Bouba Adula a.k.a. Jesus Virgo a.k.a. Unique Holmes a.k.a. Showtime a.k.a. Ching a.k.a. Jeffrey Katzman a.k.a. Burnadette Burnswell a.k.a. Kill Giuliani a.k.a. Nov is your new favorite writer. www.screamingnov.com

"...**Nathan Brown** is a classically trained young man who was taught to use his mind and think creatively, and has grown up with the audacity to do so. He's been many places, and will go many more. 'Out there things can happen, and frequently do, to people as footsy and brainy as you....'" —Nathan's Mom

Negri Cabrini is a native New Yorker of Cuban descent who grew up in Washington Heights before Giuliani, gentrification, and Starbucks. She has a BA in Spanish from CUNY.

Nicola Cairns is an Edinburgh artist born in 1975. She specializes in figurative work of detailed nudes, femme fatales, jazz musicians, and hustlers, all of which are becoming highly collectible. Cairns' style is influenced by her love of film noir, Hitchcock, and the glamour of the 40s and 50s.

COPE2 is a legendary graffiti writer who, along with his crew Kings Destroy, has been working in the streets of New York for more than 25 years. He was the main focus of the seminal graffiti movie *Kings Destroy* and has authored a photography book of his work entitled *COPE2: True Legend*.

Matthew Charles Crabe is an ex-exterminator from Budd Lake, NJ. When he is not asleep, he enjoys self-publishing booklets of his illustrations, drinking black-label beer, putting explodables in rotten food, backside smith grinds, reading comics, and getting farted on by his trophy girlfriend. Send digital food and praise to crab.blood@gmail.com.

Raphaël Dallaporta is a French independent photographer whose work has appeared in *The New York Times Magazine, Le Monde, Vogue Italia, Wallpaper**, and many other publications. His personal projects have been exhibited at Les Rencontres d'Arles, France, the Musée de l'Elysée, Switzerland, and the New York Photo Festival. www.raphaeldallaporta.com

DJ Disco Wiz (real name: Luis Cedeño) is credited as being the first Latino hip-hop DJ and the first DJ to make a "mixed plate" in 1977 along with Grandmaster Caz. The creator and founder of the Hip-Hop Meets Spoken Wordz series, Wiz is also the coauthor of *It's Just Begun: The Epic Journey of DJ Disco Wiz, Hip Hop's First Latino DJ* (Miss Rosen Editions/powerHouse Books, 2009). www.djdiscowiz.com

Deborah Dragon is the Deputy Photo Editor at *Rolling Stone*. She lives in New York and has a BFA in photography. http://gallery.me.com/dragond1

Michael Ellsberg is a professional book proposal writer and developmental editor. He is the founder of Eye Gazing Parties (www.eyegazingparties.com), which *Elle* called "New York's hottest dating trend." Born in San Francisco, he now lives in Brooklyn. www.ellsbergediting.com

New York-based painter **Ron English** is a seminal figure of the culture-jamming movement, in which artists and activists subvert existing advertisements to encourage free thought. He has replaced more than a thousand billboards with his own "subvertisements," and is the subject of the documentary *POPaganda: The Art and Crimes of Ron English*.

Derek Erdman painted a picture when he was 12 and another when he was 18. He started making them all of the time when he was 27, and now he is 34. He lives in Chicago with a cat and a roommate who is a law student. He likes candy and sleeping.

Larry Fink is a two-time National Endowment for the Arts Fellow, a Guggenheim Fellow, and a professor of photography at Bard College. The author of seven books, Fink's work has appeared in top publications including *W, GQ, Detour*, and *The New Yorker*.

Samantha Gainsborough is a kept woman. Currently residing in the lovely city of Montreal, she spends day after day reflecting on the silly things people do to get by. Gainsborough especially enjoys the occasional afternoon in Outremont indulging in Anaïs Nin erotica with like-minded friends.

Yoav Galai is a documentary photographer based in London. He is currently working on a long-term project documenting the effect of municipal and governmental negligence on the Palestinian population of east Jerusalem. www.yoavgalai.com

Jesse Gammage was born in Los Angeles and received his degree from San Diego State. He has participated in different shows in the San Diego area at venues such as Hot Monkey Love Café, Excelsior Brasserie, and a series of group showings of local art and music at The Kava Lounge on the last Sunday of every month. www.pobois.com

Harlem native **Michael A. Gonzales** has written cover stories for *Vibe, Stop Smiling, XXL, The Source*, and *Essence*. His short fiction has appeared in *Trace, ColorLines, Bronx Biannual, The Darker Mask*, and UK anthology *Tell Tales IV*. Currently striving in Brooklyn, his essay "Broken Dreams" will be reprinted in *Best African-American Essays 2009*.

Arlene Gottfried is the author of *Midnight* and *Sometimes Overwhelming* (powerhouse Books, 2002 and 2007). Her work has appeared in *The New York Times Magazine, Fortune, Life*, and *The London Independent*. She has exhibited at the Leica Gallery in New York and in Tokyo, and at the Smithsonian Institution in Washington, D.C. www.arlenegottfried.com

Griddy Grimes is a founding member of the up-and-coming hip hop group Gridlock Fam. Already featured on super-producer Nicolay's album *The Dutch Masters Vol. I* and *Here*, Gridlock Fam is currently working on *The Dutch Masters Vol. II* and a yet-to-be-titled album entirely produced by Nicolay.

Roc Herms Pont sounds German, but is Catalan. He recently discovered the world of photography, quit his job as art director at an advertising agency, and embarked on a series of personal projects. He was selected as a finalist for Lux 2007 and Descubrimientos Photoespaña 2007 and 2008.

Steven Hirsch's work has appeared in numerous publications including *Newsweek, Time*, and *The New York Times*. He has been awarded two New York Foundation for the Arts grants, and his work has been widely exhibited and is in the collections of the Museum of Modern Art and the Bibliothèque Nationale de France, among others. www.courthouseconfessions.blogspot.com, www.stevenhirsch.com

Pinch Hudson has just ghostwritten a landmark autobiography. Not that you'll ever know.

continued...

Contributors

James Hughes lives and works in the third of three apartments in his apartment building, numbered 251, which has a 11211 zip code. It's across the street from a garbage truck company named Skillman, after the avenue they share. He is currently entering his last year at the School of Visual Arts, where he studies graphic design. He likes getting mail. Send him some...

Ironically, **Idris Intifada** has no tattoos but lives in a Brooklyn domicile that houses four tattoo artists. Instead of ink to skin, he puts it to paper for magazines such as *Modern Fix* and *Mass Appeal*. When not flexing his pen game, Idris is hunting cheap energy drinks. He is currently seeking a Red Bull sponsorship.

JR is 25 and exhibits freely in the streets of the world. His work mixes Art and Act, addressing ideas of commitment, beauty, freedom, identity, and limit. His 2007 exhibit with Marco, "Face 2 Face," was the largest illegal photo exhibition ever: eight Palestinian and Israeli cities on both sides of the separation wall. He is currently working on "Women" in Sierra Leone, Liberia, Brazil, and India. www.jr-art.net

Todd Kelley is a blogger gone mad who is breaking out into the real world with his unique and captivating writing style.

John Lurie—musician, actor, producer and, most recently, exhibiting painter—emerged in the late 1970s as a saxophone player leading the avant-garde New York jazz group the Lounge Lizards. He has scored over 20 films and starred in *Stranger than Paradise, Down by Law*, and the TV series *Fishing with John*, which he also wrote and directed. He was nominated for a Grammy award for the score of *Get Shorty.*

Narayan Mahon is a photographer based in Istanbul, Turkey, where he works on photo essays and projects such as Istanbul's Illegal Literati and Lands in Limbo, about unrecognized countries. More of his work can be seen online at www.narayanmahon.com.

Brantly Martin has been working in New York City nightlife for the past eight years, hosting weekly parties at various downtown Manhattan venues. He was born in Houston and now lives in Rome.

Craig Mathis listens to A, Siege, VJ, Negative Approach, and HH. Originally hailing from south of the Mason Dixon line, he now considers Brooklyn and Las Vegas home.

Sarah Anita McNeill is an eclectic polyglot who is destined to wander continents with the occasional Gauloises, European philosophy, fountain pen, notebook, and her camera in tow. Get a peek of her travels at www.sarahanitamcneill.com.

Born and raised in uptown Manhattan's Washington Heights, **MISTA** is one of the presidents of the notorious KGKASS crue, which includes FEES, NETR, ABOT, TECH, POOMA, and YEWS. Having gone all-city with KGKASS, MISTA has hit Times Square, the George Washington Bridge, and the West Side Highway Bridge. Despite all the beef with rival crues and cops, MISTA still maintains to keep his name up and doesn't plan on stopping anytime soon.

Caleb Neelon is an artist from Cambridge, Massachusetts. He is the coauthor of *Graffiti Brasil, Street World*, and a new monograph of his own work from Gingko Press, *Caleb Neelon's Book of Awesome*. www.theartwheredreamscometrue.com

Washington, DC native **ONE9** is an artist whose paintings are based off of rhythm and negative space. He conducted a workshop at the Smithsonian Institution's Seeing Jazz Festival in 1997. ONE9's artwork has been displayed in *The Walls Belong to Us* and allhiphop.com's *Rebirth of Ill* art show. He lives and works in Brooklyn.

Minneapolis native **Mark Peterson** moved to New York in 1987 to pursue his photography career. The author of *Acts of Charity* (powerHouse Books, 2005), Peterson's work has appeared in *The New York Times Magazine, Time, Newsweek, New York*, and *People*. Peterson is represented by Redux Pictures.

Mia Petzäll began photographing in 2007, and was selected as a finalist for Descubrimentos Photoespaña in 2008. She studies with Anders Petersen, and recently completed a series called "Pregnancy." Her new project, "Instinct," is about hunting.

Joseph Rodríguez has been awarded "Pictures of the Year" by the National Press Photographers Association in 1990, 1992, 1996, and 2002. His previous books include *Still Here: Stories After Katrina, Flesh Life: Sex in Mexico, Juvenile*, and *East Side Stories: Gang Life in East L.A.* (powerHouse Books, 2008, 2006, 2003, and 1998).

Miss Rosen is Associate Publisher and Senior Vice President of Marketing and Publicity at powerHouse Books, Publisher of Miss Rosen Editions, and Editor of powerHouse Magazine. She has written for the *Telegraph, Swindle, Whitewall, Staf*, and *Jewcy*.

Joseph Rivera joined the New York City Transit Police Department in 1984, and was assigned to the elite Transit Vandal Squad in 1987. Rivera became the lead investigator of the Graffiti Habitual Offender Suppression Team (G.H.O.S.T.) within the Vandal Squad in 1998. Rivera retired in 2004.

Christophe Salet is a Paris-based freelance writer and journalist for various art and lifestyle magazines.

Ivan Sanchez is a youth advocate, motivational speaker, and the author of *Next Stop: Growing Up Wild Style in the Bronx* (Simon & Schuster, 2008) and coauthor of *It's Just Begun: The Epic Journey of DJ Disco Wiz, Hip Hop's First Latino DJ* (Miss Rosen Editions/powerHouse Books, 2009). ivan@isanchez.com

Colin Simmons moved from Minneapolis to NYC, to bust flip tricks and snap flicks. Having lived in Gotham for about a decade now, he's gained quite the photographic rep shooting for magazines like *Mass Appeal*. If you toys doubt the clout, check out his personal website, www.cadillacgypsy.com.

Lilla Szasz, born and residing in Budapest, has received awards and scholarships for her photography. She has exhibited her work internationally, and most recently was selected by Photoespaña for their Descubrimientos program. www.szaszlilla.hu

Maureen Valdes is the accidental author, bumping into a writing career by sheer luck rather than by hot pursuit. Her previous writing experience includes working for a small independent newspaper in the Sierra Nevada foothills of California.

Nathaniel Welch was born in 1966 in Miami, Florida. The author of *Jesse James and His Beautiful Machines* and *Spring Broke* (powerHouse Books, 2007 and 2004), Welch's work has appeared in *The New York Times, Newsweek, Time, GQ, Rolling Stone*, and *Interview*, among other publications.

Kurt Wilberding is a photojournalist based in New York. In addition to covering the arts and fashion for *The Wall Street Journal*, he has spent several months documenting the 2008 campaign trail. Various blogs, most notably Gothamist, Techcrunch, and Gawker, have picked up his street photography. He has a MFA from Ohio State University, and his work can be viewed at kurtwilberding.com.

Hank Willis Thomas has exhibited in galleries and museums including the Studio Museum, Harlem; Wadsworth Atheneum, Hartford; Leica Gallery, New York; and the National Portrait Gallery, Washington, D.C. Willis Thomas is the first recipient of the Aperture West Book Prize, a new annual prize awarded by the Aperture Foundation. He lives in Oakland, California.

Geordie Wood received a photography degree from Syracuse University and continued making images under *National Geographic* photographer Steve McCurry. Wood is internationally published and exhibited; his works have been featured in publications such as *The Fader, The Globe Sunday Magazine*, and *The Toronto Star*. www.geordiewood.com

"**Wooz** is my little brother. When we were much younger I would prevent him from waking up before me by any means possible. This generally involved moving a large object (usually a dresser) in front of the bedroom door, staying up all night, or waking up and locking him in. Needless to say, he is permanently ruined because of this; there is no hope at all. My advice is to stay out of his way in the morning—that or have a warm bowl of ramen and a freezer pizza prepared." —Nathan Brown

Meiko Xavier is an activist, writer, photographer, masseur, and aspiring actor/model. Currently, Xavier is working on a community education project on Youtube to provide people with information about the transgender/genderqueer experience.

Joy Yoon is a freelance writer residing in Los Angeles. yoonion@gmail.com

BOOK OF PHOTOGRAPHY BY GUY OSEARY RECORDS THE LEGENDARY ARTIST'S
GROUNDBREAKING CONFESSIONS TOUR
MADONNA
CONFESSIONS
PHOTOGRAPHY BY GUY OSEARY
pH powerHouse Books
www.powerhousebooks.com
PHOTOGRAPHY AND INTRODUCTION BY GUY OSEARY ISBN 978-1-57687-481-3 $39.95

DEAR pH MISS Rosen

I am STUMPED. This is truly my most trying LETTER TO THE EDITOR due to my inability to relate to the subject matter of being "Busted" in the sense of being busted. A topic I am unfamiliar with at this time. Growing up a Christian, Catholic, Baptist, Presbyterian, Buddhist OMM.. Korean American girl in the U.S. of A, I have dedicated myself to the pursuit of stereotypical excellence and therefore could not possibly know what BUSTED is . . .

As a child I studied hard, watched NOVA on PBS with Alan Alda, and was reading at a junior college level. Getting awards LEFT & RIGHT and trying too hard was just part of fulfilling my destiny as an overachieving Asian. With multi-cultural club and being the water individual for various sports activities and playing the flute in marching band, little time was left for meaningless shenanigans.

There were several times I can **recall** in high school where my father **thought** he busted me but was just confused by my lack of interest in adolescent mating rituals. In the 11th grade, my purity came into question when I started receiving a slew of **telephone** calls from boys in school, at all hours of the night, asking for homework help. (An early confirmation that girls are smarter than boys) In the 12th grade majority of all my study groups were comprised of girls, which prompted my father to ask "YOU, You LESBIAN?" I just yelled back, "What the ....?! You want me to get A and go to top rated medical school or what? Stop asking stupid questions." He bought my study group pizza and left me alone for the rest of the school year. Ho Hum.

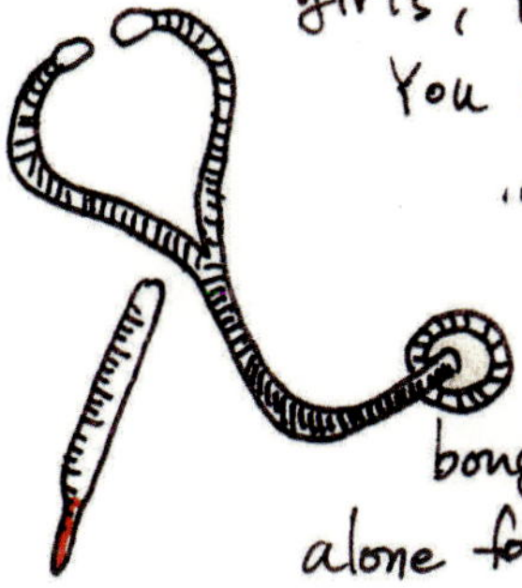

What can I say. Asian Fathers. You gotta love'em or they'll disown you.

♥ - Did I mention I love my father? - ♥

So little can be said about being busted. Though I find myself in the company of drug dealers, kleptomaniacs, people driving without auto insurance.. I still find myself in constant restraint. Never toeing the line or even looking at it. I hope to marry well and being busted never looks good on a martial resume..

A+

For Overachieving Asians Everywhere.

A+ A+

J. Yoon (윤조이)

A girl stabs her own father to death. Another shoots her stepfather. A third is a thief. They've all lost something, or never had it in the first place: parents who loved them enough, an environment that supported them in times of trouble. "Why do such people like my parents need a child?" asks Sabina with a trembling voice. "They can fuck but can't give love or tenderness." Sabina has spent three years in a detention home for adolescent girls. She was only four months old when her parents first put her in an orphanage, and spent her life shuttled from one orphanage to another.

Her father once worked as a mechanic and her mother also had a job. Then they both started drinking and the family fell apart. In 2004 Sabina and her three brothers and sisters were taken away by the government and put into a home. At that time she was the only breadwinner in the family, and she called her job "Shopping Malling," which meant stealing from shopping malls. Sabina was 10 years old when she began to steal. Then the drugs came, starting with sniffing glue and ending up on heroin. She was eventually busted for armed robbery; she and her two friends held a girl at knifepoint and ordered her to give over her wallet. The girl only had 35 bucks and a ragged cell phone.

Sabina thinks that her crime was nonsense. But she also says that she had no other choice. "My mother and father left me. Where else could I have turned?" Today she is clean and leads a normal life in the detention home. When she is free, she plans to go to a care-house with her brothers and sisters.

I have always been interested in situations in which one can experience the extremities of life. The people I photograph and their fates are not often spoken about. They live their lives, they try to survive, and the only time they come into our consciousness is if they commit a crime. Nobody cares why they did what they did. Nobody thinks there could be reasons for their behavior. I committed myself to these children to show what the beauty of life means for them, what their dreams and hopes are. And the answer is so simple. They wish to be what they are: children.

—Lilla Szasz

LILLA SZASZ

MIDNIGHT EXPRESS

BY NEGRI CABRINI

What does a single woman do when she sees a tasty morsel staring in her direction at some bullshit wedding reception she was dragooned into? Well, I can tell you what this one did—and where it led…

Gilberto, my father's long time friend, had invited us to his daughter's wedding, which was filled with white Wall Street bankers and Cubans. No in-between. Bankers bore me and the Cuban rumor mill was too much to handle. I didn't even like the bride. She was a 39-year-old virgin marrying a guy she had been dating for two minutes. The whole thing seemed desperate and I only went because it was important to my father. But now this dull reception was turning out to be a great night.

I shot him my most fabulous "come to me and I am yours" look. I nearly died as he walked over. I played it cool though. Never let them see you sweat and all. "Would you like to dance?" he asked, seductively caressing my hand as if he knew a "yes" would follow. We walked over to the dance floor and the flow was magical. I worried since salsa was not my strength but to my surprise, it worked. He knew how to lead and I surrendered to his every move. The dancing was so hot, I was completely swept away and I forgot where I was.

The song ended and suddenly reality set in. I knew every matchmaking *abuela* was keeping an eye on us. On cue my father walked over and wrapped his arms around us both. "I see you have met Gilberto's nephew, Vincent," he said with a big smile on his face. Was it official? Was my father giving us his seal of approval?

On the surface, this guy was everything I thought I always wanted: Cuban, charismatic, and sexy as hell. I knew I was in trouble.

"He is in town for spring break," continued my father.

"Spring break? How old are you?" I asked.

"Twenty," Vincent said confidently.

The plot thickened. My man wasn't even old enough to buy me a drink. And believe me, I needed to take the edge off after hearing that.

To my surprise, this boy was brilliant. He attended Brown University and was a writer. He played the guitar and could sing, was remarkably articulate and, did I mention, tasty? The conversation was stimulating; we had so much in common. We shared the comfort and passion of a language and culture. I felt as if I had known him forever. Here I was spending time with this young man…who had his shit together! I would follow his lead. I trusted him immediately. I found myself ridiculously attracted. My inhibitions soon vanished.

The following evening we went out to dinner. After spending hours together, we parked in a very romantic and secluded location. The view was to die for—too bad we spent so little time looking at it.

Vincent excited me in ways I hadn't felt in ages and he hadn't even touched me yet. The sound of his voice was sufficient to arouse me. I never thought I would be telling a young man to keep talking. This was gonna be good.

We went to the back seat and started to kiss, passionately. His little red car was beginning to fog up from the inside. I was going to keep my clothes on since we had met through our families and I was trying to be good. Then the boy went all Henry Miller on me and told me more about what he would do to me when he got my clothes off. Torture, considering we had no place to go. He unbuttoned my blouse and tried to undo my bra from behind but I stopped him. I had to take control…

I unzipped his pants and swallowed him alive. He dug his feet under the driver's seat to brace himself. I knew he loved it—because I loved it. And as I stretched out across the backseat, I worked my magic and he was completely enraptured. After a while that vein was throbbing…you know the one I mean. As I glanced up at the back of the driver's seat to breathe and focus I suddenly saw a bright green light flashing in. I wondered what it was but was not willing to stop at such a crucial moment.

Then four loud raps to the car window alerted me that we were about to have an intermission.

"WILL YOU STEP OUT OF THE AUTOMOBILE?" a voice boomed.

Good GOD! Nothing but the opening scene of *Midnight Express* could capture the way my heart was pounding. The worst part was that I knew just a few more seconds and he would have busted a nut…

"Man, I am gonna get arrested for giving Vincent the best head of his life?" I wondered just what our families would think of this one—the old Cuban rumor mill would surely have a lot to say about my oral fixations. I could see this tarnishing the illusion of "purity" I had worked so hard to project.

Vincent quickly pulled up his pants and stepped out of the car, zipper still undone. Two plainclothes officers asked him for his license and registration. "What seems to be the trouble?" he asked. While one of the officers told him we were parked in a no-trespassing zone, the other shined the light in on me and asked, "Are you alright, young lady?"

Still sitting in the car with my blouse buttons undone, my hands immediately went up as if he had pulled a gun on me. "Yes, I am here by choice!" I squealed, hoping my response would end this nerve-racking interruption. Let me tell you, every grey hair I have to date was born that night.

After all that, they let us go. We continued our tryst, but took it to a motel.

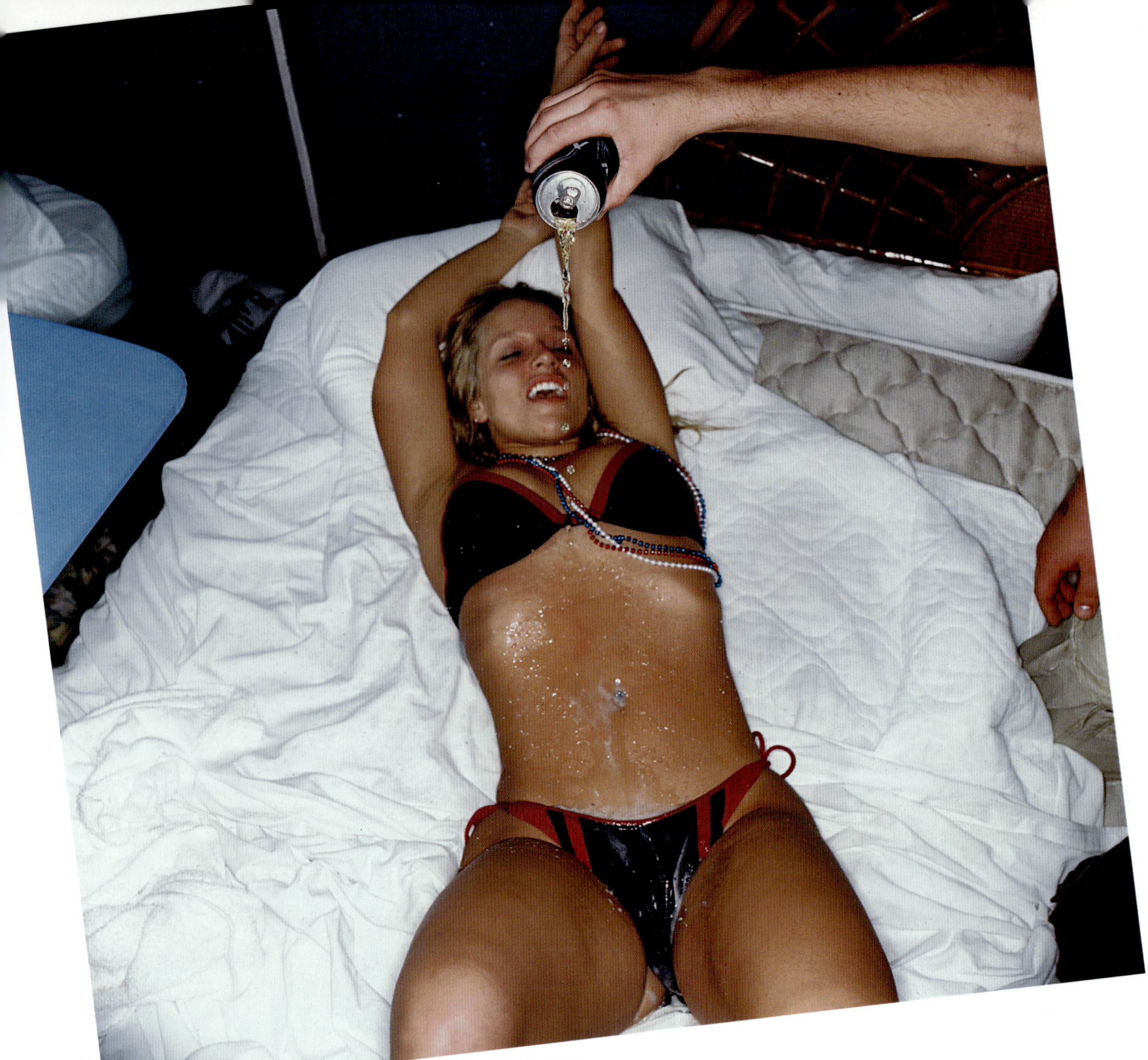

SPRING BROKE

BY NATHANIEL WELCH

PHOTOGRAPHER'S DIARY, PANAMA CITY BEACH
MARCH 19, 2000

Arrived in Panama City Beach in north Florida to a massive rainstorm. Good omen or bad? Lots of Ohio, Kentucky, et cetera plates on the way in. Everyone is in a hurry.

MARCH 19, MIDNIGHT

A kid named Jimmy drank a whole bottle of 87 proof Goldschläger and then ate something called a Hot Pocket, some sort of ham and cheese calzone thing.

I photographed him throwing up in his bed. I almost threw up as well. Very strange. It was almost like performance art. He's an education major and wants to be a schoolteacher when he graduates.

The reason I like shooting at night in the hotel rooms is the extreme nature of what happens here. The kids try to play adult because they have a hotel room for the first time without their parents. They are much drunker, more brazen, trying hard to get laid.

MARCH 20, 6PM

First full day at the beach. Weather a perfect 70 degrees and sunny. The drone of drunk hollering is already getting to me. Everybody seems to love me. "Take my picture!" has never had a more profound meaning. Last night, in a room with kids from Ohio, guys and girls were listening to old Snoop Dogg, doing nitrous, and smoking pot. I liked them the most. One girl seemed to get exactly what I was doing. She didn't laugh and yell "PARTY!" like the rest of them.

MARCH 21, 11AM

Still haven't found any interesting guys. Only the girls seem to see how pathetic it all is. And only a few girls at that. I was down at the dance party on the beach last night. DJ was spinning shit that all sounded like 2 Live Crew. Sexual tension and looseness everywhere. Lots of grinding and slobbering. This seems to be the only week of the year when girls find it just as important as the guys to get drunk and get laid. Girls totally let go. This is how men wish it always was. Hope I never have a daughter.

MARCH 21, 8PM

There's nothing bright here.

MARCH 22, 9AM

Went to sleep at 8 last night. Woke up at 3AM to

loud hooting and hollering. Last night seemed like a particularly wild night at the hotel. Late yesterday afternoon a girl on a balcony took off all her clothes and had the whole place in a boil. Cops came and took her away.

March 22, 10PM

One of the guys in the room next to mine is a 25-year-old chiropractor from Ohio. He came down with some buddies to get fucked up and meet chicks, he tells me. He had a bag of about 20 ecstasy pills. He offered to do a free adjustment to my spine, but I declined.

Talked to the fat blonde girl last night who was about to cry because her friends hooked up but she didn't. Seems impossible here for a girl not to hook up, no matter how unattractive she may be. Tonight she is in much better spirits because she made out with five guys.

March 23, 7PM

Let me explain the constant drone of drunk boys hooting. It makes no sense. It begins without warning. One does it, then others feel obliged to answer back and join in. It really gets bad about 3AM. Bars close and everyone keeps partying. It's surreal.

The kid in the wheelchair had a friend who went everywhere with him. His friend had "Sean" tattooed onto his stomach. Sean was a fuckup and a misfit. I didn't like him but was touched by the fact that he went everywhere with the guy in the wheelchair. The kid in the wheelchair was a quad and he had a squirt bottle filled with beer. Whenever he wanted a drink, he would ask Sean and Sean would bring the bottle up to his lips to drink.

March 26, 10AM

It's hard to see things as they really are. It wears you out. They're drunk, they're getting laid, they're happy. There's more, though—you just have to want to find it.

Planned on staying two more days this trip but decided to leave this morning. Can't really take it anymore. Not until next year.

ROC HERMS PONT

JR

NICOLA CAIRNS

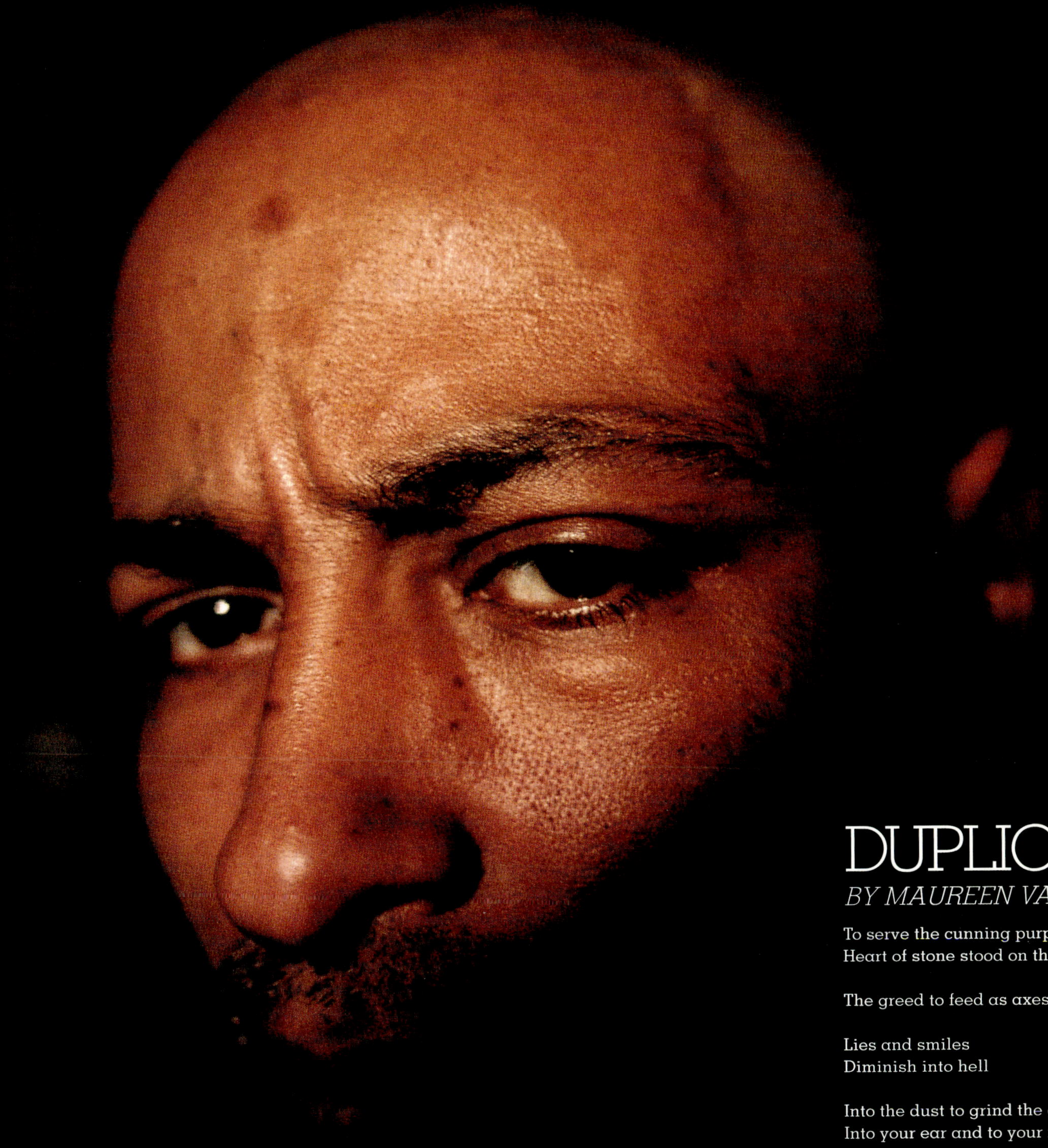

DUPLICITY

BY MAUREEN VALDES

To serve the cunning purpose full
Heart of stone stood on the river of bones

The greed to feed as axes fell

Lies and smiles
Diminish into hell

Into the dust to grind the grace
Into your ear and to your face

The one you love the great deceit
The knife will slice and clinging deep

The pound of flesh to take and take

The reaping
The bounty
The scarlet face

And when the fill has been acquired
Duplicity sate ascends retired

IN THE HEAT OF THE MOMENT: SELF-LOVE, INTERRUPTED

BY SAMANTHA GAINSBOROUGH

Thousands of people are pleasuring themselves as I write this. My older next-door neighbor is summoning the genie in his fantasyland. Your female supervisor might be working her finger magic in her darkened bedroom. That cute artsy guy from the bookstore is charming his cobra while reading some Nin. Even those folks we call "parents." Yeah, they do it too.

There is no denying solo recreation is going on around us. Hell, we're doing it ourselves. By the time I was around nine years old, I had already fully embraced such entertainment, even before I knew the name for it.

When I was five or six I used to rub myself against the floor as I lay between the couch and coffee table, watching television with my parents. It was something I did often and knew there was something personal about it, so I was discreet. But my mom brought my laundry up to my room this one day and caught me essentially humping the carpet. She asked me what I was doing and I just told her, "You should try it, Mom… it's good exercise." —*D.*

I also experimented with toys over the years during the "getting to know my body" stage. My first toy ever had to be plugged into the wall. It had a circular, flat metal head that was about the width of a dollar pancake, attached to a beige plastic handle, which had a control knob for low/medium/high positioned on it. I think I found it under my parents' sink. My mom would use it for my dad's back, but I found a better use for it.

I didn't quite know what I was doing with it, I just did it. I shared a bedroom with my older sister Vicki back then. Before I'd go to sleep at night, when she was well into a dream state (her snoring was my cue), I'd plug Metal Head into the wall socket, hide it under my sheets, and move it about in circular motions.

There was a girl I met who told me about hiding an apple between her legs when she was on mushrooms. The girl had come home after taking them earlier in the evening but she didn't start tripping until she got home. She grabbed an apple from the kitchen counter and in her twisted, silly magic mushroom paranoia, thought her mother would freak out about her eating it because maybe she'd need it for a pie the next day.

When her mother walked in, the girl yelled, "Leave me alone!" holding the apple hidden between her legs. The next morning, at their mother's instruction, her sister came to give her a pep talk explaining that pleasuring oneself was natural and okay. —*M.*

My mom opened the door to our room late one night, just when I was preparing for my private playtime. Thank goodness she was so tired, she just popped her head into the room and said, "Good night, dear," promptly closing the door behind her, not even noticing the large bump under my comforter that couldn't possibly be any of my body parts, or the cord running from the electrical socket to the right of my bed onto the floor and up underneath my sheets. I wonder what she would have said and how I would have explained. Parents are often the first people to bust us, early on in our sexual development. Sometimes, though, it's vice versa.

When I was a teenager I shared a two-story apartment with my mom. She had a Saturday-afternoon ritual of escaping to the upper level for a long bath while indulging in her favorite book. Or at least that's what I thought she was doing, until I ventured upstairs to check up on her one Saturday afternoon. I heard the swishing sounds of water and her cries coming from the bathroom. The stairs were carpeted so she had no idea that I could hear her. I went back down the stairs swiftly but quietly. —*S.*

If you walked in on someone on the highway to orgasm—at whatever age—what would you do? Pretend you don't see them? If you were the one mid-tease, would you own up to the activity? Or continue with your business? Really, who is in the more awkward position, the person indulging in the self-love principle or the one who is witness to it?

While I myself have never been caught in solo playtime, I've had my share of walking in on very private moments. While I was in college I shared an apartment with four friends. And as is the case when you share a home with friends, you inevitably start learning new things about them, often more than you want to know.

A good friend of mine was living in a one-bedroom apartment in Houston with his friend and another guy they didn't know very well. The thing was, they needed help with rent, so they let this dude move in. They had all the beds set up on opposite sides of the bedroom. They were like 19. A hostel setup. Spartan like barracks.

My friend was suddenly awoken by this jacking off sound—the friction, the motion. He rolled over and looked around and saw his roomie really into it, kind of looking at his wiener and going to town. My friend looked at him for a second—it was just light enough to see what was going on—and the guy looked up and saw that my friend was looking right at him. The dude paused, looked back at his wiener and started going to town again. My friend's like, all right then. He went back to sleep and they never talked about it. —*C.*

I kept ridiculous hours those years, usually not waking up until the afternoon. On one of those many lazy afternoons, I woke up to a very quiet apartment. No one was in the living room or kitchen or two of the bedrooms, but the door to my girlfriend P.S.'s room, which she shared with my guy friend K. (their relationship was strictly platonic), was closed. I thought maybe P.S. was taking a nap (she liked sleeping in late too), so I knocked on the door. There was no answer, so I just let myself in and got an unexpected eyeful of K., who generally was at work during the day.

The AC was on full blast so he didn't know I was even there. He was laying on his bed, partially naked with just a T-shirt on and his back was fully arched. K.'s eyes were tightly closed and he had one hand cupping his package and the other fiercely stroking away. I don't remember how long I stood there. I was a bit paralyzed, partly in fascination and partly out o

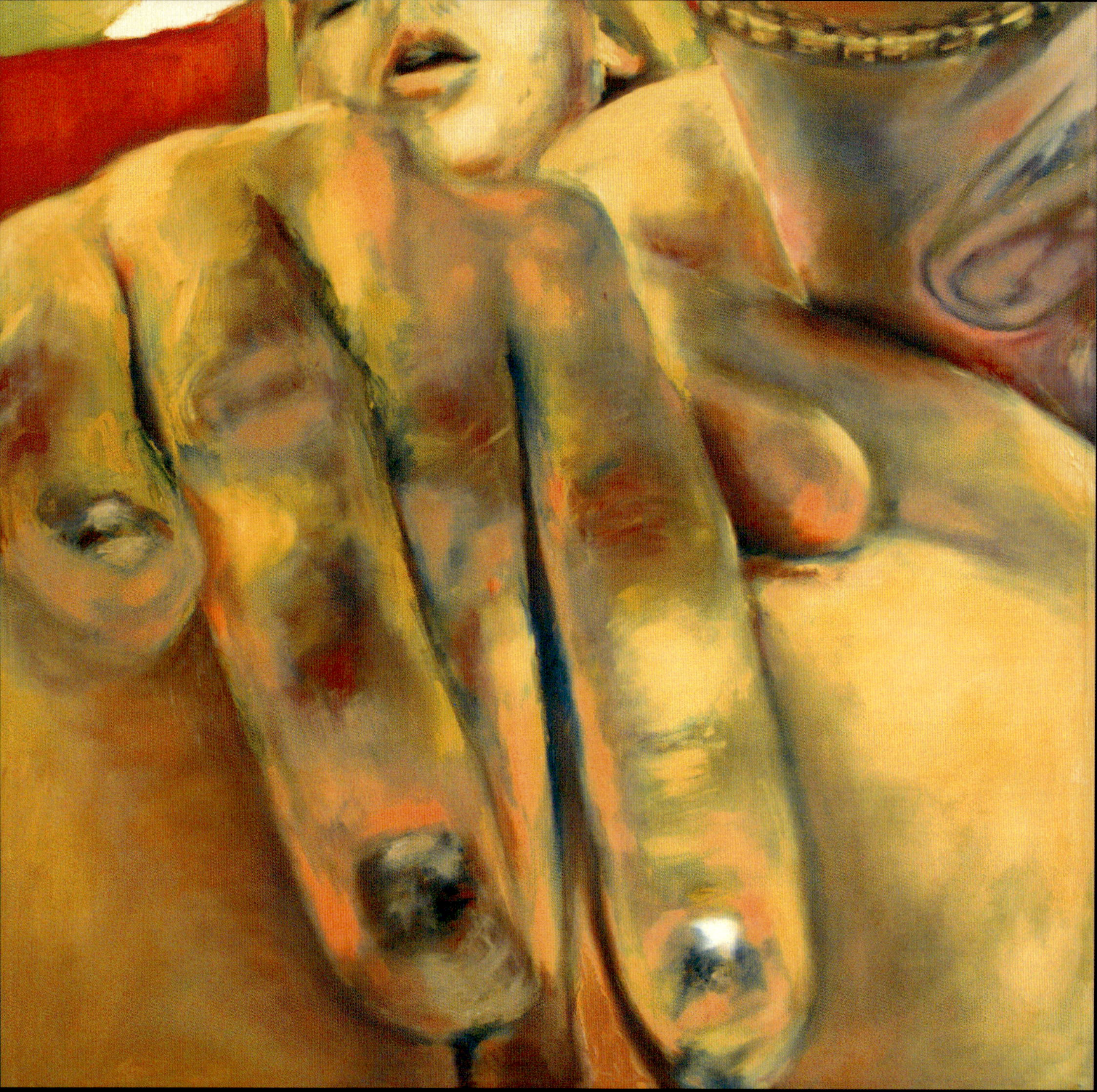

Artwork © **Jesse Gammage**

fear that if I moved he would hear me. I didn't think I could bear the embarrassment both of us would feel.

I visited my brother a few weeks ago. One day during my stay, I was hanging out watching TV in his living room. I really wanted to get it on, but I had no privacy because I didn't have my own room. But I had an urge, so not minding my brother, or the fact that it was about 7AM, I just started watching some porn online and muted my laptop. Next thing I know my brother walks in. Good thing I wasn't really naked, I was just rubbing my balls, but it was obvious I wasn't watching TV. He just turned around really quick, waited for me to realize I wasn't alone, and then asked me something. I tried to cover up my actions by telling him I was on some other website to see some hot-ass pictures, but it didn't really work. —J.

From what I remember, I just carefully closed the door and went to sit in my room down the hall for a while. With that brief glance, my whole perception of K. was changed. First of all, for a guy who was barely 5'5", K. was quite well endowed. He always did date tall, beautiful women. K. was all man in that moment, not the shy indie-rock boy most of us viewed him as. He was potent. That expression I saw on his face, there was such an intensity to it. I never let him know.

That happened more than a decade ago, but I can still recall the scene well, with K.'s bed in the far left corner of the room, his tense body on the brink of the utmost release, and the sound of that deafening 70s air conditioning unit allowing me a glimpse and then an easy escape. I still wonder who he was thinking about.

My stepfather was a violent man. We grew up in the country and he beat the crap out of me on a regular basis. When I was five years old, he used me as a target practice for his own relaxation. He chased me into the woods and shot at me with his hunting rifle. At the time, I was so small I can't tell you what I was thinking. I probably felt it was just another game, like tag or something. When I was 12, he made me swallow a live mouse.

As a teenager, I started skipping school, because I was always being picked on. When my stepfather found out, he busted my lip and put my hand through a barbed wire cow fence. He was a real sadist. I remember it was Halloween. I was sitting on the bus and saw some of these kids dressed up, the fake blood and everything. I knew I didn't even need a costume. At that point, my main goal was to escape from him.

So of course, I ran away from home. I stayed with the uncle of my best friend who lived in the city. That's how I got hooked on pills, because he worked in a pharmacy. First it was Tylenol, then Quaaludes, then "black beauties," those amphetamine diet pills. When I got to morphine, everything else came to an end. It felt so good. You keep popping the pills, 10 to 15 a day, and then you think, "Look at all these morons. They go off to their jobs while I'm having fun." It totally calms you down.

But then I had to leave, and the hustling started. I had to buy the stuff in the street. You start boosting, doing stick-ups to support your habit. It's always like, "How am I going to get the next bunch?" You eat very little, maybe one meal a day. Pineapple juice and graham crackers—that was my diet. You have people chasing you—the landlord wants the rent, you owe money to the neighbor. But you don't care. When you're on dope, your wife could be screaming, your kids could be starving, and none of that stuff gets in. The drug is more powerful. Whoever is in your way, you take them down. All you do is run after the money. It's a never-ending roller coaster.

One morning, I woke up and was withdrawing with the chills, the cramps, the horrible pains. I knew I was broke, so how was I going to get my pills? I owned one of those black leather jackets; I put it on, pulled up the collar, and went outside. It was Halloween—again! A thought flashed into my head: I'm going to rob a bank!

You remember, back then, they had these water handguns that looked like the real thing? I used one of them, black and shiny, from Chinatown. If you turned it around, you would have seen the plug! I walked into Chase Manhattan. I watched the tellers and made my pick. I handed him a piece of paper with something written on it, "Give me the money or I blow your brains on the wall" or whatever. He started stashing away the money into a bag. But you know, things don't stay quiet in a situation like this; people start staring and I had to swing the gun around. The teller froze and people started panicking. I just yelled, "Give me that fucking bag!"

I raced out and did not see that there was a camera right over the door. I could only think "I'm rich!" I left with a big smirk on my face. I had a green Ford Maverick that I stole and drove off. I parked the car and continued on a bicycle.

After that, I was on the run. I kept the money on me always. I carried it around in one of those suitcases, like a Wall Street stockbroker. I even slept with it. But I burned through that dough—you wouldn't believe it. You know, the women, the restaurants, the drugs. I went through the $20,000 in two weeks.

I saw my picture on one of those posters they put up in the subway stations. The two guys on either side of me were crossed out, "apprehended." And there I was, in between, with that fucking smirk on my face! I grew a beard, I tried to look like a flower-power hippie, did a lot of things to disguise myself. But it didn't take long. One of my enemies made a call. I wound up with 10 years.

Prison is another story. I finally had the time to get to know myself—whoever that is. I'm out now. But inside my head, it's like a bad neighborhood. I don't want to go there too often.

FICTION

THE HUNTER AND HIS PREY

BY BASTY

CLAYTON WOOD III

BY GEORDIE WOOD

Here is a friend we once knew: Clayton, the hustler and enigma, a herald from the abyss and a merchant of the illicit. At one point he was one of the most well-connected dealers in Central New York, with the clientele to prove it. Clayton's cell number was reliable and he was respected as a true businessman. When you start pushing drugs at 13 it's not only what you do, but who you are.

Yet the remains of this personality are fading. His arrest in 2004 condemned a boy struggling with identity and swept away any sense of dignity from under his feet. This image is one in a series of photographs documenting the pusher that was and the man that remains, always a hustler at heart.

DEBORAH DRAGON

Say hello to the
mediterranean
topography
staf
thenew.popculture
s.2 * staf magazine
o.0 * stafmagazine.com
o.0 * newpop.books
n.9 * so real ...
staf
@
stafmagazine
.com

Photographs © **Orkan Benli**

So the year is 2002 and I'm home in bed, resting in the crib, recovering from getting stabbed a week earlier for non-graffiti-related beef. I get a phone call from my boy Hollywood CEEP and he's telling me he's got a shitload of paint he just racked and he wants to go catch a spot (an easy, safe spot) on the 2 line.

I'm like "Damn!" I'm really fiended out to go write on something but I've still got stitches in my back and can hardly turn from side to side. Nonetheless it doesn't take much convincing before I am dressed in my bombing attire, waiting for CEEP to scoop me up. I tell CEEPER I'll hit one spot and then I'm out, back home.

He comes through and we head to the 2 line. His car is filled with white Kilz, burgundy, black Rustos, and summer squashes. Once I see the paint I am ready to do the damn thing. We go to the spot; it's a rooftop so we scheme out the area and make sure it's safe before we grab our colors and go for it. We hop over the first fence and hide in the entrance of the building. It's not looking too sweet anymore. Cars keep driving by and we still need to climb the next fence so we wait in the shadows for about an hour.

By this time we both have a bad feeling about this but we say, "Fuck it, let's go up." We climb up and pull ourselves onto the roof. It's probably only like 8:00PM, still pretty early but that's how KGK do. We're daytime hot boxes—we bomb anytime. So we finally get up to the roof, pick the first two spots, and start rocking over some toys. Around the corner from the wall, we notice an open window but we're like, "Fuck it, finish fast, and let's bounce." Just as we are finishing up our fill-ins, lights are shined at us from the rooftop above. When we look up it's the boys in blue. We both turn to each other like, "AHH, FUCK!"

We make a run towards the front of the roof where we first got up. When we get there, we spot about six cop cars. The cops look up at us and tell us to freeze. Freeze, yeah aight. We run towards the back of the roof. We look over the back ledge—it's a two- or three-story drop. There's no way to climb down. CEEP looks at me and says, "MISTA, I'm jumping."

I waste no time and jump off the roof. I land on my feet but automatically fall on my back and hit the back of my head. CEEP later tells me when he looked over, he saw me lying there motionless and thought, "Ohh shit! MISTA's dead!" Then he jumps off the side of the roof and tries to break his fall with a tree. His leg gets caught in the tree and he does a front flip to the ground. The tree rips his leg wide open but he is still able to make a run for it. He walks calmly away for three blocks, past all the police, playing it off like he's just a person from the neighborhood. He thinks he's good until some police officers notice his leg is gushing blood. They stop him, arrest him, and keep him in the cop car while they go after me. He says he was listening to the whole chase on the police radio inside the car.

I am knocked out on the ground for a couple of seconds before I jump up, catch a rush of adrenaline, and run out the back of the building. As soon as I do that, a cop comes out of

nowhere and grabs me. I hit the cop with the old football shift arm move and jet across the street. I run up this mountain of rocks, hop a fence, and run down the other side of the rock pile onto another block as two cop cars come from both sides of the street. I make a U-turn and run back up the rocks, which is where the cops that were chasing me now are. I hop another fence, end up in a bunch of backyards, and run through the backs of all these private houses. It is dark and I can't see but I don't stop. I run right through a baby pool and past about five houses until I see a shed about eight feet high. I don't know how, but I jump up there in one shot and lay up on the shed under some tree branches.

The next thing I know I hear cops and see flashing lights everywhere. I lay there quietly, still trying to catch my breath. The cops start going into the people's houses and up on their balconies. They flash their lights around and hit the spot I'm lying in. I think I'm busted but they don't see me. I can see fire trucks passing by towards the front of the houses. The next thing I hear is the cops screaming out, "OKAY, WE'RE RELEASING THE DOGS!"

Now I am thinking, "What the fuck! I was only doing graffiti and they've got a manhunt out for me!" Still, I don't give in. I lay up there for hours until the cops eventually give up. But I still don't trust getting down just yet so I close my eyes and stay there for eight hours.

When I finally jump off the shed I realize that I am so hurt, I can't even walk. Without the adrenaline high, I am in pain. I don't know how but I am able to walk two blocks, dragging my legs and looking like a monster straight out of a movie. I walk to a corner Habibi store where the owner calls me a cab. I tell the guy I just got mugged and he arranges to pay for the taxi for me.

I make it home but have to get help to go to my front steps. My feet are so swollen I can't even take off my sneakers. My feet and ankles have turned purple and all the stitches in my back are gone. I try calling CEEP but his phone is off. I know they've got him.

The next day I leave my house on some old crutches I have and there are a shitload of cops right outside, just waiting for me. It turns out they had us on camera from when we had left the building. I am charged with everything: graff, trespassing, and resisting arrest. A couple of weeks after we are released from central bookings, CEEP and I go bombing again.

Shout-outs: FREE, TAVO, RIP, ASP, NESO, PEEWEE, RENZER, KGKASS, TOTD, US, VOP, DTB, NVS, BFH, CEEP, FEES, NETR, ABOT, TECH, NOKE, GODFUCK, POOMMA, ARE, SEGE, OVAL, NAST, CROME, PUT, NEO, JAZE, DEE17, DJKASTONE, SWUE, MENOS, SANDMAN, KLEP, REL, YEWZ, LES, NOXER, JEK, SYTE, HEKTAD, TD, EVANER, CCHILD, DEKONE, FINES, DUSH, JERK, SIKER, SCO, ORE, I2LENER and all the rest of my goons!

TRACE
TRUE BEAUTY
TRACE
Art Issue
Transcultural Styles + Ideas
Art and Soul
erykah badu + Q tip
TRACE
Transcultural Styles + Ideas
BLACK GIRLS RULE!
GUEST EDITOR: IMAN
TRACE
Music Uncommon
PIMP THE SYSTEM
ERYKAH BADU
ISSN 1366-1752
JOHN LEGEND
GOAPELE
FRANK N DANK
TRACE
KERI HILSON
MICHELLE OBAMA EXCLUSIVE!
BLACK GIRLS RULE!
TRACE
THE PEACE ISSUE
FEATURING GNARLS BARKLEY
TRACE
The Art Issue
TRACE
Sporting Life
Transcultural Styles + Ideas
DIPSKATE
ROLLERBLADING, REMIXED
MODERN MOVEMENTS
NEW YORK CITY DANCERS ON THE RISE
THE FRESH FACES OF AMERICAN GANGSTER
TRACE
MUSIC'S SWEETEST COLLABORATION
COMMON + LILY ALLEN
NEW YORK CITY
CREATIVE COLLECTIVES
COPENHAGEN
A CULTURAL RENAISSANCE

IT'S JUST BEGUN

BY DJ DISCO WIZ

No matter how scared you are or aren't, you never forget your first arrest.

One day someone in the crew came back to the neighborhood with news that a department store on Fordham Road had partially burned down. The fire at Robin's meant easy access from the roof into the fashionable clothing store, so a quick plan was hatched and we took off to pull our first real heist.

As we cased the joint, we discovered that the fire had left a gaping hole in the side of the building connecting Robin's to a salsa record shop and the best they could do was patch it up with a few pieces of plywood. In a few minutes we had ripped the plywood from the wall. I was chosen to be the lookout on the roof while Tony Rome, Mico, and Josh climbed through the hole and into the store.

We had brought along a bunch of heavy-duty trash bags; the guys inside were packing the bags as fast as they could and then passing them out to me, one after another. After about a half an hour of this, as I turned to put the bag down, someone reached out and took it from me.

Oh shit! Who the fuck did I just pass that bag to? raced through my mind as I slowly turned to look.

Standing there were the detectives who had questioned us earlier about some other shit we had done in the neighborhood.

They wouldn't be asking me many questions this time—after all, I had just handed them a garbage bag full of stolen clothing. I wasn't sure whether to laugh or cry so I just shook my head in shock.

Hate Us..
COPS
ME
BOMBIN
BOMBIN' MAGAZINE
REME
FROM UNIQUE WHIPS
JAHMAL WILLIAMS
DIPSKATE
MED
A STAR IS BORN
GAIA
TELLS THE TRUTH
DCEVE
TELLS LIES
THRICE
ALL THE ELEMENTS
Available at
BARNES&NOBLE
BOOKSELLERS
Find out why at
www.bombin-magazine.com
© 2008 Bombin' Magazine. All Rights Reserved.

COPE2

It was early 1982 and I was with my boy JET FEB. He was a cool white kid from my hood—uptown, the Bronx, Mosholu Parkway. One night we decided to hit the No. 2 yards by 241st Street and White Plains Road. At the time, the No. 2 line was painted white. There was nothing but white subway cars in that yard and writers were hitting them but not that hard.

That night, JET and I started killing every subway car and panel with quick outline throw-ups. At one point, we went inside to spark a nice blunt of some buddha tye stick. Damn that shit knocked me out, but we kept killing the yard and hit around 40 cars.

The next thing I knew, there were four white dudes coming toward us, so I told JET to chill. They were trying to act like they were writers, talking about, "Hey, what you write?" I was so blunted out of my mind that I thought I was seeing shit. JET was like, "Hell no, they're cops!"

As soon as we started running, they pulled out their badges and told us, "STOP! DON'T RUN! WE HAVE YOU SURROUNDED!" I wasn't having it. JET and I split up. I don't know where he went but I made it out of the yard and went to the pizza shop, got me a slice with a soda, and started walking down 238th Street. A car rolled up right next to me and four dudes popped out. "FREEZE! DON'T MOVE!" they yelled.

I was like, "HOLY SHIT," and threw my pizza and soda at them. I ran under the 238th Street Bridge along the Bronx River and they chased me until I jumped into the river and swam across. They looked at me like I was crazy.

Once I got onto the other side I crossed the Bronx River Parkway and started walking along Woodlawn Cemetery. I was like, "Oh man, I got away!" As I was soaking wet and trying to relax, the same car with the same cops rolled up again and said, "DON'T FUCKING MOVE! WE GOT YOU NOW!" I had nowhere to run so I tried to climb over the fence of the cemetery. Just as I got to the top, two of the cops caught my leg and yanked me down like a rag doll!! They slammed me on the ground, put cuffs on me, and said, "We got you, you little fuck!"

They booked me and put me through the system. I came from a pretty poor family so I only received one-year probation and three days community service. My mom thought I was crazy but she would pick me up at the court, laugh, and say, "You have to take it easy. Stop painting those trains!" But as the years went on, she always loved me and supported me. I love you, Mom. God bless your soul...

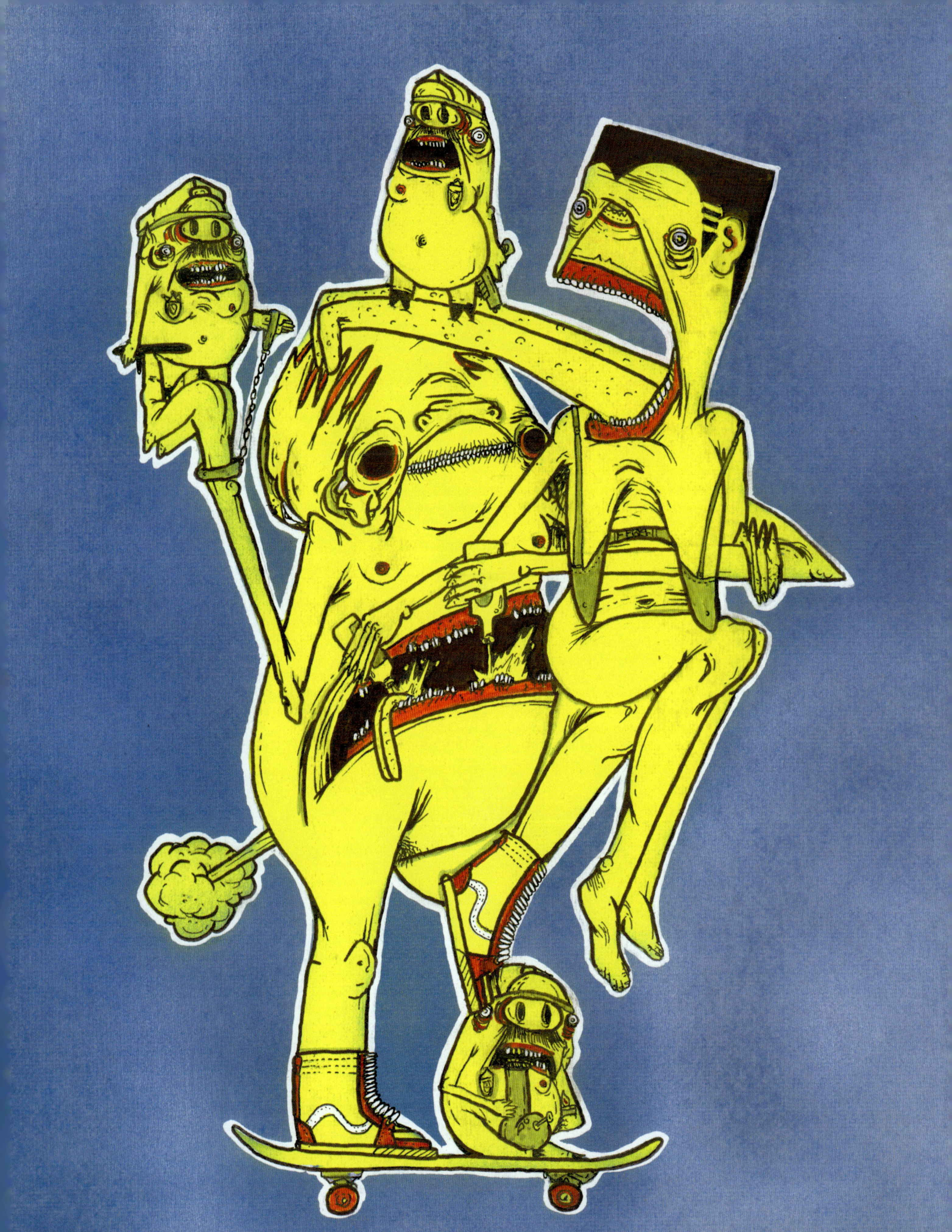

Artwork © **Matthew Charles Crabe**

BEER-COVERED PHOENIX RISES FROM ASHES, ONLY TO GET HASSLED BY THE FUZZ

BY MATTHEW CHARLES CRABE

A BRIEF HISTORY:
Raised in the middle of nowhere in the golden womb of northern New Jersey, I sought out things that would stimulate my growing young brain. It started with He-Man and the Masters of the Universe and it only got worse from there. Never able to shake the influence of poorly animated fantasy/testosterone-filled programs, I turned to comic books, Nintendo, and the least holy of the trinity, skateboarding. Bonding with other delinquents, we searched the town of Budd Lake for skate spots to destroy and trouble to get into. The combination of pain, freedom, creation, and style in skateboarding is something that I easily adapted to and will forever live by.

Life moved fast. By the end of high school I had filmed and edited two full-length skate videos, discovered what beer was all about, witnessed limbs get broken, seen people hit by cars, heard grown men cry, and tried my hand at understanding the mystic mechanisms of the teenaged girl. This pattern continued well into college, where I decided to study photography. Caught in the whirlwind of a transition of photography from a ritualistic art to a digital medium, I began to see the soul get sucked out of it and eventually tossed it aside.

With my main means of expression gone, needless to say, I was confused. I felt like a boy who had watched his puppy get burned alive on Christmas while everyone else pointed and laughed or didn't look at all. I drank a lot of beer and dated a hopeless hippie girl. The hippie girl left to become an overweight scuba diver and I was left with beer. I briefly attempted a screenwriting class, only to disappear halfway through the semester after getting run over by my own truck. After that all I did was work at a video store and drink booze, mostly by myself. This went on until I had completely drowned all that was horrible about myself and in a way, had mentally died.

Not being physically dead, I began to draw my mind back into place on a series of bar napkins. Along the way I found my desire to skateboard like a demon once again, and met a sweet woman who wasn't pregnant but would lactate once a month, and who probably saved my life. The napkins quickly evolved into paintings and before I knew it, I couldn't stop drawing. Never before encouraged to draw, I was surprised at what I could create, and was equally surprised that there were some things people hadn't been drawing. This gave me a mission, a mission that will take me until my second death to complete. A mission to ejaculate out of a pen the most ruggedly gnarly of all graphics, exposing all that is gross, depraved, and beautiful about the world.

And now, an edited list of encounters with police that I can remember:

My sister in her carseat, placed on the side of the highway after my mother got pulled over without her license.

Running from a cop after getting rolled up on at a skate spot before I realized all they do is tell you to leave.

Being told to un-duct tape Josh from a pole in front of a pharmacy where he was yelling at customers to repent. When the cop searched him, he pulled out a snot rag from his back pocket and got boogers all over his hands. If that wasn't enough, some kid with a giant head we were with was arrested for weed.

Watching a cop pull away after I asked him why you can ride a bike in the parking lot and not a skateboard.

Watching a cop throw a rock for twenty minutes at a rabid raccoon in a tree that had been living under my broken Chevy cavalier. When the raccoon fell out of the tree, the cop shot it three times in the face and it ran around the yard with blood squirting out of its head until it died.

Nasty Nat being pulled over just so the cop could ask if we had drugs and tell us there were eggs all over the side of the car.

Calling the cops on my own party.

Being pulled over by five cop cars because my headlight was out, and having my shoes searched while there were underage girls with weed in their bras in the car.

Getting into an argument with most of the police department at a party about how I wasn't a pothead, while they found my fourteen-year-old sister hiding in a closet crying.

Having the cops let me go home, giving them the finger for some reason, then getting put in the back of a car and trying to tell everyone that walked by to just open the door.

Being let go after getting caught painting a raccoon giving birth to some skeletons by the railroad tracks while wearing a pig mask on my head.

PIMP LONDON: THE GUIDE
THE DEFINITIVE GUIDE TO LONDON FOR GLOBAL STYLE HUNTERS

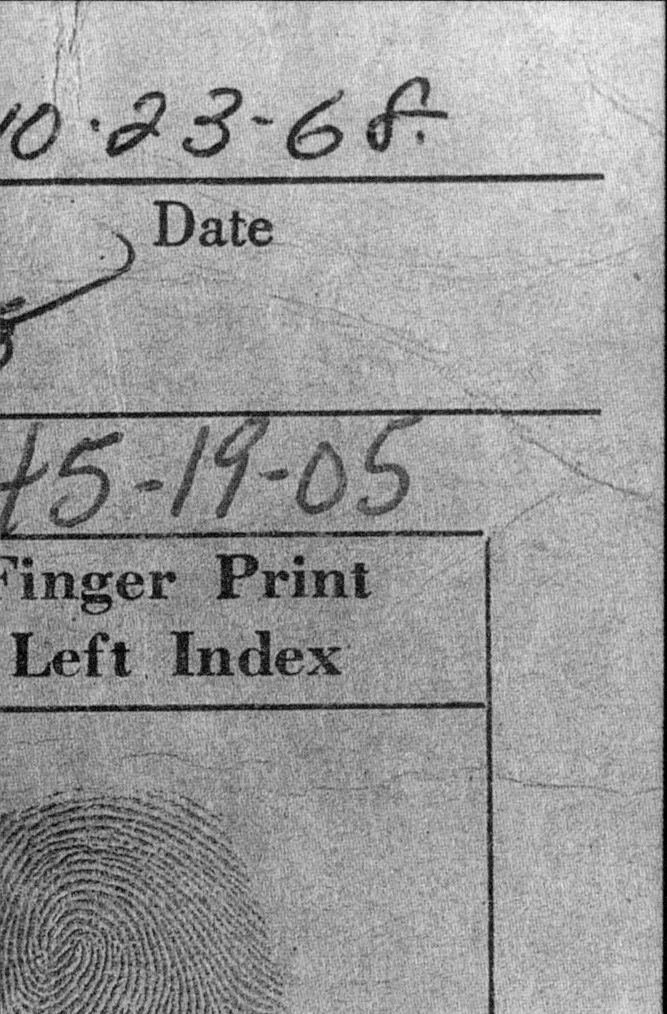

Drugs were everywhere. It was the usual stuff, starting with weed and then eventually graduating to heroin…and later to methadone. I soon became a slave to the "white lady." I became a dealer and a thief to support my habit. One night I was caught breaking into a store. The cops caught me redhanded, with one foot inside the air-conditioner duct and one foot outside. "Shit, not again!" I said to myself as I was hauled away in cuffs to night court.

Kicking the habit was cold turkey in those days: they didn't give you anything for the nightmarish pain, they just let you throw up and shit in your pants. So I just had to put my hands up and deal with my own turmoil. One day, coming back to my cell from the mess hall after breakfast, I considered jumping off

Although it was a hellhole, Rikers Island forced me to think. As I tried to deal with my problems there, I realized that prison life was not for me—although it wasn't the last time I'd be inside.

When I finally got out, I was still considered a juvenile offender, and was put on probation. I decided to get on a methadone program. I had tried many other programs, but they didn't work for me. I remember telling my mother that I was on a recovery program, and she said: "Don't lie to me! Look at you—you're high on that shit again!" The doctors had started me out at 90mg of methadone, which, until I got used to the dosage, just made me nod out and look like I was sucking on my toes.

I was hurt at first by what my mother said, but I knew it was going to take a lot of convincing for her to trust me. It didn't help my case that my stepfather was also a drug addict, and my mom had had her fill of his shit. His usual routine was this: after he got out of prison he would stay clean for a couple of weeks, and then he was right back to shooting up again. I remember coming home from school when I was 10, and seeing him nodding, with the needle sticking out of his skin as the blood ran down his arm.

My mother married my stepfather when I was 10. At first he was nice. He and my mother had two beautiful children of their own. Later on, when she found out he was doing hard drugs, she was overwhelmed. In those days, the early 1960s, you didn't talk about this stuff openly, so she carried the weight of it—she was a very strong woman. He stopped going to his job as a bricklayer, and soon the collection agencies started coming to the house. My stepfather took everything that mattered to me: my Lionel trains, my first transistor radio—all of which my real father had bought me. I hated him, and I hated myself for being stuck in this situation.

After I got adjusted to my new drug habit (methadone—much worse than any heroin), I knew that I wanted to do something with my life. I guess I had always had that desire, but didn't know how to get there. I'd thought that I could get away with stuff, but I learned that there are always consequences: getting busted, overdosing, doing time.

I slowly started to learn how to redeem myself by looking deep inside. I learned that there is a balance, a yin and yang of life.

With my photography, I have learned how to deal with these experiences.

D

1414 HAZEN STREET
EAST ELMHURST, N. Y. 11370

Block #5 (CELL NO. 3A1)

From RODRIGUEZ, JOSEPH No. 1451905

To CARMEN RODRIGUEZ Relation-Ship MOTHER

Address 807-8TH AVE APT 4 REAR

City BROOKLYN State N.Y. 11215

READ THIS
BEFORE WRITING TO INMATE

Incoming mail must contain the full name, number and block location of the inmate.

Your name and address MUST appear on the envelope.

Do not send more than 15 cents in stamps.

Send money to inmates by MONEY ORDER or REGISTERED LETTER only.

Photos not larger than 3" x 5" are permitted.

Do not mail anything whatsoever except letters, to an inmate without getting permission from the warden in advance.

DATE 11/2/68

DEAR MOM,

HI THERE! I WISH I WAS HOME WITH YOU AND MARK AND SONJA, I WAITED FOR THE BAIL BUT I SEE IT DID NOT WORK OUT. I PROMISE YOU THAT WHEN I GET OUT OF HERE THAT I WILL JOIN THE NAVY OR GO BACK TO SCHOOL BECAUSE I KNOW I PUT YOU THREW A LOT OF HARDSHIPS + TROUBLES, AND I AM GOING TO STRAIGTHEN UP AND ONE DAY YOU ARE GOING TO BE PROUD OF ME BECAUSE I DONT WANT MARK OR SONJA MAKING THE SAME MISTAKES I HAVE MADE. I WANT TO GO IN THE NAVY AND TRY TO FINISH MY SCHOOLING BECAUSE I STILL LIKE PRINTING SO MUCH THAT I TRIED TO GET IN THE SCHOOL FOR PRINTING IN HERE. ONE OF THESE DAYS I INTEND TO BE WHAT I WANT EVEN IF I HAVE TO CRAWL FOR IT AND START FROM THE BOTTOM, BECAUSE I WANT TO HELP YOU OUT AND MAYBE ONE DAY MAKE IT ON MY OWN, AND MOM LET ME TELL YOU SOMETHING DURING THIS PAST SUMMER SOME PEOPLE THAT I DONT LOVE YOU AND THE KIDS BECAUSE BOY! I LOVE YOU AND THE KIDS WITH ALL MY HEART AND RIGHT NOW I AM CRYING BECAUSE OF WHAT I HAVE DONE TO YOU + KIDS YOU CAN EVEN SEE THE TEAR DROPS ON THIS LETTER. I ALWAYS THOUGHT YOU NEVER UNDERSTOOD ME BECAUSE SOME OF THE THINGS THAT I DO YOU THINK ARE TERRIBLE BUT THEY AREN'T BUT OTHERS ARE DANGEROUS AND I APPRECIATE WHEN

DO NOT SEND CASH

Misc. 450M-426196(68) 346

YOU TELL ME WHAT THEY CAN DO TO YOU. AND IF I GET OUT
OF HERE WE ARE GOING TO TRY TO UNDERSTAND EACH OTHER
SO I WONT HAVE ANY PROBLEMS. I GOT THE MONEY YOU
SEND ME AND I BOUGHT A CARTON CIGARETTES AND COOKIES
+ CANDY BUT I HAD TO PAY BACK A COUPLE OF PACKS OF
CIGARETTES, JUST IN CASE I AM IN HERE FOR ANOTHER
COUPLES OF WEEKS OR MONTHS. MONDAY WHEN I GO TO COURT
TRY AND GIVE ME ABOUT $3.00 FOR SOAP, DEODARANT, ETC.
GETTING BACK TO THE BAIL. I WAS TOLD BY THE PATROLMAN
THAT IF YOU HAVE AN LIFE INSURANCE POLICY AND IF IT IS
ALMOST PAID UP YOU CAN BRING IT TO THE BAIL'S BONDSMEN
AND HE MIGHT ACCEPT IT. PLEASE TRY IT BECAUSE I CANT
TAKE IT IN HERE ANY LONGER. I HOPE TO SEE YOU AT COURT

WITH ALL MY LOVE

CHICKIE

P.S. I HOPE THE KIDS AND YOU ARE ALLRIGHT
GIVE THEM ALL MY LOVE

COURTHOUSE CONFESSIONS

PHOTOGRAPHS AND INTERVIEWS BY STEVEN HIRSCH

TYRONE CARTER

My name is Tyrone Carter. I'm from South Jamaica, Queens. I was walking downtown, midtown, with a friend of mine. I wasn't walking with him but I seen him and I had shook his hand, and he had made a sale to an undercover cop of some fake marijuana and the police ran up on me and him and took both of us and they said I was the lookout and he was the dealer. We gets down there and it comes to find out the marijuana was fake. I didn't sell nothing, I didn't have nothing on me but they still arrest me. I just got out on $500 bail and I'm stressed out and I'm mad but this is life, so this is what it is. Come to find out the marijuana that was supposed to be sold was Lipton tea.

MARK MARK MARK

So I got arrested for peeing and they took my phone; now I'm trying to get my phone back. I went to the precinct, they sent me to property, property sent me to back to the precinct, the precinct sent me back to property, property sent me to the courthouse, the courthouse gave me a number to call, I told them I don't have a fucking phone, I can't call a number if I don't have a phone, you guys have my phone. They let me use their phone.

Let me say on the record, I'd like to apologize to the City of New York for taking a pee. I'd like to apologize to the garbage man that took my pee away in a garbage truck. There's a reason that the garbage man gets paid more then the police. I'd like to say, fuck the police. Take off your fucking monkey suit. Let's drink a forty and swim naked in the East River. Maybe then we can be friends—until then, fuck you, fuck your life, fuck your dreams, fuck your wife. Get out of my face and give me my phone back. That's all.

Ian Jernigan

I missed two days community service from a previous crime, stealing books from Barnes and Noble. Actually, weird thing, it wasn't me that stole the books. I was there with my wife and she had taken a few Xanax. I was up cruising the sci-fi section on the fourth floor and came down to the third floor to see my wife laying on the ground passed out with a grocery bag full of books. I woke her up and said, "Stephanie, what are you doing? Come on, let's get out of here." And she's like, "No, come on, I gotta take these books." As she was walking with me she kept grabbing books off the shelf, throwing them in the bag. I kept saying, "Stephanie, stop, this is ridiculous, everyone can see you, this is so stupid." But she was so high she kept doing it, and I couldn't grab the books out of the bag fast enough. When we got to the escalator, two security guards grabbed us and we both got charged with stealing the books. She doesn't even remember any of this because she was so high. I mean, it's not the kind of thing she would do sober. We spent a couple of days in jail and then they gave us each two days of community service. But that was when my dad died, and we both went to Maryland for a month.

Janet Braha

I stole a bra. I did. 'Cause I wanted it, and I didn't have enough money. Actually, I'm not [the bra type], it was a very loose kinda bra, very comfortable, very tempting. I begged them [to let me] take it, but they wouldn't let me, they said I couldn't come back to the store for three more years. Urban Outfitters.

Sure, I would [do it again]. I had so much fun coming here to court—I met beautiful people, and I saw that it's not as bad as you think. They were all laughing, the whole time through, we were laughing, joking. Stay good, don't do bad things, you know? Do good things, don't steal.

I just needed one. That was a lacy, nice color, light blue one. I just liked it. They came to my house because I missed a court date because of the holiday, and they came to take me in. Knocked on my door and took me. Shavuot. See, now they're gonna say all these Jewish jokes about stealing. Cheapness. I'm giving my people a bad name by stealing. Yeah, I think so. I wanna do better. I wanna be a better human being.

I'm having a hard time with money. I'm having a custody battle for my children, they were taken from me by my ex-husband. And I wouldn't steal normally. But things are getting hard to keep up with. A comfortable, good, nice bra…hard to find.

NATHAN BROWN

I once knew this kid. Well, I shouldn't say "knew" since I still "know" this kid. He's my best friend. He lets me crash at his place in Queens when I am homeless, lets me piss in his toilet, lets me eat all his apples. He even once ran me over on his bike, straight across my back—I got tire track prints all across the back of my nice T-shirt. Regardless, this kid has been my boy—er, my best friend, I mean, my BFFL…for as long as I can ever remember.

His not-so-close friends called him "The Rascal," the abstract illustration of an insipid little red-haired street brat. But I know a different side of him—kind, sweet, not-so-even-tempered-on-the-rare-occasion-meaning-all-the-time, but still, a nice kid. For about seven years straight, we would just ride bikes all over town. Our bikes brought us to places no one would dream of. At age 15 outside of dive bars in Arkansas, at age 16 driving to Portland at 4 in the morning, eyes watering and hazy from cigarette smoke filling the car.

One day, we met a filmmaker and fellow bike enthusiast from North Carolina. Let's call him "The Mangler." He was even thinner than me and twice as tall. I was thin and tall, twice as much so as you might expect, or, twice as much as the normal human being, so imagine this guy…twice that. You might call him emaciated. He was a tall, skinny fucker, but he rented a box van, a Penske box van that we would throw 12 guys like us into, with couches, plywood, bikes, and lots of other contraband, and fireworks—lots and lots of fireworks. And he was a vegan, like me, so we hit it off well. We ended up taking the box van all over the United States, witnessing things people should never witness and befriending people most would never give the time of day to. Life was pretty interesting, and the food was pretty good.

On one particular trip, we decided to head to the infamous Chenga World, Ohio. We once again loaded up the truck with couches, and guys, and bikes, and plywood, and lots of other contraband, and fireworks—lots and lots of fireworks. We had a heck of a time driving there; we laughed, and some people drank, and some people smoked. I mainly sat in the back with my camera, just watching in awe, and wondering why I was chosen to be lucky enough to witness this.

When we pulled into the parking lot of the park, it was about midnight. The lights where on inside and you could see the place was packed. Bikes were flowing everywhere, people of all races, ages, and genders were laughing and drinking just like us. All of a sudden "The Rascal" piped up with, "We should lay low and surprise these losers with a fireworks show. Cool?" Everyone smiled in what was mutually considered to be silent agreement, and we unloaded the fireworks.

There were about three or four people lingering in the parking lot, doing god-knows-what in Ohio in January, and it was pretty damn cold. More and more people littered out of the overflowing skate park and into the parking lot, curious at the arrival of a strange box van with noises coming from the back. We waited; the time was not quite right. A few more people came out, then a few more, and it was perfect. "The Rascal" threw open the back door. "NOW! NOW!" he screamed.

We lit our Roman candles and jumped out. Blasts shot off everywhere, into the sky, into the skate park, on top of cars, underneath them, over buildings, at each other. It was total mayhem. I was sitting there with my camera, confused and a little scared, snapping away until—inevitably—someone called the cops.

A lone cop car idled onto the scene. The officer inside had the most confused look on his face. The parking lot looked like a war scene; Roman candles were running loose, people were yelling. "The Rascal" was in the center of it all, surrounded by smoke, with a big smile on his face.

When I saw the police, I decided to call it quits, and bolted into the park. Someone was standing at the door laughing and quickly let me in. We both peered out from the cracked door to watch the rest of the scene unfold. Like kids peeking at our parents putting presents under the tree. OK, not just like that, but we had our hands on the door, and our heads just popped out from behind it, and we were lined up vertically…so it looked the same, I guess.

The cop stopped his car, put his lights and siren on, and started beeping his horn. He must have been too terrified to get out—people were shooting Roman candles at his car for Christ's sake! Finally, he called for backup. When they arrived, the rest of our crew decided to quit and dashed inside…I would say they were scared out of their minds, that's why they stopped. I can't say it wasn't a smart move. There was still one person out there though; you could make out a figure through bellows of black smoke and police lights. It was "The Rascal"—he was armed with four Roman candles and two of them were lit, popping off at random.

Artwork © **Wooz**

Three cops jumped out of their cars and started chasing him. He ran around a car, they followed. He ran around another car, and they followed. They went back and forth, and back and forth. It was comical, like a dog chasing a cat, then the cat chasing the dog. This lasted for about ten minutes, until a cop with some sort of anti-terrorism shield confidently walked up to "The Rascal," who at this point we will call "The Offender." As Roman candle shots bounced off the cop's black armor, "The Offender" casually dropped his weapons and smiled.

The police didn't smile. Instead they grabbed "The Offender" by his shoulder and spun him around, dragging him backwards into a police cruiser. Struggling to stay on his feet, "The Offender" was a sight to see, to say the least. Black soot was all over his face, his hair was now a dirty blonde (imagine smoke rising off it), and his blue eyes were wide and vivid. Not a blink of fear was in him.

I watched as the police talked to him, towering over him as he sat in the car. All I could make out were his dirty ripped blue jeans and his black Converse sticking out from the back seat as four or five cops bounced around, jacked up on adrenaline or whatever they put in police water, shouting and pointing and screaming, their faces beet red. After a few minutes, they pulled him out by the scruff of his neck, his head just missing the top of the car door. They shoved him in our direction. He stumbled, caught himself, and then kept walking.

"The Offender" casually walked into the skate park with a smile. "I'm never getting busted," he said calmly. "I don't know why I'm not in jail, I shot a cop practically in the face…I'm just never getting busted, and it's fucking awesome…."

Photographs © **Larry Fink** / *Social Graces*

CLOSE *BY PINCH HUDSON*

Well over three years ago, I decided I was through with him. Mmm-hmm, that's right—through with him and every last one of them. I had this thought: I should go a year without sex. Collect my head and grow cobwebs. Yes. It was too easy to hold myself to a standard that proclaimed: Unattainable. I am above the likes of you.

And you. And you.

And *you*.

But it's a trap, a delusion to draw the conclusion that I could avoid the most intimate of conflicts (which is to say the wide divide between my pride and my thighs). And in this, my aging vow of abstinence is tested time and again, by a never-ending need for attention from certain men. Only at this point, I'm not the one holding out—it's them. These men. They have issues like *The New Yorker*, compelling but insurmountable troubles that prevent them from taking any steps in my direction. And I, fucked up from not getting fucked for three years, interpret this as a challenge.

Perhaps I am a Romantic, in the nineteenth-century sense of the word. For me, love is failure and ruin, as my heart holds true to the transcendent power of salvation through destruction. Which is an overwrought way of saying I love a loser, the fantasy

not being that I will save him, but that he will save himself for me. But when I awake, I am left with only the remains, my desires unsatisfied. And so the quest continues....

All of this stands in striking contrast to the status quo where women check net worth before consenting to dinner, seeking good providers for their unborn offspring. What do they call it? Husband material.

Sounds suspect to my ears.

I blanch at the idea of marriage. Marriage? A legal agreement with a man? An official ceremony binding 'til death do we part, or divorce—whichever comes first? It seems so excessive, so extraneous, so enmeshed in bureaucracy, tradition, and irrelevance. It may be best, or rather necessary, for breeders, for the perpetual procreators who require the court to secure child support because most men suck and would leave them stuck will bills. But still. No. The wedding dream was never mine.

I think you'll agree: dressing up like the cake you'll later be shoveling into each others' mouths is disturbing on several levels. And the white dress: A virgin? An angel? An innocent? An idiot? An escaped mental patient?

And the main event: Daddy hands his little girl to another man in full view of everyone he can afford to invite, relieved to know his spawn is no spinster, smiling as he pays for everything including the centerpiece someone's drunk aunt will take home later in a fit of quixotic psychosis.

This strangled gasp, this desperate grasp at codependence ends, half of the time, in estrangement, adultery, alienation of affection, separation, divorce, court, alimony, joint custody, depression, alcohol abuse, pill popping, sugar binging, weight gain, plastic surgery, reality TV appearances, new cars, hot dates, weekend getaways, formal introductions, trips with the kids, and weddings.

Because this time, it's Right.

Riiiiight.

Mind you, my parents have been happily married since 1970 so I'm not the bitter product of an environment that would inspire anti-marriage vows. Still, I never wanted a husband or a family, and I will be damned if I'll ever take another man's name. Better to be Bette Davis, who was always Miss Davis, despite her four husbands.

But I ain't trying to have no Mister Davis. That's just creepy.

When I think about it, it is clear as day. The biggest mistake I could ever make would be to enter into a contract that assigns legal terms to the most aching of intangibles:

Attraction.

"Define attraction in fifty words or less," the Notorious Hollywood Producer demanded of me one night last summer, as I sat beside him in his dining room, stuck in an overstuffed chair that threatened to swallow me whole if I so much as shifted. I had never expected this, to be in the presence of Sunset Boulevard realness, but there I was in a Golden Era Estate, planning a party for a Blockbuster Movie Director who had just released a book of photographs shot on a Kodak Disc camera kept in his back pocket for two decades. Where he got the film, I never knew.

The Notorious Hollywood Producer generously offered his esteemed estate as the site for the launch party, so I arranged a meeting in advance of the event to discuss the details. Along with my boss (who I brought along for protection), the Blockbuster Movie Director's Excellent Executive Assistant, and the Paranoid Party Planner Hired by the Blockbuster Movie Director's Blow Job Bitch, I was introduced to the Notorious Hollywood Producer by his Unctuous Estate Manager in the legendary screening room, which overlooked an elegant oval pool to one side, and a tennis court (where the Kennedys played tennis a thousand moons ago) to the other.

NHP, as he shall now be known, could not stop speaking, spouting, espousing his wisdom and slightly deranged logic that employed three dozen women to call two thousand bookstores across the United States and inquire about the availability of his autobiography in their shop.

"And you know what?" NHP hectored like Rush Limbaugh after one OxyContin too many. "Ninety percent of the stores never heard of my book. *Never heard of it!*"

He couldn't be stopped. Except by the forces of nature. There was a brief pause caused when NHP directed, "Wait here one moment," and then walked toward the bathroom. The moment the door closed, the Unctuous Estate Manager asserted himself, ordering everyone, "Let's go. Now."

He led us back to the main house and into the dining room, where we seated ourselves in voluminous chairs around a massive yet intimate table. Then we began to discuss the logistics of the launch.

"Why did you leave?" NHP asked from the doorway at which he appeared unobserved, like a lizard leaping lithely, ever so quietly. At the sound of his voice, we arose—in unison—from those distended chairs and stared stupidly as he shuffled slowly past, shouting the name of his houseboy, who came running.

"A double Cosmo," NHP commanded over his shoulder while he took his seat at the head of the table, to my immediate right.

Returning to his post as Director of Everything, NHP hijacked the conversation, taking us on a scenic tour through his toxin-addled, stroke-rattled mind, a mind that served alternately to exalt and ravage his life. It reminded me of the time I inadvertently ate an eighth of mushrooms; tripping my ass off at Palladium on a Sunday evening, I recalled offering seven questions to the universe, each answer unlocking a secret of the world. But after I came down I could only remember the last question: "Why can't it always be like this?" To which the Universe replied: "Because then nothing would ever get done."

Much in the same way, my brain could not contain the intricate maze that unfolded before me as NHP disclosed the meaning of life. But one moment was forever burned in my memory.

"I have written a book, yet it has taken me years to answer this question," NHP asserted dramatically, then stood—sort of. With great effort, he was freed from his chair. He took the glass in hand and asked, "How would you define attraction in fifty words or less?"

Looking adoringly at his drink, NHP took a sip and paused to savor the sweet swill of liquor as it ran across his pickled tongue. Then he turned to me.

"You!" he demanded as he stood above me, dressed in a Puma tracksuit that unzipped to his navel, with the requisite gold medallion hanging in a garden of hair that grew in abundance for a septuagenarian. Holding the Cosmo a few inches from me, the slight tremor in his arm incited a mildly exciting tension inherent in the possibility of baptism from this man. I looked up expectantly but was disappointed to see the glass had been returned to the table. In awe, yet unable to submit to his strange game of domination, I answered his question with an eyebrow arched, a smile snarked.

"Purrr."

Infuriated with my singular response to his dinner party polemic, he seethed, *"That's not a word!"*

Unflapped by his comeback, I responded with the utmost authority, "Yes, it is. It has three Rs."

Put off, he summoned all of his being to disengage from my loophole, and turned to my boss, who sat to my immediate left. "You!" NHP barked.

"It's when two people feel a physical, spiritual, emotional, and intellectual pull towards each other created by mutual curiosity and desire," my boss said with self-satisfactory relish.

"Wrong!" NHP steamed before turning his attention to the next victim.

The Notorious Hollywood Producer had several wives, one of whom left him a week after the party, claiming he was having an affair with a tartlet a third of his age. More reasons to avoid matrimony I do not need.

So what is attractive when one doesn't require a long-term investment, when one is not looking for husband material, a product for fatherhood, or someone to cosign a mortgage application?

Damned if I know.

SEX
PHOTO: GABRIELLE PIEDMONTE
wwt
tHE APPAREL BRAND
WORSHIPWORtHY.com
tHE ONLINE MAGAZINE
WW the Agency
tHE CREATIVE AGENCY
WORSHIPWORtHY

SPORTS ★ ★ ★ ★ FINAL

SPECIAL 16-PAGE SECTION WRAPS MAIN NEWS

DAILY NEWS

50¢ www.nydailynews.com **NEW YORK'S HOMETOWN NEWSPAPER** Tuesday, June 11, 2002

THE LAST DON

JOHN GOTTI

1940

2002

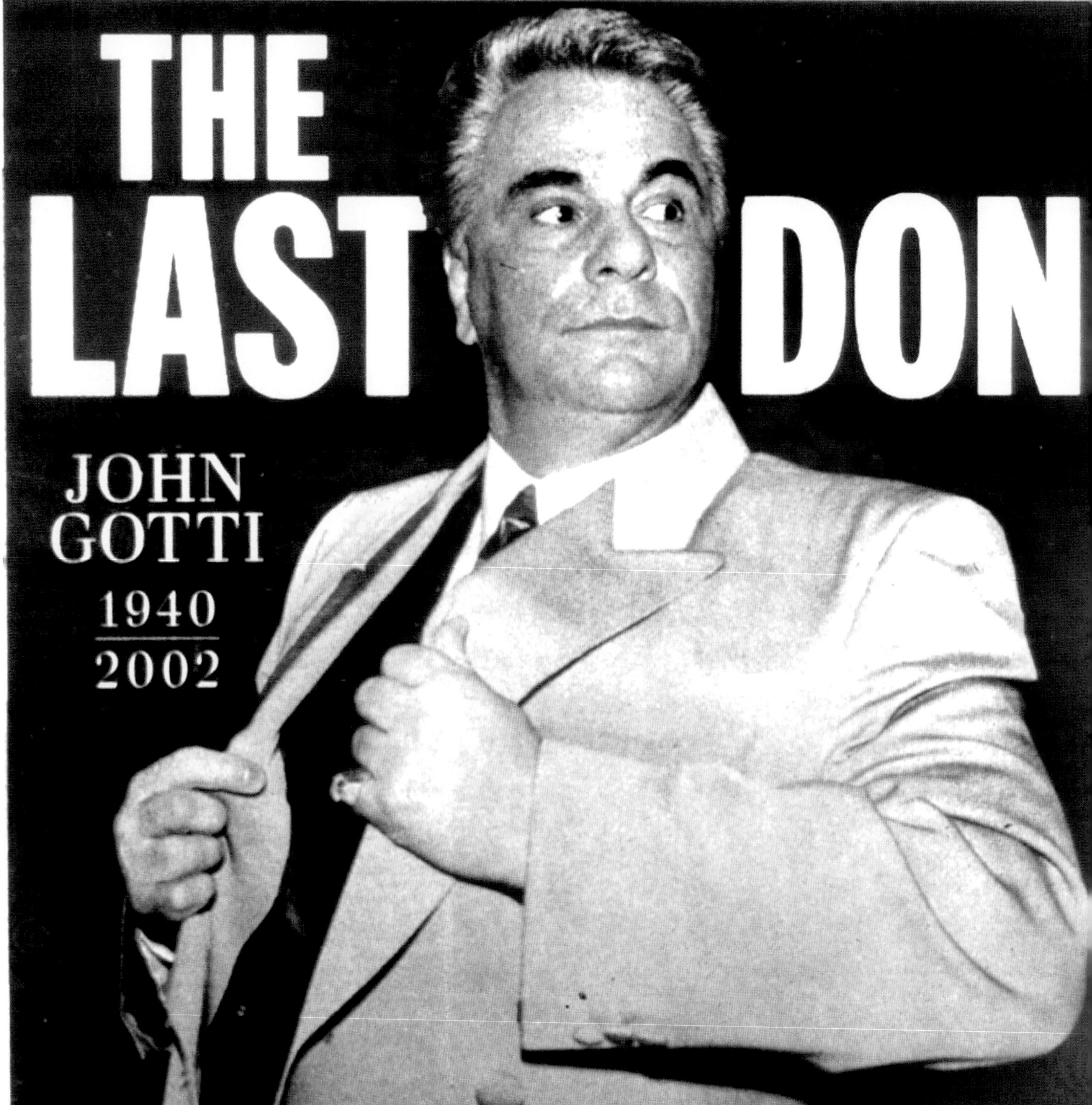

FEDS FOIL DIRTY BOMB TERRORIST PLOT

SEE MAIN PAPER FOR FULL COVERAGE

ALL MY IDOLS DIE IN JAIL

BY IVAN SANCHEZ

New York can trace its criminal genius back to La Cosa Nostra and so can I, although mine was more of a criminal stupidity—there was nothing genius about it.

It's said that Carlo Gambino was enticed into a life of crime in the early 1900s after witnessing the Honored Society, including members of his own family, walk around in his native town of Caccamo in Palermo, Sicily. The members of the criminal enterprise dressed in the finest suits and hats of the day and were so respected, feared, and powerful that they ran their town without interference from the police or military. Gambino so idolized the Honored Society that when Benito Mussolini came to shut them down he fled to New York to continue in the traditions of his mentors. The Gambino name will forever be synonymous with organized crime at its best, pre-Sammy the Bull. Even John Gotti, who would later become the head of the Gambino crime family, started his career idolizing the flashy made men he saw driving around in Cadillacs in his South Bronx neighborhood.

Apparently, as men, we are all doomed to adore, worship, and admire the fast buck, and will risk our freedoms and our lives to acquire it. By the time I was coming up in the Kingsbridge Section of the Bronx the mob had settled in nicely, owning a chain of video stores called Top Video. The stores were used as fronts; they ran videos out the front door and numbers out the back. The first time I saw a desk covered with stacks of Benjamins I knew right then and there that I to had figure out a way to make that my reality one day.

My idols were these guys named Big Al and Big Tony who ran the operation. These guys had no problem operating within the stereotypes of the mob, driving long-ass Cadillacs and wearing big diamond pinky rings. The rings were always prominently displayed when they talked with their hands, making sure to extend the pinky for emphasis on what they were saying. I can honestly say I stole from the mob and lived to tell about it. Not because I was smart enough to not get busted, but most likely because I was a stupid fifteen-year-old who was cut some serious slack by Al.

SIZE, a neighborhood graffiti writer I worked with, put me onto memorizing the prices of renting three or four videos to customers. Once you knew the exact prices, you'd be able to ring up one video, while verbally charging the customer for the three or four they were actually renting. This, of course, was before digital display cash registers. I'd keep track of the overages in my head and at the end of the day I'd simply take the money out of the register when no one was looking. It wasn't exactly safecracking, but it was still stealing from the mob and I was risking a serious beatdown at the very least if caught.

The scam worked beautifully for about a year, until Al walked into the store one day and told me to "Get the fuck out." I tried to ask him what was going on and he shot me a look that warned, "If I have to repeat myself I might not be able to stop myself from beating you all over this store and shoving movie after movie up your ass." The jig was up—I was busted and I'd have to find another criminal in the neighborhood to idolize.

Shortly after the incident, my hardworking single mother had grown weary of the drugs, violence, and crime the Bronx had to offer. She decided to try to relocate my siblings and me to Virginia Beach. My brother and I ended up closing the U-Haul truck on her dreams and staying behind for a life of crime. With my mother out of the way I was free to dive in head first and I continued to turn to criminal mentors in the neighborhood, much for the same reason Gambino and Gotti had idolized their mentors. They had money and they looked good as a result of it.

One of my mentors was a young Puerto Rican guy from my building named Pedro "P-Funk" Gonzalez. P-Funk had been selling weed in the neighborhood since he was 13 or 14 years old. The kid always looked fly, rocking the latest hip-hop attire. He introduced more styles to my neighborhood than anyone I can remember and he eventually wore gold chains so big he'd make Biz Markie and Big Daddy Kane tuck their shit.

When P asked me if I wanted to turn my apartment into a stash house I don't believe I hesitated for a second. With the promise of paying all my bills, keeping food in the refrigerator and money in my pocket, I had just landed a nice deal for myself. It was 1989 and in the beginning everything was going smoothly. One of the guys running the stash house went by the name Chippy. The first time he ever knocked on the door he looked like someone who was coming to kill me. But in addition to being menacing, he also had a real swagger about him and eventually I became comfortable around him, even with hundreds of pounds of weed in the apartment and the .357 Magnum I slept with under my pillow. The fucking gun was bigger than me and I wasn't sure I'd be able to shoot the shit if anyone ever broke into the stash house.

Chippy had a connection with the rapper Slick Rick's cousin and we'd spend countless hours smoking blunts and plotting a robbery of the rapper when he was out of town on tour. It was no big secret that Slick Rick rocked hundreds of thousands of dollars in jewelry and kept jewels, money, and guns in a safe in his house. I often allowed myself to envision what I'd look like rocking some of Slick Rick's jewelry around my neck. I was five-foot-nothing and weighed about 140 pounds soaking wet. I would have looked crazy with a gold dookie rope chain dangling from my neck. Crazy fresh, that is....

One weekend my mother came to New York on an impromptu trip and begged me to come to Virginia Beach with her for a visit. She wanted to see if I liked it so she could try to keep me there. And although at first I refused her requests, seeing my mother cry made me give in and take the drive back to Virginia with her. When I returned a few days later I knocked on the apartment door to find that no one was there

Photograph © **New York *Daily News***

to let me in. I walked to the other side of the building and banged on P-Funk's door to ask for the spare key. When he opened the door he told me to come in and sit down.

I figured the stash house had been broken into or busted, but it was worse. He told me that Chippy had tried to rob Slick Rick that weekend and had been shot to death—apparently by someone in Rick's camp. Although it was never proved who killed my boy Chippy, Slick Rick would eventually be arrested for shooting his cousin for trying to set him up.

With Chippy gone, new management was brought in to watch over the stash house. This time it was a six-foot-plus Jamaican behemoth known as Barney. His eyes were bloodshot yellow and if you looked into them you'd see nothing at all.

P-Funk eventually told me to take over the corner sales of weed on the block. We didn't really have a weed spot in the neighborhood, even though the entire crew chain-smoked blunts like our lives counted on it. Apartment 5F was a stash house and P didn't want people stopping up there all the time looking for small quantities of weed when we only dealt in pounds. He showed me how to bag weed to make it look like there was more in the bag than there really was. He also showed me how to soak it in water and dry it on the radiator so that it would puff up and look like more.

I had all the expertise and supply I needed but I made for a horrible drug dealer. If I took 20 bags of weed downstairs to sell, I sold ten and smoked ten with my boys. I spent my re-up money on the latest Nike Air Maxes and then had to take my weed on credit instead of flipping my profit back into my product. Truthfully, I just didn't give a fuck at the time. For me it was just something to do to watch the days go by. Eventually I started stealing drugs from the stash house, much the same way I stole money from mob video. When I was busted by Barney and threatened I simply fled to another part of the Bronx—all the way to the other side of West Kingsbridge and Bailey Avenue—to live with my cousins.

P-Funk eventually branched out into selling coke, crack, and heroin in addition to weed. I never had any beef with P but the relationship was strained because I had stolen from his people. But every time I saw him over the years his cars and jewelry continued to be upgraded. I felt more jealousy for P than anyone I ever grew up with, especially when he came out with the all-white Toyota Land Cruiser, then the Range Rover, then the 600 Benz with the TVs in it.

The one thing that never seemed to happen to P, though, was that he never got busted. I mean never. I still had people like my brother-in-law working with P and they'd tell me how he'd become a millionaire over the years. I always felt like if I had done the right thing by him maybe that could've been me, right there with him, making more money than I could count. At times my jealousy turned to hate, especially after I heard he had put a contract on my brother-in-law, who was caught up in his criminal drug enterprise.

One day I stopped by the old neighborhood to see if there was any truth to the rumors that Pedro told his goons they could feel free to kill me too if I was with my brother-in-law when they caught him. It was several years after hearing rumors in the streets that P had expanded his empire to include murder. But being the hardheaded nut job I was, I rolled up on the block like I had it like that. Interesting enough, P seemed sincerely happy to see me after 10 years. I told him if he ever wanted to invest his money in legal business to give me a call, and of course I never heard from him again.

Over the years I often thought about P-Funk and counted the years he'd been successfully running the streets and making money without getting arrested. Lord knows that life only allows you five years if you're lucky. By 2008, P had been running on the wrong side of the law for 23 years. I often thought about whether or not he'd ever get caught. It was the jealousy in me that kept me wondering, and then one day the phone call came: "P's been busted."

My brother-in-law Jorge called to ask me if I had read the New York *Daily News*. The story was dated May 7, 2008, and the headline read, "Bronx Drug King Gets a 35-Year Term." The story went on about "Pedro (Pete) Gonzalez, 37, who rose from a street corner crack and pot dealer with his brother David, 33, in their Kingsbridge neighborhood to running a citywide multikilo cocaine and heroin ring."

I guess they didn't know him as P-Funk, they knew him as Pete. And he'd now be spending the next 35 years of his life in a federal penitentiary as a result of his 23-year run on the streets. My man P, the person I stole my first bike with and learned how to bag my first nickel bag with, had been sentenced to spend the rest of his life in prison. All the jealousy went right out the window. All the times I wanted to be like him seemed ridiculous, and all the times I idolized him and his cars, his jewelry and his money, meant nothing.

Another one of my idols was going to prison. P-Funk was busted, and all I felt for him now was pity. He'd most likely die in jail, like John Gotti, my other idol.

Photograph © **Ivan Sanchez**

BALLS AGAINST THE WALL: THE ELECTRO-PUNK BAND JUGGERNUT VS. THE NYPD

BY IDRIS INTIFADA

"What do you mean I can't bring my sword into the bar? That's bullshit!" Aric sang with so much conviction that even the ice-vaginaed Judge Judy would have to rule in favor of his right to bear arms. Unfortunately, *Aric vs. The Local Drinking Hole* never made it to CourtTV, so instead he pleaded his case for the fair treatment of blade carriers, in song format, on a crowded Williamsburg street corner. As the electro-punk duo JuggerNut, with multi-instrumentalist Cliff dressed as a gigantic scrotum (with fake pubic hair, bulging veins, and all), Aric vented his self-proclaimed "struggles with the inner conflicts and legalities of having a sword at the bar." The band unleashed a beer-can-smashing, trash-can-launching public performance and passersby quickly crowded the scene. With all eyes fixed on the man-sized testicle playing an electric maraca and the vocalist wilding like the Incredible Hulk after a prescription pill binge, the impromptu audience wondered what forces had incited this musical rampage.

It all began when Aric, after years of intercontinental traveling, moved into the Bowery's infamous Sunshine Hotel. "That flophouse is super crazy and interesting; there's all these junkies living there," he describes. "Cockroaches are all over you all day long." Down the street from the hotel, Aric gave musical birth to the half-gonad, half-comic book superhero JuggerNut, on stage at the Bowery Poetry Club. Originating as a sketch in his notebook, at the club Aric would bring the JuggerNut character to life with his solo act. In the fall of 2007, Cliff caught the one-man tribute to the cartoon super-genitalia and just had to get down with the Nut. "We decided to both work on JuggerNut as an entity and as a cosmic idea," says Cliff. "The next thing you know, I'm in the nut suit and in the band."

Some compare JuggerNut's distorted sound to the 70s synthesizer band Suicide, but in actuality, the band plays their own unique brand of electro-punk they call "sexual comedy." Songs like "A Woman's Ass" and "Cockroaches in My Cocaine" perfectly capture the pleasures and pains of everyday life to a drumbeat more infectious than hepatitis pee on a public toilet seat. To conjure the adventurous superhero spirit of the JuggerNut, the band adds a performance art element to their music. But don't get it twisted, Aric and Cliff aren't two sissy mimes acting in Central Park. While playing their song "Stun Gun," they shoot off live electric tasers as they storm the music venue. Every so often an audience member gets electrocuted but a tiny shock seems insignificant when compared to the larger idea of JuggerNut.

Sadly, the NYPD doesn't think stun guns are so insignificant and the zappers have put JuggerNut at odds with the law. After a show, the band got a ride home in a friend's Thunderbird. "My friend had a few drinks in him and we drove to this stoplight. He doesn't feel like sitting at it, so he pulls right through it. He sees this cop and he just takes off! We run 20 blocks of red lights and stop signs, just flooring it. Two cop cars cut us off from in front and behind. They come out with guns pulled and shove our friend into the cop car," recounts Cliff.

After the high-speed chase, "They searched the trunk and our stun guns were back there. I also had a knife on me, and two Xanax. They arrested me for everything but the Xanax. Actually, I think they took the Xanax," Aric conjectures. "Then, all of a sudden, the cops start firing the stun guns off. They were all laughing and having fun. It was a pretty cool scene in the dark. Later they asked where I got the stun guns because cops aren't allowed to carry them in the city. I was all drunk, screaming at them, 'Fuck you, we pay your bills!'"

For JuggerNut, bad luck doesn't come in triples but in twos like testes, and a month before the stun gun incident, Aric had another run in with the boys in blue. "I'm half-Swedish, so I don't know if it's a Viking thing or what, but I like to bring my sword to the bar and spin it around," he explains. "I brought my sword out that night in my belt, and walked outside the bar with an open Corona. The cops swooped down on me. I had thirteen Klonopins and three Xanax and it wasn't even a question if I was getting arrested."

Photographs © **Colin Simmons**

Magic Cigarettes

by joy yoon

Often, I find myself visiting my father.. This was one incident...

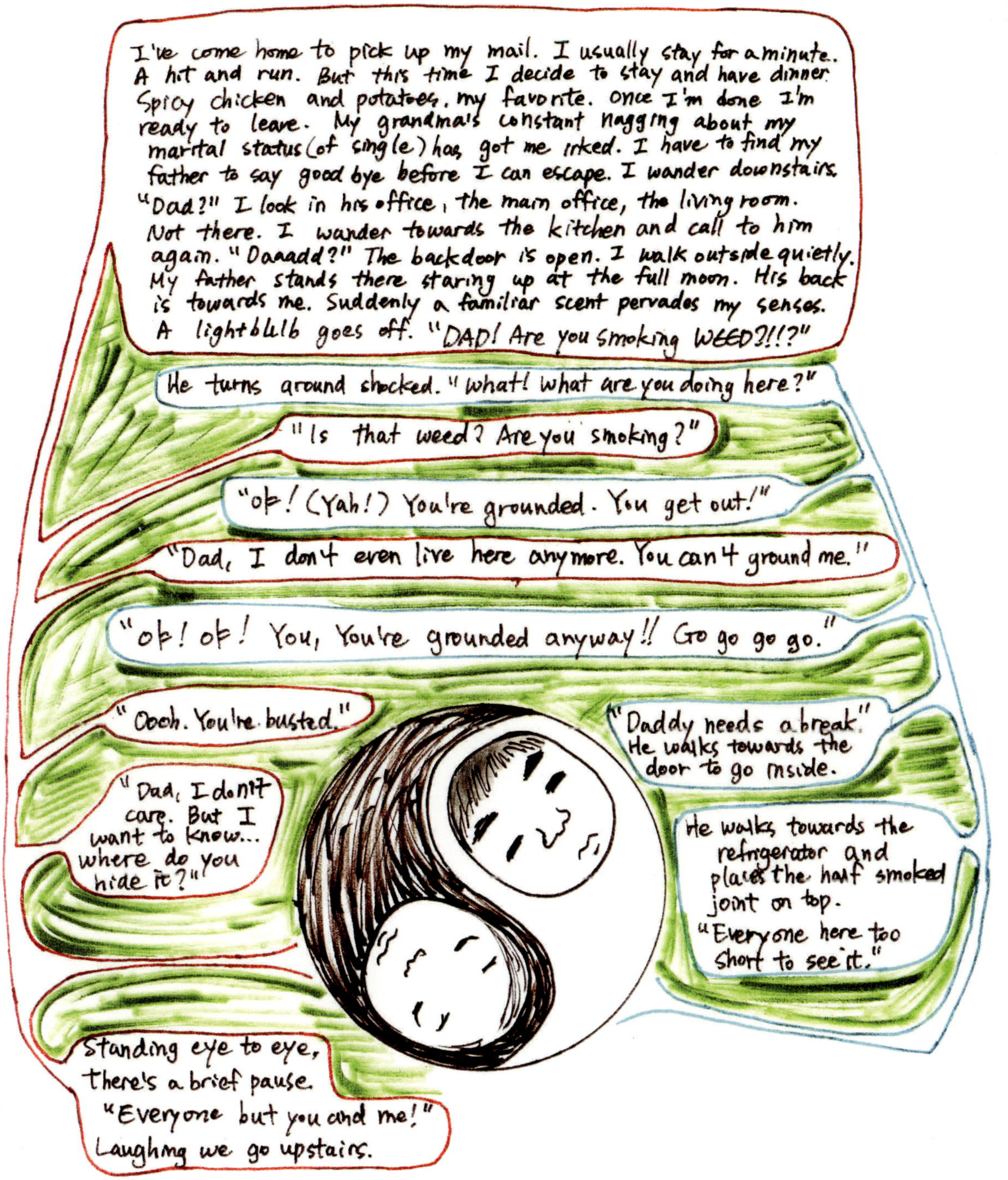
I've come home to pick up my mail. I usually stay for a minute. A hit and run. But this time I decide to stay and have dinner. Spicy chicken and potatoes, my favorite. Once I'm done I'm ready to leave. My grandma's constant nagging about my marital status (of single) has got me irked. I have to find my father to say good bye before I can escape. I wander downstairs. "Dad?" I look in his office, the main office, the living room. Not there. I wander towards the kitchen and call to him again. "Daaadd?" The backdoor is open. I walk outside quietly. My father stands there staring up at the full moon. His back is towards me. Suddenly a familiar scent pervades my senses. A lightbulb goes off. "DAD! Are you smoking WEED?!!?"
He turns around shocked. "What! What are you doing here?"
"Is that weed? Are you smoking?"
"Ok! (Yah!) You're grounded. You get out!"
"Dad, I don't even live here anymore. You can't ground me."
"Ok! Ok! You, You're grounded anyway!! Go go go go."
"Oooh. You're busted."
"Daddy needs a break." He walks towards the door to go inside.
"Dad, I don't care. But I want to know... where do you hide it?"
He walks towards the refrigerator and places the half smoked joint on top. "Everyone here too short to see it."
Standing eye to eye, there's a brief pause. "Everyone but you and me!" Laughing we go upstairs.

PHOTOGRAPHS AND TEXT BY CALEB NEELON

Artwork © **Alex Lukas**

Dan Haga and Dan Ayers are Urban Atrophy, a team of Baltimore-based explorers looking for the perfectly abandoned. With a take-nothing-but-pictures-and-leave-nothing-but-footprints ethic, their missions have taken them over and under fences surrounding disused psychiatric hospitals, missile silos, jails, and churches, among other places. There's a small community of people that are serious urban explorers, following the same ethic—no stealing, no vandalism, leave it the way you found it, and, just as important, don't tell the world where your good spots are and get them blown. All of these are important ethics with one very tangible goal in mind: not getting caught.

When it comes to sharing his favorite locations, Dan Haga is necessarily cagey, but once a place has been demolished, he can open up a bit. And so it was that he was willing to share the story of Baltimore's Westport Power Plant. —Caleb Neelon

I find a lot of locations that catch my interest by browsing aerial images of Baltimore city. Over time I've gotten the eye for it. This building was huge and had all the classic signs of abandonment: rough-looking roof, overgrown grass, fencing, and sealed windows and doors. The next step is to figure out what it is—or what it was. After figuring out an address, things like tax records can tell a lot: its previous owners, descriptions of the property—things like that. You don't want to wander into an abandoned place without knowing what hazards to watch out for—there's a reason these things are fenced up. I also like to like to research the history and find historic photos so that I can compare what it looked like new to what it might look like now.

What I was looking at was Baltimore's Westport Power Plant, located right on the harbor. When it was built in 1906 this was the largest concrete-reinforced power station in the world. The plant generated power by burning natural gas and fuel oil to produce steam, and also had a combustion turbine used for burning natural gas. It fully closed sometime in the 1990s.

My next step was to scout the place out. I drove around it several times. Baltimore is good at rough neighborhoods, and the plant was in one. Looking at it from the front, I could see activity inside the perimeter fence, as well as an occupied guard shack. The fencing was top-notch and all windows and doors were not simply boarded up but sealed with welded steel plates. Access from the front was obviously impossible. They really didn't want visitors.

We plotted a way in around the back. There were a few fences to hop, but that wasn't a problem because they were totally hidden from view. But there's always some surprise, and in this case it came when I tried to cross a patch of what looked like shallow, gray mud. It turned out that there was thigh-deep, disgustingly dirty water underneath the mud. I was carrying a couple grand in camera gear, and it was a below-freezing day. It really sucked.

That's still no reason to stop. While not recognizable in the film, Westport was reportedly used in the filming of the movie *12 Monkeys*. One of the first peculiar things we found was a *12 Monkeys* logo stenciled on the wall. That was on the outside, though. We wanted in. It took forever to find a place to squeeze inside; access ended up being an old conduit hole that required us to climb high up a wall and pull ourselves through a narrow hole that most wouldn't fit through.

UNDER (FRIENDLY) FIRE

PHOTOGRAPHY AND TEXT BY YOAV GALAI

This incident was considered marginal even on the day it happened, August 12, 2006. One hundred fifty-five Israeli soldiers were injured and 24 killed on that day in the Lebanese town of Ainata, and the incident didn't even make the news.

I was sent by my newspaper to Haifa to help cover the missile attacks when I received a phone call from the picture editor asking me if I wanted to join an incursion into Lebanon. I immediately said yes and drove my bike to the border.

On our second day in Lebanon, while we were setting up a room to receive wounded soldiers who had been run over by a tank the night before, we were fired at with what we thought were RPGs.

These photos were published as a double-page spread the day I returned to Israel. I licensed them to the Associated Press so that they also appeared in *Paris Match* and other international newspapers. They were also used by "friends of the IDF" in fundraisers (with my permission—it meant that my friends who were injured got to attend those events all over the world).

For these photos, I was to receive an Israeli Press Photography Award. During the ceremony, one of the soldiers who were injured in that incident told me that it was actually caused by friendly fire. I was shocked. Until that time, I had thought that the Hezbollah were responsible for this destruction. But I realized soon after that it made no difference.

I was invited to a few TV and radio interviews along with the same soldier to talk about the award. We both wanted to talk about the incident but were informed by the producers that we should not mention that the injuries were caused by friendly fire.

I did manage to exhibit the photos in Los Angeles, in the UCLA Dortort Center. The show was curated by Mrs. Perla Karney and funded by Hillel. The "friendly fire" caption was included and it didn't matter one bit; everyone who came understood that it is a part of the reality of war.

The reactions were very powerful. Both Israeli and American veterans told me it brought back memories of their war and younger people were surprised to find boyish faces springing out of the familiar green uniforms.

After the exhibition closed, an Israeli NGO wanted to bring it to Israel in order to help raise awareness for PTSD (post traumatic stress disorder). We got as far as checking out the venue, but my insistence on including the fact that it was a friendly fire incident killed the show.

The prints are still in Los Angeles.

RUNNING WITH ABBIE

BY PATTI ASTOR

In 1968, at age 18, I took off from Cincinnati, Ohio, for Barnard College, New York City. From nearly the first day I stepped on campus, I had become a member of SDS (Students for a Democratic Society). By the time this story took place, I had already dropped out of school to go anti-war full time and had been arrested three times, including one charge of "felony riot."

By this time, Weatherman had been happening, but there was a group of us who passed on their heavy trip and formed the Running Dogs, a tongue-in-cheek reference to Mao Tse Tung's scathing disdain for "imperialism and its running dogs." We had no problem with fighting in the streets, but skipped the indoctrination routine and still went out to party after the demos.

Rennie Davis, one of the Chicago Seven, organized the final massive demonstration against the war, which was held on May Day, 1972. The theme: "If the Government Won't Stop the War, We'll Stop the Government." The plan was to demand a "People's Peace Treaty" between the citizens of the United States and the citizens of Vietnam.

The whole extravaganza began on Sunday in Washington, D.C. with a rock concert. Then everyone pitched their tents in "Algonquin Peace Park" on the banks of the Potomac River. We were 10,000 strong. At 6:00AM we were awakened by Chinook helicopters whirring low over the encampment. Just like our Vietnamese confreres on TV, we scrambled out bewilderedly. Military voices issuing from loudspeakers mounted on the 'copters warned:

"A state of emergency has been declared. Your presence is not legal here. You must evacuate at once or there will be a sweep of the area. Repeat, we are beginning a sweep of the area."

Well shit, TPF (Tactical Police Force, the robin's egg blue-helmeted cops with riot shields, clubs, et cetera, which was what we usually faced in NYC) were one thing but this was ridiculous. I stood in the light of dawn, transfixed by the machine guns mounted on the fuselages. They were a horrible shade of green and I was close enough to see detail. I remember thinking, "This is what it's all about."

So began May Day, May 1, 1972, when the Woodstock Nation made its last stand.

The legions for peace migrated. We all had assignments for which major traffic points to disrupt. "Stop the Government"—remember? If they can't get to work, they've been stopped. We deployed past a beat-up panel truck hastily painted in huge letters with "DON'T GO." I am proud to say very few people pulled out at this point.

Running Dogs and NYC Weathermen, ass-kickers that we were, were assigned to DuPont Circle, a major location. Because we were in D.C., we were equipped with homemade gas masks. In Washington, you had to expect gas (not as many pedestrians would be affected). At DuPont Circle we pushed over the police barricades and blocked the street with our bodies, effectively bringing traffic to a halt. One kid in our legion was struck by a car and ended up with a broken leg.

Busloads of riot squad cops arrived. The cops deployed and made a running charge. We split up into affinity groups (usually four or five people who fought together all the time and would stick together defensively no matter what) to book it up the side streets. A couple of guys joined in with our group of four Running Dogs as we tore off down a street and turned into an alley, and suddenly, we'd escaped! There were six of us standing there catching our breath in this totally deserted and eerily quiet suburban alley. We couldn't believe it when one of the two new guys turned out to be the storied yippie leader Abbie Hoffman. Count on Abbie to be where the action was! He smiled and said, "Let's all sit down, join hands, and sing 'We Shall Overcome.'"

Everyone cracked up and we ran out to hit the streets again.

(Years later, I saw Abbie for the last time at the U.S. Open, at the famous Borg vs. McEnroe final in 1981. He was sitting right behind me! It was after he had resurfaced from the underground as a fugitive. I told him that story and he got a big kick out of it. Continually hounded by the law and beset with financial difficulties, he would commit suicide in 1989.)

The survivors of the morning's skirmish began to converge at the college in Georgetown—a necessary regrouping from a morale standpoint, but tactically it was suicide. Jerry Rubin (pre-networking) delivered a fiery exhortation in a red, yellow, and blue classic National Liberation Front star shirt. Then the cops cordoned off Georgetown U and gassed the shit out of the whole area while phalanxes of riot police performed a pincer movement toward the center. We would charge the cops and then they would charge back with more gas. As piles of the noxious creamy powder heaped up in the streets, we fell gasping and crying. Guerrilla medics performed battlefield clearings of our eyes and we ran on.

Holed up in Georgetown, we narrowly escaped arrest by slipping into a dorm and hiding in a frat-boy closet while we listened to the pigs shake the place down. After they left we watched the whole thing on TV in the basement rec room. Choppers flew across the Washington Monument and troops deployed onto the lawn. It looked just like the TV war coverage in the rice fields.

Over 3,000 people were arrested that day, Abbie eventually among them. Everyone was held in a stadium. Abbie, in a move that became legendary, escaped by hiding in a garbage can and going out with the trash.

This demo was the last really strong show of force by the anti-war movement. Many dropped out when they were no longer threatened by the draft, but the real problem was caused by the disintegration of the central SDS leadership. Most of them became Weathermen and went on to increasingly nutty things

Abbie Hoffman harangues Columbia students and others at statue of Alma Mater. He urged them to disobey all laws. Later about 500 stormed through campus, breaking windows, after university turned down demands for bail money for Black Panthers. 3/13/1970 Photograph by Tom Cunningham © **New York *Daily News***

like holding up armored cars, et cetera, which did not lead to public support. I have always wondered if the whole Weatherman thing was an FBI- or CIA-incited plot designed to split and disarm what was a very powerful voice for change. Someday when I have the nerve I would like to request to see my FBI file.

The mainstream movement came to an end and soon after, happily, so did the war. My generation, whether G.I. or protester, was left with a legacy from that war: a deep trust in our fellow freedom fighters and an overwhelming disillusion in the system—The Establishment. For four years my life consisted of tear gas, mounted police, and beatings—nothing to equal a firefight, but I think all would agree our youth went to that war. Everyday my thoughts were of my USA brothers over there getting their heads blown off for no good reason. I would do it all again without question.

Beat poet Gregory Corso said, "The revolution has happened and you'll never get them back."

LIVING INSIDE OUT

BY TODD SOLOMON

He hides what he doesn't want you to see. He keeps the darkness hidden to shelter you from knowing the real him, for if you knew the real him, he fears you wouldn't stick around for very long.

He's been running from who he really is for years, but no one can see his tracks in the dustless path he has left. He runs from who he is as he tries to be what everyone else needs him to be.

She doesn't talk much, but when she does it's laced with sarcasm and doubt. Her confidence barely exists; she always thinks everyone can see straight through to all the secrets that she hides. It's one lie after another.

When she was growing up she didn't want to wear dresses and bows in her hair like all the other girls. She wanted to run and play with the boys in her baggy jeans and Karate Kid *sweatshirt. She'd play in the dirt and the mud without a care in the world about how dirty her clothes were going to get. She would spend hours outside digging in the mud, looking for worms. If the weather had made the ground dusty enough, she would take all her Hot Wheels outside and make dirt roads in the middle of the backyard, creating her own* Smokey and the Bandit *scenes.*

She creates a new life that she can't fully live. She hides behind photos of someone she longs to resemble. She doesn't understand why it happened, but everyone else comes up with a reason, declaring it medical or calling it a choice she herself made. "Dare to be different," they say. But when you dare to be different, who picks up the pieces of the hearts you shatter along the way?

He fell in love with the most amazing woman in the world. They can talk about anything or nothing and it always feels like home. He feels safe with her. She makes him feel things he never thought he'd feel, but he feels afraid and guilty because she doesn't know his secret. He fears she will leave him if she ever learns the truth.

They are one person sharing the same body, struggling to find a place in the world where they belong. No one can see this person for who they are instead of what they appear to be. They can't imagine the struggle. The doubt. The pain. The fear. The reality of being transgender. This isn't a choice.

When you are transgender, your mind and your body are not in agreement over your gender. You are trapped in the wrong body, forever in a battle with yourself, with society as a whole, and forever fearing the casualties of the war of the people you love most. You live in a costume you can't stand and wish you'd never gotten to begin with, but in reality it's your own skin you aren't comfortable in. Many transgender people cannot stand in front of a mirror without feeling hate or recrimination. They look for ways to hide those telltale signs.

She stood in front of the mirror after a shower while she brushed her teeth. Her breasts were staring her in the face almost as if they were mocking her. She clenched her teeth and held a firm grasp on the towel lying on the counter. It was all she could do not to pick up a razor and cut them off herself.

Hormone therapy is often a crucial part of living transgender. For female-to-male transitions, testosterone treatment enables the growth of facial hair, squares up the jaw line, and helps build up muscles while redistributing fat cells. Many find it very hard to pass without being on testosterone, and to have someone identify you by your birth gender can be a devastating blow.

They had stopped at a small café downtown to eat lunch during their shopping excursion. He was feeling confident that he was passing well that day, but when the waiter came up to take their order he greeted them with, "Good afternoon, ladies." His mind went a million different directions. "He knows I'm a girl! I really thought I was passing well today. I didn't shave my legs—but he thinks I'm just some hippie lesbian out with her classy girlfriend. I just want to go back home."

Most people hear the word "transgender" or "transsexual" and automatically imagine something perverse but like anyone else on this earth, these people want to find love and give love. Most of all they want to be seen for who they are, instead of what they are; to be accepted. They wish that people could understand that their circumstances don't make them freaks.

I can't remember the last time I was able to walk outside without having someone stop dead in their tracks with that look of confusion on their face. They study my every move, like I'm going to falter in some way that's going to give them a clear sign of who I really am. Does it matter?

Some people think that being trans has something to do with sexuality, but it doesn't. Many transgender people are asked why they can't just be gay, but it's deeper than that. It's not about sexual preference, but physical and psychological identity. It's about how you see yourself as a whole person.

There's nothing easy about this. There's nothing I can do to make it easy on everyone. If I don't transition I'm not sure how long I'll be able to stand the pain, but if I do I don't know if I'll be able to stand the people I love not accepting me. The only thing that is going to change is my outside appearance. I'll always be the same person.

You live in fear because you don't want people to find out what you've hidden so well. They continue to dig, for no other reason than that they feel they need to know. They are people that don't know you. They don't know what you like, what you don't like, or what you feel. All they know about you is what they see and they swear it isn't all that it appears to be.

Every time my phone rings, or I get a text or an email or a knock on my door, I'm afraid that they've found out, afraid my world will no longer exist because they won't see it for what it is. I've had to lie to feel like I belong, but they aren't going to understand that. I'm living in my own personal hell, with no one to save me but myself.

Photograph © **Meiko Xavier**

Wooz

Story & stuff By Joy Yoon
JOY
AS A CHILD I GREW UP OBSESSED WITH KUNG FU. I WAS AMAZED BY DAVID CARRADINE, JEAN-CLAUDE VAN DAMME AND MY HERO, BRUCE LEE. BUT THAT ALL CHANGED THE DAY...
BRUCE LEE GOT ME ARRESTED.

I look good today. I'm wearing my new overalls and my favorite fuchsia Lacoste t-shirt. Today I plan to strut. On the block where my parents work there's a florist, hair salon and drycleaners nearby. I'm not allowed to wander. So I make my rounds. 

At the florist its boutonnière making. At the hairsalon, its Spanish with Maria and talking about Julio Iglesias. Then its on to the lawn at the drycleaners. I am examining grass today. I need to concentrate. All of the sudden an old white man is standing in front of me. "What are you doing? Where are your parents? Who's watching you?" I'm busy so I answer quickly. "Playing. My dad's in Korea & my mom's at work. No one." After a few more questions, he handcuffs me to his car as he goes inside to pick up his dry cleaning. He thinks I've been abandoned. He won't listen. Back at the station I'm questioned more. All of the sudden I'm stopped in my tracks. "What's your name little boy?" **BOY?!** Clearly I'm a well dressed girl. "I'm a girl!" They accuse me of lying. "Girls don't have haircuts like that." I've been insulted. "Bruce Lee has this haircut and he's a kung fu master." Apparently no girls like Bruce Lee. These cops are insane. We argue for almost an hour. Finally they decide to take me to my mom.

I'm escorted into my parents shop where my mom is talking to my aunt and uncle. She stops when she sees the officers. "Excuse me ma'am, we found your daughter wandering the streets." Liars. She looks at them. "That's my daughter." They stare at her, at each other, then at me. I glare with my hands on my hips and yell. "See! I told you I was a girl!" They sheepishly tip their hats, mumble an apology and leave. Everyone starts laughing at Joy the Boy.

I go to the bathroom and stare at myself. How could I be mistaken for a boy? It must be the hair. My Bruce Lee haircut caused all this. Our relationship would never be the same after that day. But at least I didn't get busted, stupid cops!! ><

HIDDEN

INTERVIEW WITH RAPHAËL DALLAPORTA BY CHRISTOPHE SALET

"We have seized unprecedented amounts of illegal drugs and have denied drug traffickers and terrorists the profits they need to conduct their deadly work."
—George W. Bush,
National Drug Control Strategy 2008 annual report

A few months before this triumphant announcement, an investigation posted on the National Public Radio website gave a slightly different slant on the situation. Entitled "War on Drugs Hasn't Stemmed Flow Into U.S.," the report included testimonies from both anti-drug professionals and convicted smugglers saying basically the same things: drugs are plentiful as ever. As long as there's a demand, people will figure out a way to get them onto the market.

You may think that former President Bush's quote is fitting for someone who is prompt to see victories in the most desperate situations. Yet the American authorities are not the only ones that want you to believe that it is possible to put an end to drug trafficking. "We'll win the War on Drugs" is a mantra used by governments in all countries where illegal drugs are a social problem, and it is mainly affluent countries that are concerned.

In a series called *Hidden*, photographer Raphael Dallaporta documented the incredibly inventive methods used to smuggle drugs into France. Though ingenious, these methods were not flawless, for their objects of transport sometimes wound up in the hands of French customs officers. Most of these seizures were touted in trumpet-blowing press releases and got generous news coverage. But what do they really represent when compared to all the rest that slips through the net? When Dallaporta's series was later published in the French newspaper *Liberation*, the article simply concluded: "In France, customs seized more than 80 tons of narcotics in 2004. It had no effect on the market."

Christophe Salet: ***How did you get the idea to photograph these objects for your* Hidden *series?***
Raphael Dallaporta: The idea came from Clementine Mercier, a photo editor at *Liberation*. She had already published my photo series on landmines and knew my work. One day, she forwarded me a press dispatch from AFP (Agence France Presse) about a drug seizure, along with a message saying, "Could be an interesting subject for a photo documentary." The attached picture featured a Santa Claus made entirely out of cocaine.

So it was a commissioned work? Yes, but the pictures were not originally intended to illustrate a report, not immediately at least. She was interested in developing a photo documentary that could be published at a later date as part of a special report on drugs or something. It was all quite unusual. The interesting thing was that a newspaper was willing to publish a series like that: generally, if there are no people in the picture, they don't see the point of printing it.

Isn't it weird to talk about drugs by showing objects that appear completely unrelated? Not really. If you think about it, on the one hand, you have people who would do anything to smuggle drugs, absolutely anything. And on the other hand, you have customs officers who claim to have such strong intuition that they are able to detect the most sophisticated tricks. To me, it's all a bit of a joke, with each side taking the other for a fool. I figured we'd play around with that idea and just go ahead and shoot the objects. Of course, it's ultimately about people but their story is told through the objects. The story is not about how the people got there, but how the drugs got there.

So how did you work together? The hardest thing was finding a way to access the objects. We started by signing up for the customs' press alert service: as soon as they made a seizure, their press office would contact us and I would go there immediately with my camera. It was kind of a thrill, really. But each time, the drugs had been concealed in ordinary suitcases. So there wasn't much to see...apart from cocaine, of course. Which was funny until you noticed the guy handcuffed to his seat. Then you realized that for him, life was over. But hey... this is drug trafficking. I guess it's part of the game.

What did you do then? I began to talk to the customs officers. I often had to wait until they stopped questioning the guy before I could take any pictures. And during that time, other officers watched over me to make sure I didn't steal the dope. So I asked them if there was a place where they stored more unusual objects, you know, like trophies. That's when they told me about the Customs Officers Training College in La Rochelle where they send all these various "concealment methods," as they are known in the official jargon. I went there with one of the newspaper's drug journalists, Arnaud Aubron, a.k.a. "Mr. Drug," the one most favorable to legalization. We reached the school in a rented car and the first thing we saw was the dogs; they have these huge kennels with around 50 dogs that are trained to detect all kinds of stuff. Then the journalist started to freak out and emptied everything out of his pockets into the glove compartment. That was awesome!

How long did you stay? Two days. On the first day, they took us on a guided tour. Most of the stuff was kept in display cases. But some were so big they were stored elsewhere—hydraulic pumps, a truck, a car. We selected about 30 items, but it's not like there were hundreds to choose from anyway. The customs guys see these things as trophies. They really are super proud. But when you add it all up, it's no big deal. All together they can't even weigh more than a ton, apart from the truck of course. It's kind of ridiculous.

How did you choose to shoot them? I felt it was important not to give them too much dignity, as they have no aesthetic value. Most of them are even kitsch. Even if I shot them with a 4x5 view camera, I based my technique on the way people photograph objects to sell on eBay. I just put them on the floor and kept the boundary line between the floor and the wall in the

background. This line appears in every single picture and links them altogether. If I had used a completely neutral background, they would just have looked like any old everyday objects, and my aim was to show that each item had its own history and had made its own particular journey, which ended at the customs house.

What did you know about the circumstances under which they were seized? Each time I photographed a new object, the journalist would ask the customs officers exactly how it had been seized. Usually they do searches based on leads they get from informers, though they would never tell you that: they always talk about their intuition. Some objects travel routes that immediately arouse suspicions. For instance, they had seized some sugar cane that had been sent from Jamaica via Amsterdam. Or some drills that had come from Miami via London (why would you send a drill when it's something you can find in any country?). Sometimes they damage goods and don't find anything. Some fairground people had their bumper cars entirely stripped down because informers had said they contained narcotics. It took them four days to dismantle everything and in the end there was nothing.

How did you get on with the customs officers? Like I said, I couldn't help thinking that what they were showing us with so much pride was nothing compared with what actually gets through. At the beginning we managed to keep a straight face and play along with them. The journalist was really into the investigation at first but soon began to crack: when he was told about some Islamic prayer beads made out of hash and began to smell them, he couldn't help saying it was really good stuff. And he didn't stop there. They were telling us the story of a fire extinguisher filled with heroin that they had found in the trunk of a car. They probably wouldn't have searched that far if they hadn't found a small sample under the driver's seat. At that point, I heard a voice say, "What a moron!"

What did you like most about this project? The same with my personal projects, it's about showing things that are usually kept hidden. I did this series in 2004–2005, right after my work on landmines, which was also on objects that you never see. After that, I worked on a project called *Domestic Slavery*: I shot the fronts of buildings or houses where cases of domestic slavery had been reported and gone to trial. So you can find a common thread through each of these projects: when you look at the pictures, you can't really tell what's going on until you read the captions that come with them. In a way you get fooled by the pictures in the same way as the objects they feature have already fooled other people. In the case of drugs, the drug has been found but the object still works as a lure.

You're not afraid that people simply miss the point? I believe that there's lot to learn from going around a problem instead of always tackling it head-on. That way, you can see all the implications, everything that goes with it. Too often when photography deals with social issues, it tends to move people to pity. I could have shown the victims of landmines, but the more you show the victims, the more you forget that what causes so much damage is something that is man-made. Sure, it's important to show civilian victims but the fight against landmines is not only humanitarian, it's also political, economic, and ecological. My approach might not be so immediately understandable but I believe that the awareness it raises is more profound and maybe more durable. This being said, *Hidden* is a far less committed work than *Landmines* or *Domestic Slavery*. It's more of a parody. I found the subject fun.

Does that mean you don't consider drugs to be a serious issue? Of course I do. Drugs are a real curse. But again, I didn't choose to talk about people who are dying from an overdose or who are trying to get out of it. I'm talking about the people who try to fight drug trafficking. What most influences my work is the relationship I have with the professionals I have to deal with. For the *Landmines* series, my relationship with the mine-clearing experts was essential. For *Domestic Slavery*, I worked very closely with the director of the Committee against Modern Slavery. In both cases, they are people I truly admire and who do an incredible job. The customs officers I met were...complete jerks. I mean, they are nice enough people but I wouldn't say that they work real hard to sort things out. Look, what time is it now? [He looks at his watch.] 3PM? They're probably just coming back from lunch, which they started at noon at the bar next to their cafeteria. That's what I saw during those two days. And you could tell it wasn't special treatment on our behalf. If anything, they were holding back. Believe me, two days there was more than enough. Stay a full month and you'd be an alcoholic.

TAUNUS
638
B.Z.
1. Borgplaatje uittrekken
2. Druktoets krachtig indrukken
4. Na ieder gebruik hervullen

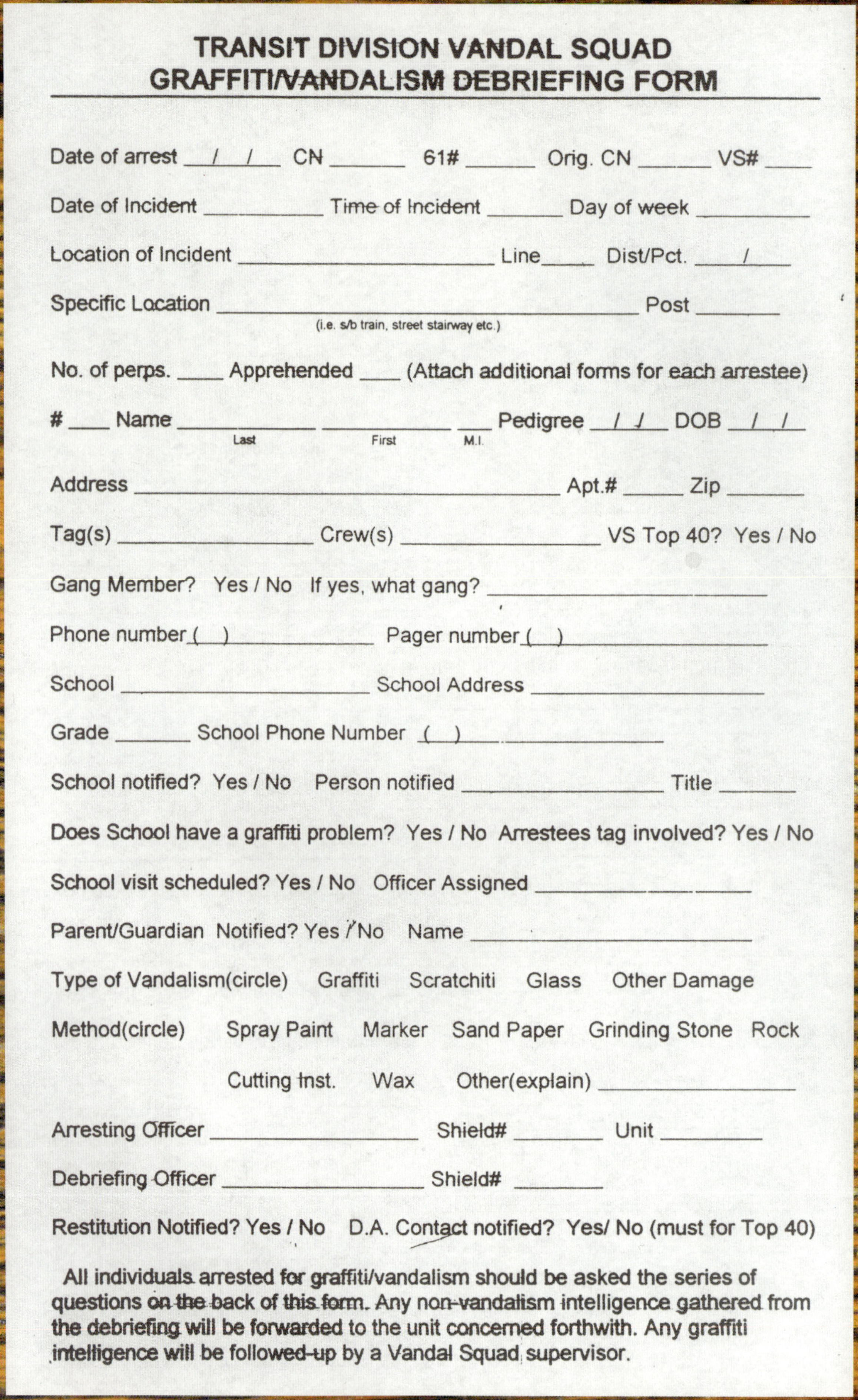

TRANSIT DIVISION VANDAL SQUAD
GRAFFITI/VANDALISM DEBRIEFING FORM

Date of arrest __/__/__ CN ______ 61# ______ Orig. CN ______ VS# ____

Date of Incident __________ Time of Incident _______ Day of week __________

Location of Incident ______________________ Line____ Dist/Pct. ____/____

Specific Location ____________________________________ Post _______
(i.e. s/b train, street stairway etc.)

No. of perps. ____ Apprehended ____ (Attach additional forms for each arrestee)

____ Name ____________ ___________ ___ Pedigree __/__/__ DOB __/__/__
Last First M.I.

Address ____________________________________ Apt.# _____ Zip _______

Tag(s) _________________ Crew(s) _________________ VS Top 40? Yes / No

Gang Member? Yes / No If yes, what gang? ________________________

Phone number ( ) _____________ Pager number ( ) _______________

School _____________________ School Address ____________________

Grade ______ School Phone Number ( ) ______________

School notified? Yes / No Person notified _________________ Title _______

Does School have a graffiti problem? Yes / No Arrestees tag involved? Yes / No

School visit scheduled? Yes / No Officer Assigned _________________

Parent/Guardian Notified? Yes / No Name ________________________

Type of Vandalism(circle) Graffiti Scratchiti Glass Other Damage

Method(circle) Spray Paint Marker Sand Paper Grinding Stone Rock

Cutting Inst. Wax Other(explain) _________________

Arresting Officer __________________ Shield# _______ Unit _________

Debriefing Officer _________________ Shield# _______

Restitution Notified? Yes / No D.A. Contact notified? Yes/ No (must for Top 40)

All individuals arrested for graffiti/vandalism should be asked the series of questions on the back of this form. Any non-vandalism intelligence gathered from the debriefing will be forwarded to the unit concerned forthwith. Any graffiti intelligence will be followed-up by a Vandal Squad supervisor.

V VANDAL SQUAD: INSIDE THE NEW YORK CITY TRANSIT POLICE DEPARTMENT, 1984–2004 BY JOSEPH RIVERA

GRAFFITI/VANDALISM DEBRIEFING FORM Page 2

All persons arrested for graffiti vandalism should be asked this series of questions. Questions may be phrased in any manner which facilitates obtaining the maximum amount of useful intelligence.

What other tag names do you use (have you used)? ____________________

What other graffiti crews are you a member of? ____________________

Do you know any other graffiti vandals? Yes ___ No ___ If yes, list their tag names and real names-____________________

____

____

How did (do) you obtain graffiti instruments? Buy ____ Steal ____ Other _____

Where did (do) you obtain these graffiti instruments? ____________________

____

____

Do you have any knowledge of any graffiti related crimes (past/present/future)?

____

____

____

____

____

____

____

Do you have any knowledge of any other related crimes (past/present/future)?

____

____

____

____

____

____

____

Any other intelligence information-

Note: A Vandal Squad supervisor **must** be notified of any intelligence information gathered during the debriefing. **DO NOT DELAY.**

BACK TO THE BEGINNING

BY GRIDDY GRIMES

Jay was a lowly mail clerk at a Fortune 500.
On his break he used to sneak out back and get blunted.
Just an ex-drug dealer trying to better himself
But struggling to find a legal way to generate wealth.
His boys were always trying to lure him back to the streets
But he promised his girl he'd stick to the 9-to-5 beat.
So he kept his head down and kept doing his job.
Then the next day he's in the break room and met Rob.
He was a real young dude he'd always see when he's leaving,
Rob told him that he works downstairs in receiving.
After three or four months of talking and just building
Rob said he had a sure fire scam to make a killing.
He said he could access and manipulate the system
And orders for laptops that arrived could go missing.
Jay thought that it was too good to be true and too risky
But told him that he'd do it if they split it 50-50.
So Rob started simple and slow with one a week
And Jay would bag it up and flip it out on the street.
Now it's three months in and they bumped it up to five.
Each week they were bagging 3 G's on the side.
He had to lie and tell his girl his hard work had paid off
And they promoted him and now he's the mailroom's top boss.
So they finally got to make the move out of his mother's basement.
Found a really nice crib near work and straight laced it.
Now six months in and they were getting too greedy
'Cause free money was coming to these kids too easy.
The next day he's doing what he does in the mailroom,
No clue he'd be behind bars needing bail soon.
Four cops burst in: "Everybody freeze!"
Got Jay on his knees
And threw the cuffs on wit ease.
Later he found out Rob ratted him out.
When they caught him Jay's name was the first thing out his mouth
He was given three years but will likely do one,
But Rob was neva charged cause he's the CEO's son...

Artwork © **Hank Willis Thomas**

3-piece suit: $250
new socks: $2
gold chain: $400
Picking the perfect casket for your son:
MasterCard

PITCH BLACKNESS

BY HANK WILLIS THOMAS

Brands have become so omnipresent that their impact and control is, for lack of better word, absolute. The Pavlovian response by consumers who are slavishly compelled to get the next must-have item (while they drown in a mountain of debt) will be one of the legacies of our generation. In this climate, NY-based artist Hank Willis Thomas has created the *BRANDED* project. The project consists of images that utilize well-known logos to draw correlations between the slave trade of America's past with the sophisticated advertising campaigns of today. In this piece, Hank speaks about his inspirations for the project and some of the goals he hopes to accomplish by exposing these images and themes to the public.

—Craig Mathis

BRANDED was inspired by a book I read about Michael Jordan and the global economy. In the book there was a quote by Stanley Crouch where he referred to a *Sports Illustrated* article in the 1990s. To paraphrase, it said if teenage girls in Ohio had pictures of a half-naked black man on their wall in 1960 there would have been hell to pay. That quote put an image in my mind of Michael Jordan being born in the early 60s, during the peak of the Civil Rights movement, and what kind of impact it would have had.

I also started thinking about slaves being physically branded as a sign of ownership in the eighteenth and nineteenth centuries. And eventually how in the twentieth century the descendants of slaves have become the fuel in marketing industries that their forebears provided the physical labor to build. In those terms, additional themes and correlations were developed; the cotton fields and slave auctions during the times of slavery in America compared with the football fields, basketball courts, and NFL and NCAA drafts of today.

Also the realization that for most of our generation, we are as much informed by popular culture as we are reading books. Personally, MTV and BET have probably had as much influence on me as reading *Crime and Punishment*. So realizing the power that advertising wields has informed my work to a significant degree. One of my biggest goals is to try to use the language of advertising to talk about things that advertising can't talk about.

While it may appear that the motive for my work is to "bust" a particular company or campaign, I choose to avoid making it a personal attack on a corporation specifically or a person specifically. As an artist (and as a participant in the overall system), I strive to make work that is uplifting while still remaining critical, and at the same time not holier-than-thou. It's part of my mission to make a comment on society, to speak back to the culture as a whole in my own way but also with a wink and smile.

Artwork © **Hank Willis Thomas**

RON ENGLISH

It was a Thursday night in Manhattan. I was feeling pretty respectable when I turned the corner onto 14th Street. I had suited up and gone to dinner with my businessman friend Lenny at an upscale uptown restaurant. We had consumed a nice bottle of wine and were in fine spirits. I wanted to take Lenny to Barmacy, one of my favorite downtown bars, to let him experience a little life on my side of 14th street. I also wanted to show off a couple of my recent billboard liberations. A few doors down from and six feet above the bar was my latest creation, "Think Different" with Charles Manson, and on the billboard on the opposite side of the building was a giant vagina pasted over a Nissan Minivan Ad, courtesy of the Billboard Liberation Front. The caption on the billboard had originally read "Don't Have Kids Without One." I had changed "Kids" to "Sex."

As Lenny gazed up at Charles Manson, I slapped a sticker on a friend's door as a sort of howdy. I would have rung his bell, but I knew he turned in early and didn't drink. We continued on to the bar, where Deb, the owner, was sitting outside checking IDs. She smiled and waved as we approached. As we closed in on our destination, I realized I was being flanked by six guys and a woman. Although I didn't recognize them, I figured we must have been on pretty friendly terms at one time, because they were being pretty aggressive with their affections. After they had me on the ground in handcuffs, it occurred to me why I hadn't recognized them. They were undercover officers.

As they dragged me to my feet, I thought about the giant vagina looming six stories above. I didn't look up at it, but I felt it there. Deb was screaming, "You can't arrest him! He's a famous artist!" I was pretty sure that wouldn't endear me to the officers. I couldn't believe I'd finally gotten caught. Everyone knows that returning to the scene of the crime is a criminal's dumbest move. As for the vagina, when the BLF showed up in town with that thing, my first thought was, I don't want to help put THAT up. But, of course, I did.

As I watched the one of the male officers rifle through my bag, I inquired to another about the reason for my arrest. He replied, "We saw you put up that sticker."

The other cop removed three more stickers from my bag and announced, "He's in possession of three more!"

I said, rather relieved, "Well, take off these cuffs, give me my ticket and let me go have a drink with my friend."

But that wasn't the way it was going down. They informed me I was going to the police station. So I was figuring this was some sort of desk ticket. The station was what, seven blocks away? So maybe I was out of commission for one or two hours. I told Lenny to wait for me at the bar. I'd be back in a round or two.

A police car arrived on the scene, and they sandwiched me in the backseat between a young male cop and the female cop. I tried to be friendly. I was in a pretty good mood. The female cop starts badgering me. "You're going to get somebody killed!"

"I don't understand," I said. "What do you mean, I'm going to get somebody killed?"

"Have you been drinking?" she said.

"Well, I split a bottle of wine at dinner with my friend."

"That's right, you're drunk. You're going to get into a car and run over someone. What if it's a kid, What if you kill a kid? Do you have kids?"

"I have a couple kids but I don't have a car. I take cabs."

"What the hell kind of father are you, driving around drunk?"

I looked at the young cop on my right. "What is she talking about? I don't even have a car!"

He rolled his eyes as the car made a left onto Broadway. Where the hell were we going? The station was on East Fifth Street, wasn't it?

"It's scum like you, that's the problem with this world. You won't stop till you've killed someone. Probably a kid."

I asked the guy where we were going. "We have to make a stop," he replied. We crossed Canal Street.

The car pulled up in front of the station by the Holland Tunnel. The driver got out and went inside. My two cop friends were silent. I asked the younger cop why we were here. He shrugged his shoulders.

I asked him how many people he usually arrested in a night. "Usually one a night."

"One a night? Do you usually use seven cops?"

"It varies."

"The other five cops, are they out arresting someone now?"

"No, you're tonight's collar."

"Guess being an undercover cop's not the adventure one would think."

"I haven't been doing it that long."

"Well, you must have some good stories. You should come to dinner at our house. My wife's a great cook. You could tell us some good stories. I could tell you some good stories."

Our driver returned.

Our next stop was the Tombs. I recognized it from the rat cage sculptures out front.

The cops removed my cuffs and locked me in a small cell. After a while I got a roommate. He thought I was his lawyer. I guess it was my suit. I gave him a little legal advice. Then some other cops took me to be fingerprinted. My young cop was there.

"I thought you would be back on the streets," I said.

"No, I stay with my collar."

"I'm not getting out real soon."

"You'll see a judge tomorrow."

"For putting up a sticker?"

"For committing a crime."

"I'm sure I'm going to get a lot of respect from the other prisoners—stickerboy, menace to society. I don't suppose you'd tack on aggravated assault or murder, you know, so I can get a little respect."

The cop finally smiled. "I suppose that could be arranged."

"Or you know what would really be helpful? What if you went back and peeled off the sticker? Then I wouldn't be a criminal at all. I could take you to the bar. I could buy you drinks all night. It would be more fun than this."

"You know what? I am going to do something for you. I'm going to come upstairs with you."

I wasn't sure what he was talking about.

My new cop friend led me up the stairs to a hall lined with other prisoners and stood next to me. "Do you have any money on you?"

"I think I have around three hundred dollars," I said.

"Take it out and put it in your shoe. If you have more than one hundred cash you're a drug dealer and they will take your money."

I followed his instructions. We were all ordered to stand with our hands on the wall. We were relieved of our belongings. Then we were instructed to pull our pants to our knees and put our hands on the wall. My cop said, "Just keep your hands on the wall." The cop who had been giving the orders walked into an office and returned with a latex glove on his right hand. Not the kind doctors use; the kind you do your dishes with. Then he shoved his index finger up the ass of the first guy in the line, then the second guy, then the third guy. He wasn't changing gloves or even fingers. I looked around at my cop. He rolled his eyes. Latex Glove Man got to me and my cop friend said, "He's okay," and he moved on to the next guy.

"You're my new best friend," I said.

"No problem."

Next stop was a nice big cell with maybe 40 prisoners, all black, and one phone. The guy using it was pretty big and pretty engaged in his conversation. Didn't look like I'd be making a call anytime soon. Thankfully there was some bench left and I got a seat. I sat quietly for a couple hours until they called my name. I was led to a counter, where I peered over at a guy doing some paper work. He finally looked up and read me my charges. There were three misdemeanors, illegal advertising, graffiti, something else, and then there was a felony, endangering the public.

"What the fuck is this?" I said.

"What the fuck are you worried about?" He answers. "You're a white guy. They don't do nothin' to white guys."

"I put up one sticker and I was in possession of three more, for this I get a felony?"

"Get this white boy out of my face!"

The population in my cell had doubled upon my return, but I still got a seat. Same big guy was on the phone. Evidently he was having a quiet night watching TV with his woman. "What's happening now, baby?" There were a few stray conversations going on, but I chose to indulge myself in a little quiet self-pity. That was until I smelled something familiar. I looked up to see two young guys being shoved into our cell. They sat farther down on the bench, so I got up and walked over to introduce myself. Some of the black guys got irate.

"Fuck you, man! You sit in here and you don't talk to nobody until a couple white boys show up!" I realized that probably did seem a little wrong, so I answered the criticism.

"I didn't talk to him because he's white, I talked to him because of the way he smells. Don't tell me you can't smell him."

There was a general agreement in the cell that the white boy did have a nasty odor. "Spray paint," I explained. "I knew he was a graffiti artist when I smelled him. I'm a graffiti artist too." Everyone thought that was pretty funny.

An hour later a few of us were moved to another cell, including the graffiti artist and his sidekick, who, it turned out, didn't know the artist; he just happened to be preaching to him while he was tagging, and then cops showed up and extended his preaching opportunity into the Tombs. I didn't sit next to my white boy buddies in the new cell. They turned out to be a bit duller than I had imagined they would be. I was kind of watching the young evangelist out of the corner of my eye to see if he might take advantage of the captive audience and try his hand at saving a few sinners. He didn't, but graffiti boy actually decided to give an impromptu lecture on the oppression of the black man. He stood up on his bench and, and to the amusement of the other prisoners, off he went for about a minute until he slumped back in his seat. All eyes in the cell were still on him and I was getting a little worried they would find other ways to make him the entertainment, so I started a diatribe about the coming revolution.

I have been in jail about thirty times, and I know that a little inspired revolutionary banter is a great ice-breaker. The revolution is to prisoners what the Second Coming is to Christians. Like the riots in L.A., it will be sparked by a certain event and it will come down fast. There will exist in the first two to three hours, before the authorities regain control, the opportunity to seize power. You have to always carry with you the awareness of who has to be taken out and where they are, and what facilities have to be seized, i.e. television and radio stations. Hell may freeze over while we wait for Jesus to return from the dead, but revolutions really happen. My little rant was well received.

The next day, they brought a couple hundred of us up to a big hall, where we all met for two minutes each with a state-appointed lawyer. The meeting took place through a small barred window. My lawyer laid out my charges and the deal: eight months sentence deferred if I stayed out of trouble for eight months. One slip and the eight months were automatic. In the

Artwork © **Ron English**

meantime, four months of community service. I told her I didn't want to do any community service. She asked me what exactly I had done. I explained that I had put up one sticker and had been in possession of three more stickers that I had intended on putting up in the bathroom of the bar, a custom encouraged by the owner. She said she would try to get me out of the community service.

We were less than a minute with the judge. He gave me eight months probation, four months community service. My lawyer said, "He doesn't want any community service."

The judge said without looking up, "Eight months probation, no community service."

After a couple more hours in the big holding cell, they took me down to claims, where they returned

my wallet and keys. I pointed out that my driver's license was missing, and they told me to come back on Tuesday. I figured the DMV would be more fun.

Outside the Tombs it was dark again. I stopped to admire the rat cage sculptures when one of my cellmates approached. "We were lucky," he said.

"How so?" I enquired.

"Dumb fucks picked up tonight stay until Tuesday."

"Tuesday?" I was puzzled.

"Monday's a holiday", he explained. "They picke up two hundred last night. They'll pick up twice tha tonight, and twice that Saturday night. You think las night was crowded? Tonight's fucked."

Viva Revolution.

GO HOME
N WURK-
TANNY!
YOURE NOT
WANTED
HERE!

FICTION

BOOM FOR REAL

BY MICHAEL A. GONZALES

After a 48-hour binge of booze, broads, and cocaine, crashing back to the bleakness of ~~early~~-morning reality felt like slow death. Leaning against the sticky counter inside Starbucks on 125th Street, I cursed the bright lights overhead.

Clad in a crumpled black linen suit and wrinkled white shirt, I looked like one of those filthy-faced "drunkies" who stumbled down the broken boulevards of Harlem. Standing at the crowded counter were a few priests from Rice High School around the corner. A young soul sister lingered a few feet away.

The pretty woman glared with brown eyes as though she knew me. With a smooth, Hershey-hued face, full lips, and a shoulder-length weave, she seemed familiar. "Just another funky phantom," I reasoned, as the aftershocks of alcohol caused me to shiver. When you've been partying in Harlem dives and after-hours spots as long as I have, the city is full of familiar strangers.

Sugar crunched underfoot as I walked to a seat in the back of the store. A few feet away, a goth girl clad in scratched Doc Martens, torn black stockings, and a black dress left a coffee-stained *Daily News* on the table. Snatching the tabloid, I glanced at a few stories. Turning it over, I saw the front of the newspaper read July 13, 2007.

It was 30 years to the day since the blackout of '77, and the paper was filled with retrospective stories. On the bottom of page 49, a small obit caught my attention:

Arthur "Grandmaster Dynamite" Isaacs, Hip-Hop Legend, Killed at 45

NEW YORK—Arthur Isaacs, best recognized by the moniker Grandmaster Dynamite, was found dead yesterday after being brutally beaten inside the subway tunnels at the 145th and Broadway station. He was 45.

A pioneer in the world of rap music, Arthur Isaacs produced singles for various artists including Kool Keith, KRS-One, and Starr MC, with whom he constructed the landmark album *Black Noise in Riot City* in 1988. According to sources, Mr. Isaacs had lived in the subway station for several years. At press time, this could not be confirmed.

Thinking about Dynamite, I closed my eyes. Behind shut eyelids, old school b-boys boogied to the drumbeat that blared throughout the Battlegrounds. A "yes-yes-y'all" hip-hop park on Amsterdam and 152nd, it was at the Battlegrounds playground (which had gotten its nickname from the basketball boys who had played there since the 50s) where Grandmaster Dynamite had rhythmically reigned supreme on the wheels of steel in the summer of '77.

A few years before I heard of Herc, Flash, or Bambaataa (the only thing I knew about da Boogie Down Bronx was the zoo), Dynamite introduced our neighborhood to the funky beats that would later be called hip-hop. Alongside his partner and "uptown verbalizer" Starr MC, who "rocked the mic till the twilight," Dynamite transformed our hood into a paradise of bass-heavy delights.

Hooking up his megablast system to a nearby lamppost, he hypnotically spun delirious discs that included the James Brown funk anthem "The Payback" and the jungle boogie of Incredible Bongo Band's "Apache." Looking at the blissful faces of the crowd, I realized that this "get down" music was the soundtrack of new revolution.

Malcolm X and MLK were dead, and Motown no longer meant dancing in the streets. Post-revolutionary folks, tired of being busted and disgusted, just wanted to bop to the beat. Stax was gone, the *Soul Train* had derailed, and black noise outlaws like Grandmaster Dynamite hijacked the tracks.

***

It had been a year since I'd last seen Dynamite. Like many disposable heroes, D. had become another lost loser drifting on the edge of hip-hop's murky memory. A shadowy figure begging for change on a crowded Times Square block, the once-upon-a-time ghetto superstar had been sadly reduced to a footnote in our age of hip-hop superstars swimming in golden pools of Cristal.

Sadly, Dynamite was a zooted zombie past his prime; frail and sickly, he held tightly to the frayed McDonald's coffee cup. Reeking of piss and broken dreams, he stood in the middle of the sidewalk and depended on the kindness of New Yorkers. "Change for a brother," he mumbled.

Staring with lobotomized eyes, he shook the dirty cup in front of frightened tourists. Since the bustle of the city rarely paused for any bum's dead president hustle, most people passed Dynamite as though he were invisible. Disgusted, he inched closer to the shadows of a former porn emporium, hoping to catch at least one sucker willing "to help a brother out."

Dressed in tattered Nikes, a ripped *London Calling* t-shirt, and soiled jeans, Dynamite was a fallen monument of a man who had once meant as much to New York City as the Empire State Building. Yet, like Grandmaster Flowers and DJ Hollywood before him, Dynamite had become just another deferred dream of hip-hop's past.

Glancing at his grubby fingers and cracked nails, I almost wept seeing those once prized-hands that had moved many crowds with the perfect beat. Those once-steady hands that had held thousands of dollars and millions of crack pipes; those used-to-be-pretty brown hands were as battered as the squalid sidewalks that had become his home; those hands that were once smooth as the sea, precise as the pyramids, and specific as science, had become skeletal and brittle.

Looking at Dynamite was like facing my own worst fears. Witnessing small signs of madness everyday, I wondered how close to the edge one might totter before falling off into a vortex of delusion. How long did it take before your demons finally swallowed you whole?

A passing car blasted Puff Daddy's capitalistic anthem "All About the Benjamins." As I shoved three twenties into his cup, our eyes met. "You know who I used to be, don't you, brother?" Dynamite droned in a lithium tone, as though he might not be so sure who he was himself.

***

Of course, I knew who Dynamite was, because at one time I wanted nothing more than to follow in his sonic footprints. Fact was, I still had recurring dreams about the chaotic night of the infamous '77 blackout, when I almost got busted looting Fleischman's Electronics simply because I too wanted to be a grandmaster.

That sweltering July night, I was standing in the middle of Broadway dressed in husky jeans, a stained T-shirt, and Pro-Keds when the streetlights began to blink. Moments later, the entire city was shrouded in darkness. For a split second, everybody was still as mannequins in Macy's windows. A heartbeat later, savage voices screamed, *"Blackout! Vic the stores!"*

Fires were set, bottles were thrown, cars were overturned, and the taste of smoke was bitter on my tongue. "When you ain't got nothing, you ain't got nothing to lose," I thought as a gang of pipe-wielding hoodrats ripped the gates from the A&P supermarket. A band of wild boys broke the windows of Mr. Fleishman's stereo store (a Holocaust survivor, he had blue numbers tattooed on his wrinkled left forearm), and I just couldn't resist the temptation.

Since first seeing Dynamite spin at the Battlegrounds a few weeks before, I had been staring at a gleaming set of Technics SL-1200s through the shop window. That night, over the anguished screams of crazed citizens, I scrambled across the shattered glass and grabbed a brand new turntable from the shelf. Mama hadn't raised no thieves, but I wanted to "get mine." A black boy all of fifteen, I figured the world owed me something.

"Animals!" Mr. Fleishman screamed from the sidewalk as he watched black, brown, and beige-skinned men he had thought of as friends and neighbors destroy what was left of his world. "Animals!"

Running with the stereo in my meaty arms, I trekked across the sea of debris that littered Broadway until a crazy cop fired a few warning shots over my head. I dropped the turntable on the ground and it shattered on impact. The police officer laughed as I zoomed back to my apartment building a few blocks away, the brutal ghetto music of shattered glass and gunfire exploding in my ears.

***

Outside the Starbucks, a backfiring truck jolted me awake. The newspaper was still open to Grandmaster Dynamite's obituary. "Fleishman's never reopened and I never became a DJ," I muttered bitterly.

Finally leaving the coffee shop, I drifted down Seventh Avenue. Stopping for a minute, I leaned against a tarnished Cadillac. Pulling a crumpled cigarette pack from my jacket, I cupped my hands to keep the flame from blowing out. Staring at a rustic rooftop across the street, I saw a vision of Grandmaster Dynamite hovering over the concrete gargoyles affixed to a red-bricked apartment building.

Dressed in a vintage Troop jacket, divine Black Power red, black, and green angel wings swelled from his back. With a peaceful gaze on his sanctified face, the turntable master slowly ascended towards heaven. If I was going to believe in demons, I thought it best to make room for a few angels too.

IT WAS AN ACCIDENT!

KEIJI ANDO

A Fine Example of Art **by John Lurie** (powerHouse Books)
A compendium of indie musician, actor, and producer John Lurie's foray into the visual arts, *A Fine Example of Art* weaves images of anthropomorphized animals into a dream world of domestic memes, a synthesis at once hilarious and profound.

FUN! The True Story of Patti Astor
(Miss Rosen Editions/powerHouse Books)
A visually charged autobiography of the former 60s radical turned 70s underground film star turned 80s art world doyenne. Written with all the flair of a woman who named herself after a street in the Village, *FUN!* is everything the name promises and more! Fun is coming in 2010.

It's Just Begun: The Epic Journey of DJ Disco Wiz, Hip-Hop's First Latino DJ **by Ivan Sanchez and Luis "DJ Disco Wiz" Cedeno**
(Miss Rosen Editions/powerHouse Books)
Recounts the rocky path DJ Disco Wiz traveled from the streets of the South Bronx to the New York penal system, and how he found his redemption in the love of hip-hop culture. Coming 2009.

Juvenile **by Joseph Rodríguez** (powerHouse Books)
Himself a former juvenile offender, photographer Joseph Rodriguez follows the lives of a dozen kids from arrest, trial, and incarceration to release, probation, employment, and the search for redemption and renewal.

Midnight **by Arlene Gottfried** (powerHouse Books)
A chronicle of madness, compassion, and constancy, this series of simple portraits offers a glimpse into the life of a man harried and haunted by his own illusive demons and grants us insight into the crucible of his friend and documenter who refuses to let him go.

Next Stop: Growing Up Wild-Style in the Bronx **by Ivan Sanchez** (Touchstone Fireside)
A portrayal of the Boogie Down Bronx from the inside, this memoir details with both nostalgia and brutal honesty a turbulent life growing up in the birthplace of hip-hop.

Novurkistan **by Loucious Broadway** (Xlibris)
Review from Amazon.com: "yea yea word up in the book he said, 'I don't care if I sell one or one million but I think you should cuz If I only sell one then out comes the hood' but I don't know if y'all heard that cuz you're spending your money on cheap perfume to mask the funk that comes from within, when you should be reading books that can do it for you."

New York Photo Festival 2008 Catalog
The official catalog of the inaugural New York Photo Festival: The Future of Contemporary Photography, curated by Martin Parr, Lesley A. Martin, Tim Barber, and Kathy Ryan, and featuring work by *Busted* contributor Raphaël Dallaporta.

It's All Good **by Boogie** (Miss Rosen Editions/powerHouse Books)
A gripping look at the predators and prey in the drug game today, this book features crackheads, junkies, and gangsters living in the projects of Bedford-Stuyvesant, Bushwick, and Queensbridge, New York, a place where pathos prevails, tension is always evident, and escape is one rock, one shot, one glock away.

Hank Willis Thomas: Pitch Blackness (Aperture)
With his characteristic pointedness and dark humor, Willis Thomas shows in *Pitch Blackness* why he is considered one of today's most compelling emerging artists.

Social Graces **by Larry Fink** (powerHouse Books)
In juxtaposing images of New York's blue-blooded social elite against those of staid, working class Pennsylvanians at home, Fink provides powerful commentary on the contemporary American cultural and economic divide. First published in 1984, this study's stark relevance remains today.

Spring Broke **by Nathaniel Welch** (powerHouse Books)
A document of the depraved, alcohol-fueled annual rite of passage for the American adolescent, spring break. In capturing this teenage dissolution, Welch pushes below the superficial veneer of abandon and wanton sexuality, evoking an unanticipated pathos.

Vandal Squad: Inside the New York Transit Police Department, 1984-2004 **by Joseph Rivera** (Miss Rosen Editions/powerHouse Books)
The only book on graffiti told from the perspective of law enforcement, recounting the days and nights spent in pursuit of some of New York City's most notorious vandals.

Photograph © **Boogie** / *It's All Good*

This is

Fine Art
photographic
prints
from the
archive of
The Daily News

now

INTERNATIONAL EDITION
THE ART NEWSPAPER

anubis

BAKED